Genetic Complexity of Hormone Sensitive Cancers

Genetic Complexity of Hormone Sensitive Cancers

Editors

Jyotsna Batra
Rupert Ecker

MDPI • Basel • Beijing • Wuhan • Barcelona • Belgrade • Manchester • Tokyo • Cluj • Tianjin

Editors
Jyotsna Batra
Queensland University of Technology
Australia

Rupert Ecker
TissueGnostics GmbH
Austria

Editorial Office
MDPI
St. Alban-Anlage 66
4052 Basel, Switzerland

This is a reprint of articles from the Special Issue published online in the open access journal *Genes* (ISSN 2073-4425) (available at: https://www.mdpi.com/journal/genes/special_issues/Hormone_Cancers).

For citation purposes, cite each article independently as indicated on the article page online and as indicated below:

LastName, A.A.; LastName, B.B.; LastName, C.C. Article Title. *Journal Name* **Year**, *Volume Number*, Page Range.

ISBN 978-3-0365-7116-4 (Hbk)
ISBN 978-3-0365-7117-1 (PDF)

Cover image courtesy of TissueGnostics GmbH

Contents

About the Editors

Jyotsna Batra

Associate Professor Jyotsna Batra, PhD, is a Laboratory Head at School of Biomedical Sciences and an Advance Queensland Industry Research Fellow at the Centre for Genomics and Personalised Health at Queensland University of Technology, Australia. Dr Batra leads a research group working on the molecular genetics of cancer. Her current research focus is to identify cancer risk-associated genetic variants and to understand their molecular consequences on cancer initiation and progression. She aims to develop better biomarkers to detect cancer early and to identify genetic biomarkers that can distinguish slow growing diseases from very aggressive diseases at an early stage, so that better therapeutic interventions can be performed. Dr Batra has contributed to >200 research articles, including those in high-impact journals such as *Nature Genetics*. Dr Batra has received several poster and oral prizes for her research work. She has also been a finalist for the prestigious ASMR and Women in Technology (WiT) Awards and won the Queensland Young Tall Poppy Award (2019) and Cure Cancer Researcher of the Year Award (2018). She has received >AUD 8 million in funding, including international funding from the US DoD Idea Development grant.

Rupert Ecker

Rupert Ecker, PhD, is an Adjunct Professor at Queensland University of Technology (QUT) and has studied cell biology at University of Vienna. He is an entrepreneur, holder of multiple patents and co-founder of the TissueGnostics group in Austria, Romania, the USA, and the Asian Pacific. He is Chief Executive Officer of TissueGnostics GLOBAL. His current research focus is on artificial intelligence (AI)-based decision support systems for next-generation histopathology/tissue cytometry. He aims to develop ISO 13485- and in-vitro-diagnostics-compliant image cytometry software for biomedical research, as well as clinical diagnosis. Dr. Ecker is an inventor of more than 10 patents and has contributed to >50 research articles and book chapters. Together with his collaborators, he won the science2business Award 2011 of the Austrian Chamber of Commerce.

TISSUEGNOSTICS
PRECISION THAT INSPIRES

Preface to "Genetic Complexity of Hormone Sensitive Cancers"

Cancer is a multi-faceted phenomenon and requires a concerted action across different disciplines to define a roadmap for precision medicine to treat cancer, from genomics, bioinformatics, biochemistry, proteomics, cellular and tumor biology, pharmacology, oncology, hematology, radiology, and surgery to relapse monitoring and palliative care. In this Special Issue, we focus on the initial segments of this roadmap by elucidating the 'Genetic Complexity of Hormone-Sensitive Cancers', and the book chapters provide insights into selected topics and technologies at the current frontiers of cancer research.

As scientists working in the field of biomedical research, novel and innovative techniques and technologies are appearing at an increasing pace. These are exciting times for cancer research. Technologies that simultaneously determine the expression levels of large groups of genes and link the genotype with cellular phenotypes and functions are essential for understanding regulatory networks, not only those made up of proteins (enzymes, hormones and the signal transduction cascades they constitute), but also genetic regulation.

Today, completely novel technologies and instruments are revolutionizing our view of biological entities, e.g., genes, and also, novel extensions to and revisions of existing and well-established technologies such as genetic probes and bar-coding techniques, which add a whole new dimension to microscopy as they allow us to visualize DNA sequences and mRNA in situ. It is the combination of well-established technologies with novel approaches, for example, the identification of newly spliced isoforms of key oncogenes, that invites new perspectives on old questions and allows us to connect that which previously appeared to be separate or even contradictory.

Only 2% of genetic material belongs to defined genes, while 98%, the evolutionary detritus, were considered to be unused. There is emerging evidence, however, that this genetic material is neither unused nor detritus, but rather, it exerts regulatory functions, both physiological and pathological ones. In Chapters 3 and 5, the authors present data on long non-coding RNA (lncRNA), which code for small peptides referred to as micropeptides (miPEP). To find coding information in "non-coding gene sequences" and to identify huge regulatory networks, which until now, have been elusive, was completely unexpected, calling into question everything we had assumed about intracellular regulatory networks and our genetics. To analogize the work of cosmologists who detected that the vast majority of mass in the universe is dark matter, we term these 98% of the genetic material of unknown information and function, ***Genomic Dark Matter***. Hence, we are close to understanding of how genes and proteins are intertwined, going far beyond the classical model of a transcription-based unilateral information flow from genes to proteins, and we might even witness a paradigm shift over the coming years. The regulation of genes and proteins through small, "transient regulatory units" that may act simultaneously with lncRNAs through their transcription products, miPEPs, is a revolutionary and refreshing new concept and we might soon observe the formation of new sub-disciplines in genetic science.

The editors of this Special Issue wanted to create a unique forum for the international biomedical research community to present their research and review articles on the genetic complexity of cancer. It is notable that the guest editors and contributors are based all over the globe. We express our gratitude to the contributing authors from multiple institutions in nine countries. While the chapters in this book cannot claim completeness in any way, they represent an attractive collection across different aspects of hormone-dependent cancers and will hopefully stimulate further targeted research.

The first chapter is a review article on the state of the art of computational pathology, the next step of digital pathology. Over the previous two decades, the automated slide scanning and digital representation of tissue sections for histopathological examination have become mainstream, and many commercial options have become available. The terminology generally used to describe this technological approach is Digital Pathology. This term is actually misleading in that the "act of pathology" is not digital. Pathology is the "ability" to recognize relevant structures in histological sections of human tissue and deduct diagnostic and prognostic conclusions. In Digital Pathology, human experts look at digital representations of tissue samples on monitors, rather than through the oculars of a microscope, but the pathological analysis and interpretation as such is still performed visually. The next step of development aims to introduce a digital workflow into the actual digital analysis by using contemporary tools such as tissue cytometry and spatial biology and in situ genomics. These novel approaches are also referred to as Next-Generation Digital Histopathology.

Chapter 2 is a review on the Genetic Complexity of Prostate Cancer (PCa), its various pathways and mechanisms, its genetically distinct subsets, and the extraordinary complexity of genomic alterations in this form of cancer. The authors provide an overview of genetic changes that can occur during carcinogenesis and tumor progression and reflect on potential therapeutic approaches.

An intriguingly novel research topic—the regulatory function of long non-coding RNAs (lncRNAs)—and its emerging importance in prostate and breast cancer is reviewed in Chapter 3. Recently, germline genetic variations associated with cancer risk have been correlated with lncRNA expression and/or function. In addition, single nucleotide polymorphisms (SNPs) may occur within cancer-associated lncRNA and are correlated with cancer risk and may have potential as therapeutic targets for cancer treatment. As some of these lncRNA have a tissue-specific expression profile, they require further investigation to explore their potential as biomarkers for specific cancers.

The authors of Chapter 4 present their research work on the gene signatures fundamental to controlling cancer progression and their importance for the development of biomarkers and diagnostic tools. Two sets of overlapping signatures linking hormone signaling, cell cycle progression, and control pathways are discussed for prostate cancer, as these pathways may act together to promote aggressive cancer development.

Chapter 5 scrutinizes the alternative splicing of Iroquois-class homeodomain protein 4 (IRX4), a homeobox transcription factor, which has been implicated in PCa as a tumor suppressor through genome-wide association studies (GWAS). Additionally, tumor growth, metastasis, and therapy resistance benefit from aberrant splicing. In the study presented, twelve IRX4 transcripts in PCa cell lines were investigated, including seven novel transcripts, demonstrating unique expression profiles between androgen-responsive and non-responsive cell lines. These IRX4 isoforms might induce distinct functional programming that could contribute to PCa hallmarks, thus providing novel insights into diagnostic, prognostic, and therapeutic significance in PCa management.

Chapter 6 provides an in-depth comparison between six prostate cancer cell lines and tumor tissues with respect to gene-expression levels of androgen, insulin, estrogen, and oxysterol signaling. Major differences between the PCa tissue and cell lines and between different cell lines are described, and these elucidate the importance and limitations of cell lines for understanding PCa formation and progression. This systematic characterization provides a solid basis for the PCa research community to choose the appropriate cell line model for any hormone pathway of interest.

Acquired resistance to cyclin-dependent kinases 4 and 6 (CDK4/6) inhibition in estrogen receptor-positive (ER+) breast cancer remains a significant clinical challenge. Chapter 7 investigates differentially expressed genes associated with an acquired resistance to palbociclib in ER+ breast cancer.

Chapter 8 reviews alternative splicing in gynecological cancers, which is also induced by human papillomavirus and may cause cervical cancer, the fourth most common cancer on a global scale and the most common cancer in developing countries with rapidly increasing mortality rates. Serine/arginine-rich (SR) proteins and heterogeneous ribonucleoproteins (hnRNPs) have prominent roles in modulating alternative splicing. Aberrant splicing events in cancer-related genes lead to chemo- and radio-resistance, and the authors reflect on the role of alternative splicing events and splicing variants in cervical cancer as potential biomarkers for diagnosis, prognosis, and the development of novel drug targets.

Bladder carcinoma (BC), another hormone-dependent form of cancer with increasing incidence and mortality rates, is the subject of a study presented in Chapter 9. The study presented here utilizes a global proteomics approach to identify differentially expressed proteins in bladder cancer cell lines and identified hundreds of proteins that overexpressed or downregulated several BC cell lines. which were compared with a normal human urothelial cell line. It identified UDP-N-acetylhexosamine pyrophosphorylase (UAP1) as a promising therapeutic target for bladder cancer.

The last chapter reviews the current research on head and neck cancers with a focus on alternative splicing. These forms of cancer originate from the mouth, nasal cavity, throat, sinuses, and salivary glands, and the growing incidence rate makes them to sixth most common cancers worldwide. Such as other cancer forms discussed in previous chapters, in head and neck squamous cell carcinomas (HNSCC), human papillomavirus (HPV) is a major risk factor, followed by tobacco use and alcohol consumption. The genetic basis of the development and progression of HNSCC includes aberrant non-coding RNA levels. Different protein isoforms, the alternate methylation of proteins, and changes in the transcription of non-coding RNAs (ncRNA) can be used as diagnostic or prognostic markers and targets for the development of new therapeutic agents.

Recent methodological advancements in genetics, cell biology, biochemistry, bioinformatics, and information technology allow us to perform tests, experiments, and in situ investigations that the previous generation of life scientists might have not even dreamt of. Currently available commercial solutions not only permit us to identify specific genes in cells within their native tissue environment, but detect genomic signatures in situ and on the single cell level. Tools for spatial biology, also referred to as contextual tissue cytometry, allow us to simultaneously quantify a continuously increasing number of phenotypic/proteomic markers and relate them with the genetic profile of specific cells. These novel conceptual and experimental approaches are essential aspects of today's research endeavors and aim to successfully establish a roadmap towards Precision Cancer Medicine. As these chapters present such advancements, we hope this book may contribute towards achieving this bold aim, and we wish to inspire and enlighten all of our readersin order for them to further their own research.

Jyotsna Batra and Rupert Ecker
Editors

genes

MDPI

Review

Next-Generation Digital Histopathology of the Tumor Microenvironment

Felicitas Mungenast [1,2,*], Achala Fernando [3,4], Robert Nica [2], Bogdan Boghiu [5], Bianca Lungu [5], Jyotsna Batra [3,4] and Rupert C. Ecker [2,3,4,*]

1 Institute of Pathophysiology and Allergy Research, Center for Pathophysiology, Infectiology and Immunology, Medical University of Vienna, 1090 Vienna, Austria
2 TissueGnostics GmbH, 1020 Vienna, Austria; robert.nica@tissuegnostics.com
3 Translational Research Institute, 37 Kent Street, Woolloongabba, QLD 4102, Australia; achala.vitharanage@hdr.qut.edu.au (A.F.); jyotsna.batra@qut.edu.au (J.B.)
4 School of Biomedical Sciences, Faculty of Health, Queensland University of Technology, Brisbane, QLD 4059, Australia
5 TissueGnostics SRL, 700028 Iasi, Romania; bogdan.boghiu@tissuegnostics.com (B.B.); bianca.plosnita@tissuegnostics.com (B.L.)
* Correspondence: felicitas.mungenast@tissuegnostics.com (F.M.); rupert.ecker@tissuegnostics.com (R.C.E.)

Abstract: Progress in cancer research is substantially dependent on innovative technologies that permit a concerted analysis of the tumor microenvironment and the cellular phenotypes resulting from somatic mutations and post-translational modifications. In view of a large number of genes, multiplied by differential splicing as well as post-translational protein modifications, the ability to identify and quantify the actual phenotypes of individual cell populations in situ, i.e., in their tissue environment, has become a prerequisite for understanding tumorigenesis and cancer progression. The need for quantitative analyses has led to a renaissance of optical instruments and imaging techniques. With the emergence of precision medicine, automated analysis of a constantly increasing number of cellular markers and their measurement in spatial context have become increasingly necessary to understand the molecular mechanisms that lead to different pathways of disease progression in individual patients. In this review, we summarize the joint effort that academia and industry have undertaken to establish methods and protocols for molecular profiling and immunophenotyping of cancer tissues for next-generation digital histopathology—which is characterized by the use of whole-slide imaging (brightfield, widefield fluorescence, confocal, multispectral, and/or multiplexing technologies) combined with state-of-the-art image cytometry and advanced methods for machine and deep learning.

Keywords: next-generation digital histopathology; tissue cytometry; multiplexing; RNA ISH; cancer; tumor immune microenvironment; tumor microenvironment

Citation: Mungenast, F.; Fernando, A.; Nica, R.; Boghiu, B.; Lungu, B.; Batra, J.; Ecker, R.C. Next-Generation Digital Histopathology of the Tumor Microenvironment. *Genes* **2021**, *12*, 538. https://doi.org/10.3390/genes12040538

Academic Editor: Gael Roue

Received: 11 March 2021
Accepted: 1 April 2021
Published: 7 April 2021

1. Introduction

Cancer is a crucial global health challenge. The incidence of new cancer cases is predicted to increase by around 70% over the coming two decades [1]. Due to the idea that cancer originates from a deranged genome, exploring the genomic, transcriptomic, and proteomic nature of cancer is vital for understanding and utilizing remedies for the treatment of cancer [2]. The tumor cells and the surrounding microenvironment, which includes various types of immune cells, signaling cells and molecules, fibroblasts, and the extracellular matrix comprised of adjacent blood vessels, are highly interdependent compartments. We have started to understand the complex interplay between each of these compartments, with the tumor itself shaping and directing its surroundings—the tumor microenvironment (TME)—while at the same time obtaining signals from this microenvironment for further progression [3]. It is often believed that cancer seeds are germinated primarily in an appropriate microenvironment [4]. Precise localization of molecular indicators by spatial immunophenotyping techniques inside the microenvironment concedes an additional comprehensive analysis of the tumor to foresee its progression and therapy response [5–7].

In the past few years, in-depth profiling of cancer cells/tissues has determined the cancer genome, the transcriptome, and the proteome as powerful sources of diagnostic, prognostic, and predictive markers/biomarkers [8,9]. In this regard, spatially mapped cellular gene expression has appeared as a critical method to understand the localization and complicated multicellular interactions of DNA, RNA, and proteins within cells located in the tumor as well as in the TME [10,11]. Interrogation of the tumor cellular organization context at single cell level with the cell's interactions with neighbor cells helps towards a better understanding of the heterogeneity of the TME between individuals as well as within the same tumor sample [12,13]. Thus, a need arises for multi-omics approaches where many DNA, RNA, splice variants and protein targets can be visualized by various staining techniques in situ. This dictates the need to quantify stained tissue sections, in terms of intensity, presence (expression levels), and/or spatial distribution in an unbiased, objective, fast, and automated way. Next-generation digital pathology is able to fulfil these requirements in research as well as in clinics. Even though the term "digital pathology" has been used for decades, its practical definition is still limited to digitizing samples. The actual analysis in digital pathology is still performed visually—by pathologists looking on a monitor rather than through a microscope's oculars. Converting immunohistochemistry (IHC), immunofluorescence (IF), or RNA in situ hybridization (RNA ISH) stained markers within tissues sections into digitized images is a prerequisite, but for pathology to become really "digital" and automated, further processing and extraction of quantitative data, termed as image cytometry, is required. Several commercial systems are available [14] that offer specialized software solutions utilizing image cytometry, but are methodically focused on the analysis of histological sections and are thus referred to as tissue cytometry [15,16]. Due to the constant evolvement and the increasing reliability of these systems, two commercially available whole slide imagers are approved by the 'U.S. Food and Drug Administration' (FDA) and can be used for clinical approaches [17]. To evaluate the reliability of these next-generation digital pathology platforms in clinics in terms of prognostication and patient management, Nagpal et al. conducted a comprehensive study using prostatectomy specimens. They established a deep convolutional neural network addressing Gleason scoring, which was trained by pathologists on 912 hematoxylin and eosin (HE) stained tissue slides. Next, they compared the classification of 29 additional pathologists with the results of the deep learning-based system. As an outcome of the study, the deep learning-based Gleason classification system showed a significantly higher sensitivity and specificity than 9 out of 10 pathologists [18]. Further studies that used deep convolutional/deep learning/machine learning networks for cancer tissue classification/TME on HE cancer samples for follow-up alignment with clinicopathological parameters are those by Jiao et al. [19], Kwak et al. [20], and Bidal et al. [21] on colon cancer samples, Mittal et al. on breast cancer [22], Wang et al. on lung adenocarcinoma [23], and Diao et al. on skin cutaneous melanoma, stomach adenocarcinoma, breast cancer, lung adenocarcinoma, and lung squamous cell carcinoma [24].

All the above-mentioned studies are good examples that show that tissue cytometry may provide the methodological basis for next-generation digital pathology, which is the state-of-the-art technology to use and constitutes an enabling factor for precision medicine in clinics as well as in research. Within this review, we are going one step further by addressing the concepts of next-generation digital pathology using imaging-based tissue cytometry, in combination with multiplexing and RNA ISH technologies, as an emerging and central method within precision diagnostics, and discussing various applications.

2. Multiplexing Techniques as Useful Tools for High-Content Phenotyping

To achieve high-content phenotyping, optionally in combination with applying genetic markers for well-defined DNA loci as well as total RNA or specific mRNA measurements, the importance of multiplexing staining techniques continues to increase in research and clinics, especially for the purpose of determining the complex immune and tumor microenvironment status in patients suffering from cancer, graft versus host disease, and other pathological conditions related to immune responses [25]. In clinics the assessment of various immune cell markers as well as immune cell populations is required for prognosis, diagnosis, and selecting the therapeutic intervention strategy. Conventional IHC/IF staining techniques are restricted by the number of markers which can be detected at once within one tissue section. This problem was bypassed by staining consecutive tissue sections, with the main limitation being that high-dimensional co-expression analysis is not possible and very precious information is lost [14]. However, in recent years the ability of multiplexing, in terms of visualizing a high number of markers at one time within a sample, has evolved and thereby represents a powerful tool for investigating complex molecular/functional processes and interactions within cells as well as in the complex native tissue environment. In this section, we discuss various immunohistochemistry and immunofluorescence multiplexing techniques.

IHC-based multiplexing methods: Conventional IHC staining usually used in pathology only allows the detection of one marker per tissue section, and therefore no co-expression analysis is possible. With IHC multiplexing techniques the number of stained markers per tissue sections can be increased drastically, which leads to more detailed staining of patient tissue, especially important for clinical applications in respect to diagnostics and prognosis [26]. Previously published multiplexing methods based on IHC are "multiplexed immunohistochemical consecutive staining on single slide" (MICSSS) [27] and "Sequential Immunoperoxidase Labelling and Erasing Method" (SIMPLE) [28]. These two techniques use the chemical property alcohol solubility of the peroxidase substrate 3-amino-9-ethylcarbazole (AEC). The protocol is similar to conventional IHC but includes after image acquisition the removal of AEC with organic solvent-based destaining buffer, and the restaining with new antibodies targeting other markers of interest. Thereby, MICSSS and SIMPLE enable multiple staining rounds. As a final step, the images taken after each staining round are overlaid and sometimes even transferred into a pseudo-color IF-like image. The advantages of these two staining techniques are that they allow co-expression analysis and there are no limitations in terms of antibody species (same antibody origin species can be used for each marker), which is a limitation in conventional staining IHC techniques. However, MICSSS and SIMPLE allow only one marker at each staining round and therefore are limited in number of markers (accordingly to published data, up to 5–10 in total) and are highly time intensive [28,29].

IF-based multiplexing methods: Multiplexing methods based on immunofluorescence are much more common and comprise many advantages over IHC-based multiplexing methods. With IF multiplexing techniques, conventional immunofluorescence staining/imaging can be extended from around 6 to up to 60 markers. Published IF multiplexing techniques include "TSA Opal multiplex immunohistochemistry" (Opal mIHC, PerkinElmer, Waltham, MA, USA) [30], "in silico multiplexing workflow" [31], "tissue-based cyclic immunofluorescence" (t-Cycif), MultiOmyx (MxIF) and "multi-epitope-ligand cartography" (MELC) technology as well as DNA barcoding-based techniques such as "CO detection by InDEXing" (CODEX, Akoya Biosciences ,Marlborough, MA, USA) and GeoMx® (NanoString, Seattle, WA, USA). IF-based methods are much more effective and faster than the IHC-based methods, given that more than one marker can be stained simultaneously in each staining round [14,31–35]. The Opal mIHC technique is based on sequential staining rounds, and the secondary antibodies are tagged with tyramide signal amplification system (TSA)-conjugated fluorescence molecules. Heat-treated stripping of the tissues in between the staining rounds removes the primary and secondary antibodies but not the TSA-conjugated fluorescence molecules. After multiple staining rounds, the slides can be acquired. There is no limitation in the number of different antibody species but there is a restriction in the number of fluorochromes [30]. Blenman et al. established a workflow for multiplexing that includes multiple staining rounds of the tissue, whole-slide imaging with the tissue cytometer TissueFAXS PLUS (TissueGnostics, Vienna, Austria), dye inactivation by chemical bleaching after each acquisition step, as well as merging the images from all staining rounds and quantitative analysis of the stained markers/cell populations with StrataQuest software (TissueGnostics) [31]. A similar strategy is used by the t-Cycif and the MxIF techniques [32,33]. One big advantage of these chemical bleaching-based methods is that they substantially reduce autofluorescence of the tissue after each acquisition step [36]. However, chemical bleaching-based technologies are still time consuming; for a staining protocol of 30 markers, approximately 1–2 weeks are needed. One main disadvantage of the repeated chemical-based bleaching steps for fluorochrome removal after each staining/imaging round is that the preservation of cell and tissue integrity cannot be guaranteed. Lin et al. reported that after 10 staining rounds, a loss of 2–46% of the cells within various tissue types was observed [31–33]. Another technology used for multiplexing is MELC, which is based on fully automated and repeated rounds of multiple marker IF staining, imaging as well as chemical and photobleaching (at the excitation wavelength) of the fluorochromes on a tissue section. The main limitation of the MELC technology is that the photobleaching and imaging step can be only applied to one microscopic field of view [35]. A rather innovative and novel technology able to deal with a very high number of different target antigens is the DNA barcoding-based method CODEX. A cocktail of up to 50 unique oligo-DNA (barcodes) conjugated antibodies specific for the target markers is applied at once on the tissue section. Next, the barcodes are detected by highly specific dye-labeled reporters, which are barcode-complementary oligonucleotides labeled with fluorochromes. Multiple rounds of staining, imaging, and removing of the reporters allow high-dimensional phenotyping [34]. Similar technology is used by GeoMx® (NanoString, Seattle, WA, USA), which is also based on oligonucleotide tags (barcodes) in combination with microscopic imaging to identify a high number of markers (proteins, mRNA, miRNA, etc.) in one hybridization reaction [26]. A summary of the above-mentioned staining methods is provided in Table 1.

The enhanced number of stained markers offered by several multiplexing methods also increases the necessity of appropriate next-generation digital pathology platforms that provide fully automated acquisition of the stained tissue sections as well as computer-assisted/digital high-content phenotypic analysis and high-dimensional data mining.

Table 1. IHC and IF multiplexing techniques.

Method	Process	Advantages	Disadvantages	References
MICSSS (IHC)	Multiple staining rounds; AEC removal AEC with organic solvent-based destaining buffer; imaging	• No limitation by the number of different antibody species • No company-specific reagents/devices are needed	• Time intensive • Limited to 10 staining rounds	[29]
SIMPLE (IHC)	Multiple staining rounds; AEC removal with organic solvent-based destaining buffer; imaging	• No limitation by the number of different antibody species • No company-specific reagents/devices are needed	• Time intensive • Limited to 5 rounds of staining without loss of tissue antigenicity	[28]
Opal mIHC (IF)	sequential staining with AB tagged with TSA conjungated fluorescence molecules, AB removal by heat-treated antibody stripping; imaging	• No limitation by the number of different antibody species • up to 7 markers	• Time intensive • limited by the number of fluorochromes	[30]
In silico multiplexing workflow (IF)	Multiple staining rounds; Dye inactivation by bleaching with alkaline solution + H_2O_2; imaging	• No limitation by the number of different antibody species • No company-specific reagents/devices are needed	• Each round of staining may take at least 24 h depending on the antibodies and the tissue	[31]
t-Cycif (IF)	Multiple staining rounds (like MxIF); bleaching by hydrogen peroxide, intense light and high pH; imaging	• Background noise decreases with cycle number due to multiple rounds of fluorophore bleaching • No limitation by the number of different antibody species • No company-specific reagents/devices are needed	• Relatively slow (each cycle 6–8 h, most time consuming is the imaging) • after 10 cycles, 2–45% loss of cells	[32]
MxIF (IF)	Multiple staining rounds; Alkaline oxidation chemistry was developed that eliminates cyanine-based dye fluorescence within 15 min; imaging	• No limitation by the number of different antibody species • No company-specific reagents/devices are needed • Removal of fluorescence dye within 15 min • Up to 60 biomarkers	• relatively slow due to scanning times	[33]

Table 1. *Cont.*

Method	Process	Advantages	Disadvantages	References
MELC (IF)	Multiple automatic staining rounds; during each cycle the sample is incubated with one or more tags and imaged before bleaching by soft multi-wavelength excitation	• Automated cycles of fluorescent staining, imaging and photobleaching • No limitation by the number of different antibody species	• Bleaching/acquisition can be applied only to one field of view • Special devices are needed	[35]
CODEX (IF)	Antibodies conjugated to a CODEX barcode; visualized by the binding of highly specific corresponding dye-labeled CODEX reporter	• No limitation by the number of different antibody species -> no secondary antibodies • Fast, each round of extension and bleaching (10 min) • Up to 35 rounds with 3 markers	• Special devices and reagents are needed	[34]
NanoString (IF)	Antibodies conjugated to a barcode; visualized by the binding of highly specific corresponding dye-labeled reporter	• Up to 40 markers • No autofluorescence and spectral overlap	• Limited number of regions of interest • Special devices and reagents are needed	[26]

AB, antibody; AEC, 3-amino-9-ethylcarbazole; CODEX, co detection by indexing; IF, immunofluorescence; IHC, immunohistochemistry; MELC, multi-epitope-ligand cartography; MICSSS, multiplexed immunohistochemical consecutive staining on a single slide; SIMPLE, sequential immunoperoxidase labelling and erasing method; TSA, tyramide signal amplification system; t-Cycif, tissue-based cyclic immunofluorescence.

3. Advanced Imaging for Digital Pathology

The first step in a tissue cytometry/next-generation digital pathology workflow includes whole slide scanning or at least acquisition of a region of interest of the stained slide. The second and even more important step comes with the subsequent computer-assisted quantitative image analysis. Next-generation digital pathology technology aims to guide the workflow away from visual observation with a standard microscope and subjective estimations, which are funneled into scoring schemes describing marker expression with "+/++/+++", to a fully automated computerized platform for the detection and numerical quantification of stained markers in defined cell subpopulations in relation to specific histological structures. Not only are these platforms providing a fast analysis of markers, but they also seek accurate, unbiased, reproducible, and standardized results. These platforms are already well integrated and used in various fields of research [37,38]. Additionally, in 2017 the FDA approved the first next-generation digital pathology program (Philips IntelliSite; PIPS) as a clinical digital diagnostics tool in routine diagnosis [39].

Several whole slide imaging platforms (with or without image analysis software) are commercially available in various configurations (e.g., TissueGnostics, Akoya Biosciences, Leica Biosystems, Hamamatsu, Zeiss, 3DHistech, PerkinElmer, Roche, Philips). As the name already indicates, these scanners are able to acquire whole slides instead of only individual captures of fields of view, and thereby provide complete composite digitized images of slides in high resolution. The technology used is image acquisition by either tile scanning or line scanning with a follow-up stitching of the images [14,40]. Depending on the specific next-generation digital pathology platform configuration, these scanners are able to perform whole slide imaging in different imaging modes such as brightfield, widefield fluorescence, confocal, structured illumination, multiplexing, and/or multispectral. The hardware components are usually the following: microscopy stand (upright, inverted) or boxed system without a phototube, cameras (color and/or monochrome), light sources for fluorescence and/or brightfield mode, multiple filter sets for multicolor fluorescence imaging (may include single-, dual-, and/or multi-band filters), high-quality objective lenses for acquisition with different magnifications (1× to 100×), motorized slide scanning stage or high-throughput slide loading systems. Some platforms offer objective auto-oiling for high magnifications and/or provide a slide bar-code reader for higher efficiency. A powerful computer workstation and high-resolution computer monitors for the viewing of the digitized slide as well as for the potential follow-up image analysis [37,41] are mandatory. For controlling all the individual components, for digitized slide viewing and data management, the platforms also include highly functional slide imaging software [40]. In some instances, the platforms also offer or are equipped with image analysis solutions for quantitative analysis. Such quantification is not stoichiometric, and hence does not provide chemical concentration of markers, but is rather based on comparison with negative controls, which is referred to as cytometric.

The basic Theory of Scales of Measurements defines four different types of scales—nominal, ordinal, interval, and ratio [42], the first one being referred to as qualitative, and the other three being accepted as quantitative. In all three scales, systematic, observer-independent measurements of well-defined attributes of objects can be performed, resulting in numerical values that allow for comparison of the objects under investigation as well as statistical evaluation of the assigned attributes.

Most slide scanners today provide area and distance measurements in metric values, whose measurements fall into the ratio type of scales. The amount of any given molecular marker expressed in certain cells, however, is usually determined as a relational value (e.g., mean relative fluorescence or optical density in brightfield microscopy) rather than an absolute value (e.g., μmol or nanogram). Hence, such measurements belong to the interval type of scale. Such measurements do permit comparative measurements of more and less, but it is not possible to draw conclusions by building the ratio between two values. If a cell or cell population expresses a certain molecule at a mean relative fluorescence of 7000 and another cell or cell population exhibits a value of 14,000, in the scope of a cytometric measurement it is safe to state that "the second cell/population contains more of that molecule than the first" and that "the mean relative fluorescence increases from 7000 to 14,000" in a comparison of these two entities, but it cannot be concluded that the amount of molecules doubles. This is similar to our daily temperature readings in Celsius or Fahrenheit: 30 °C is not "twice as hot as 15 °C". Cytometric measurements belong to the interval scale and are thus to be considered quantitative.

Image analysis options can provide unlimited applications depending on the platform and among others may enable basic single cell analysis, dot detection, cellular co-expression as well as subcellular co-localization analysis, meta structure detection, multiplexed high-content phenotyping, proximity measurements, structural tracing (e.g., neurites and/or axons), particle and/or single cell tracking, as well as the analysis of spatial relationships for next-generation digital pathology [41,43]. A representative example of high-dimensional data analysis and the power of these platforms is shown in Figure 1.

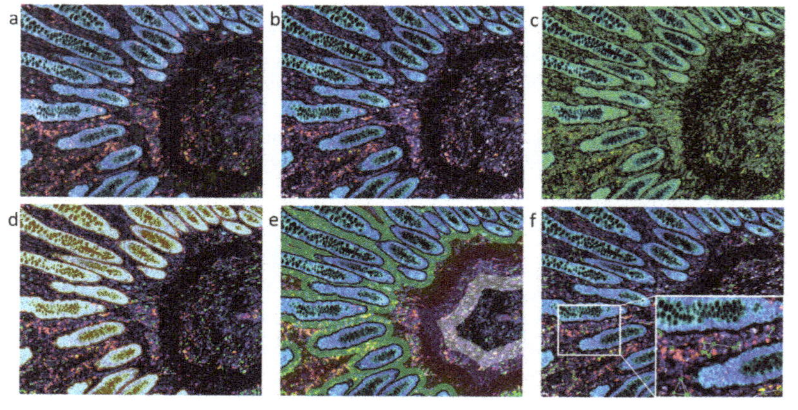

Figure 1. A representative example of high-dimensional automated tissue cytometry shown on a colon sample stained for seven markers. (**a**) Original multicolor immunofluorescence image data set acquired by a multispectral imaging technology. Nuclei stained by 4′,6-diamidino-2-phenylindole (DAPI) in blue; immune markers/immune checkpoint markers CD4 in green/PD-L1 in yellow/PD1 in red/CD68 in pink/CD8 in orange; pan-cytokeratin marker in turquoise. As this raw data image contains overlapping emission signals from the fluorochromes, the colors appear mixed. (**b**) Image with clearly separated fluorescent signals obtained by a mathematical procedure referred to as spectral unmixing. (**c**) Nuclei detection, highlighted by the green contour mask shown in overlay to the original image. (**d**) Metastructure detection of epithelial cells, highlighted in orange overlay. (**e**) Proximity measurements in relation to detected metastructures with various distance zones highlighted by different colors. (**f**) Analysis of spatial connections among and between single cells of a specific cellular phenotype highlighted by a green mask and white connecting lines. The images were provided by and analyzed using TissueGnostics' image cytometry solution StrataQuest.

4. Role of Machine Learning

A big fundamental improvement step in recent years elevating the next-generation digital pathology approach is the integration of artificial intelligence (AI) algorithms for pattern recognition into the image analysis/image cytometry process [44]. Over the past few years, these AI tools have become more robust, and with only minimal user input can be applied to automatically detect objects such as nuclei and specific structures as well as for the classification of various anatomical tissue entities within an entire digitized slide [44,45].

Understanding molecular and cellular interdependencies quickly leads to complex questions, which require the elaboration of extensive algorithms and enormous amounts of computing power to get to an answer. While machine learning has been known to have great potential in this field for many decades, in the recent past it has advanced greatly in its practical use due to the availability of powerful computer technology, in particular parallel computing on multiple CPUs and/or CPU cores as well as due to new software tools, programming languages, and advanced machine learning techniques, which have made the technologies much easier to use without the requirement of advanced theoretical knowledge [44,45]. By engaging state-of-the-art technologies, computer scientists and engineers try to generate models that can provide answers to the complex problems given by nature. Machine learning's power resides in its robustness in generating customized models designed to solve (very) specific problems. [44]. Machine learning models are generated by learning on examples consisting in observations. Current techniques of machine learning comprise supervised, unsupervised, transfer, federated, and reinforcement learning [45].

In the case of supervised machine learning, the observations are tagged to a class by a human expert, and therefore the model efficiency is strictly related to the quality of the training data set used. An optimal training data set should cover a wide enough range of variability expected in the real-world data, for example a well annotated slide. Failing to do so can increase the possibility of misclassifications. [46].

Unsupervised learning refers to a machine learning method where the algorithm learns from examples without being able to refer to predefined target values or classes (untagged/unlabeled data). The algorithm tries to identify patterns by creating an internal representation of the data and looks for density probabilities (e.g., clustering analysis). This method is suited to search for patterns which are not obvious, or are difficult to identify even for/by the human eye [47].

The transfer machine learning method is characterized by the fact that an already trained algorithm can be used to answer different, but related questions. It means an existing trained model can be adapted/tweaked to solve new tasks without the need to train a new model from scratch [48].

In the federated machine learning method, the algorithm learns from data spread/stored on multiple devices. Federated machine learning is similar to distributed learning, but the focus is on training on heterogeneous data and not on parallelization. No training data information is shared between the devices as part of the learning process [49].

Another machine learning method, enforcement learning, is based on an algorithm that needs to take optimal decisions based on the new data presented and the cumulated experience (knowledge). The learning process is continuous; each decision taken by the algorithm is labeled using a system with rewards and punishments. The aim is to solve the task by maximizing the cumulative positive feedback [50]. Precisely, the step-by-step development in machine learning aims towards a human-like learning, in the sense that humans learn from existing experience even in unrelated sectors and can transfer knowledge to new arising tasks rather than start from the basics, which is, however, still the case in machine learning.

Due to the versatile range of applications of next-generation digital pathology discussed in the following section, these platforms (with or without AI) can be seen as a crucial part of precision medicine by providing a solid and fully automated tool for the gaining of novel information on the pathology of specific diseases, identification of novel predictive and prognostic biomarkers, as well as targets for therapy [37,38,44].

5. Current Applications of Next-Generation Digital Pathology

5.1. RNA In Situ Hybridization (ISH)

In clinical settings, a routinely used method to measure RNA is real-time PCR [51]. However, this grind-and-bind technique is unable to visualize the individual cell signals in their original context, and is prone to becoming contaminated by unintended cell and tissue types and masking the different cellular subpopulations and phenotypes in the heterogenous TME [6,52]. Next-generation sequencing and single-cell sequencing technologies can detect RNA expression at the single cell level, but dissociation from their native setting deprives the data related to their spatial relationship [53]. With the latest developments in RNA ISH, multiple approaches came into play such as non-isotopic fluorescently labeled ISH (fluorescence in situ hybridization—FISH) and biotin or hapten labeled nucleic acid probes (chromogenic in situ hybridization—CISH) to gather spatial data [52,54–57]. These methods opened a new data dimension, supporting localization and quantitation of target RNA in single cells to detect precise RNA expression in specific cell types [52,58]. However, these techniques only allow a restricted number of labels to be integrated into the probes, leading to reduced sensitivity of expression for most of the genes [52]. Due to a high possibility of cross-hybridization and non-specific binding in complicated tumors, the signal-to-noise ratio is constrained, and extreme technical complication limits the performance of these methods [52,58]. In Figure 2, a representative example of the automated quantitative analysis of FISH and RNA ISH is shown.

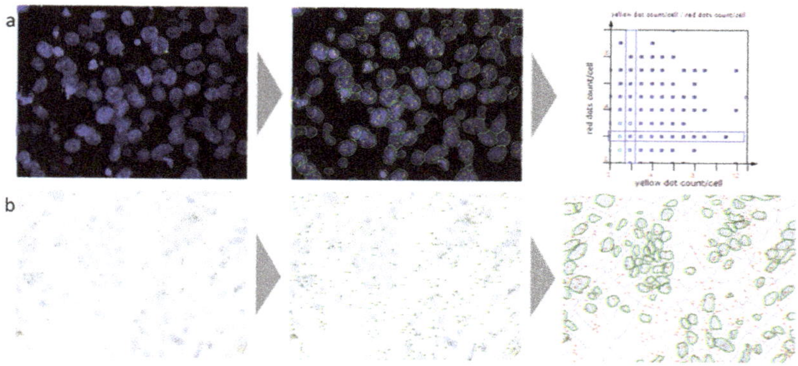

Figure 2. A representative example of automated analysis of fluorescence in situ hybridization (FISH) and RNA in situ hybridization (ISH) stained cells using a next-generation digital pathology platform. (**a**) FISH staining (blue, nuclei stained for 4′,6-diamidino-2-phenylindole (DAPI); red and yellow dots, FISH probes); on the left the original image is shown, in the middle the corresponding analyzed image including cell and dot detection mask, and on the right the analyzed data visualized in a scattergram. (**b**) RNAscope staining (blue, nuclei stained for hematoxylin; brown, RNAscope staining); on the left the original image is shown, in the middle the original image overlaid with the detected dot mask, and on the right the original image overlaid with the nuclei mask, the cellular mask, and the identified dot mask. Both images were provided by and analyzed using TissueGnostics' image cytometry solution StrataQuest.

RNAscope by Advanced Cell Diagnostics Inc., Hayward, CA (ACD) has presented the most pragmatic method that overcomes these limitations of traditional RNA ISH by a unique probe design and an advanced signal amplification system [52,59]. This technology excels due to its specificity, sensitivity, low turnaround time, and robustness in a wide range of applications across various disciplines including infectious diseases, neuroscience, cell or gene therapy, and single-cell transcriptomic profiling in cancer [52,60–64]. In the TME, RNAscope has prominent advantages such as spatially mapping a cell atlas [65,66], visualizing and characterizing gene signatures and generating the immune landscape, and even identification of novel cell subtypes [67,68], classifying and identifying highly heterogeneous and immunotherapeutic cell types [69,70], and identification and characterization of a gene signature of stem cells [71–73] and circulating tumor cells [74,75] as well as analyzing or predicting their response to drug treatments [76,77]. Compared with a one-probe RNA ISH hybridization system, the possibility of nonspecific amplification in RNAscope is considerably low since it implies a double-probe independent hybridization system and improves the sensitivity and the signal-to-noise ratio, allowing better quantification of RNA expression [52,78].

The RNAscope method allows robust detection of mRNA, long non-coding as well as microRNAs [57,79–82], and multiple gene transcripts generated by alternative splicing [83,84] simultaneously in fresh-fixed, fresh-frozen, and formalin-fixed paraffin-embedded (FFPE) clinical specimens, revealing the full potential of RNA [85]. For example, the expression of a majority of androgen receptor (AR) splice variants other than the full-length AR variant remains unclear in prostate cancer progression. RNAscope has been proposed to be a capable technique for detecting expression and localization of splice variants by designing probes specifically to target distinct splice variants. For example, AR and AR-V7 expression have been detected in FFPE prostate tumors by RNAscope where AR expression was found to be 3-fold higher in primary tumor cells compared with benign glands, while AR-V7 expression was higher in metastatic castration-resistant prostate cancer than in primary prostatic tissues [84].

Emerging new therapeutic strategies broadly target both cellular and non-cellular components of the TME more than ever, by various therapies such as immune checkpoint blockade therapy, dendritic cell vaccination, and antiangiogenic therapy [86]. Detection of RNA targets in the TME that are involved in tumor immunotherapy with the RNAscope assay can facilitate these therapies predominantly. RNAscope applications enable the determination of localization of specific immune cell types (i.e., cytotoxic lymphocytes and regulatory T cells) in the TME [87], spatial relationships between different cell types in the TME [88], and immune activation state and function of tumor-infiltrating immune cells in the TME [89,90]. For example, Monte et al., using RNAscope assay, reported that infiltrating basophils in the TME regulate tumor-promoting Th2 inflammation and reduce survival in pancreatic cancer patients [89]. Besides, this technique is an attractive strategy to determine cell type-specific expression of immune checkpoint markers [91] and differentiate activated CAR+ T cells from endogenous T cells [5]. RNAscope's aptitude to precisely identify the cellular sources of secreted proteins (e.g., cytokines and chemokines) is a distinct benefit since although the mRNA will always localize in the cells of origin, secreted proteins tend to dilute and diffuse in the intercellular space [67,87,92]. Besides, RNAscope provides valuable information on the differentiation of paracrine and autocrine signaling, which aids in the classification of subtypes of several cancers [93]. A dual gene analysis approach with RNAscope has been utilized for simultaneous detection of CD44+ cells and PD-L1 in head and neck squamous cell carcinoma, which found that CD44+ in the TME induces expression of PD-L1, thus subsequently suppressing T cell-mediated immunity in the TME [94]. The localization and quantification of multi-RNA from several genes simultaneously by RNAscope provide greater time saving and significant results from a single feasible technique. However, rapid mRNA translation and RNA degradation in cells can affect RNAscope applications, and thus BaseScope, a subfield of RNAscope, has been recommended for short RNA targets of 50–300 nucleotides [95]. Instead of using 20 probe pairs, BaseScope utilizes short 1–6 probe pairs to target small regions of RNA more effectively. Thus BaseScope is a successful method to determine the expression and quantification of small nucleolar RNAs (snoRNAs), microRNAs, and the RNAs which have a high potential of degradation and transient expression in the TME [95].

The newest approach of RNAscope, in combination with IHC and called dual RNAscope ISH/IHC, has proven to offer an ideal platform to generate more reliable data that can be used to study gene expression signatures at the RNA and protein level with spatial and single-cell resolution in complex TME [5]. This allows correlation of both RNA and protein expression in a single slide, simultaneously validating antibody specificity [78,96–98]. For example, combined detection of HPV RNA by RNAscope and Cdc2 protein expression by IHC has been useful to predict the prognosis of oropharyngeal squamous cell carcinoma patients. Even more, the results conclude that the sensitivity of RNAscope was higher than that of PCR reverse dot hybridization [98]. The automated RNAscope is a significant advancement over manual RNAscope and improves the clinical advantage by allowing more samples to be analyzed in a standardized way simultaneously with less time, less inter-user variability, and less manpower in an observer-independent manner [86]. The method has proven consistent and provides reproducible results in quantifying transcript levels. Overall, the spatial resolution presented by the RNAscope method brings a novel dimension to precise localization of target RNA in single cells and allows localization and quantitation of RNA expression in specific cell types in the TME [86].

5.2. Assessment of the Tumor Immune Microenvironment

One of the most promising fields in biomarker and therapy target detection in oncology is dedicated to the exploration of the patient-specific immune contexture in situ with conventional and multiplexing IF and IHC staining techniques in combination with automated quantification [14].

One prominent approach for immune cell assessment within a particular tumor tissue, colorectal cancer (CRC), was developed by the group of Galon et al., where they successfully established a patient stratification strategy based on the detection/identification of T cell populations within the tumor core and the invasive margin named Immunoscore (ratio of the markers CD3 and CD45RO, CD3 and CD8, or CD8 and CD45RO). It is currently undergoing evaluation/implementation as a routine parameter for prognostic and predictive diagnosis in clinics for colon cancer [99,100]. To demonstrate its power the group of Pages et al. conducted a large-scale study, where his group assessed the Immunoscore by using a digital pathology method of a large patient cohort (n = 2681 CRC patients), aligned it with clinical pathological data, and thereby was able to show the power of the Immunoscore in the prognosis of survival prediction and treatment response in CRC patients [101]. In order to provide a representative (yet not complete) overview of recent applications, Table 2 shows further examples of studies using conventional and/or multiplexing IF and/or IHC staining techniques in which next-generation digital pathology was the central method for the quantification of various immune cell markers/populations in different cancer types and aligned with clinicopathological parameters.

Table 2. Studies using next-generation digital pathology for the assessment of the tumor immune microenvironment.

Cancer Type	Markers	Scanner/Microscope	Quantification System	Reference
Breast cancer	CD4, CD8, Foxp3	Olympus BX51 (Olympus, Tokyo, Japan)	UTHSCSA Image Tool (University of Texas Health Science Center at San Antonio, San Antonio, TX, USA)	[102]
Breast cancer	CD4, CD8, CD3, CD20, FOXP3, CD68	Leica SCN400 F (Leica Biosystems Inc., Richmond, IL, USA)	ImageJ software (NIH, Bethesda, MD, USA)	[103]
Breast cancer	PD-L1	Aperio AT2 Scanner (Leica Biosystems Inc., Richmond, IL, USA)	QuPath (University of Edinburgh, Edinburgh, UK)	[104]
Breast cancer	CD8	ScanScope XT (Aperio Technologies, Vista, CA, USA)	HALO (Indica Labs, Albuquerque, NM, USA)	[105]
Breast cancer	CD3, CD20, Foxp3	NanoZoomer (Hamamatsu Photonics, Hamamatsu City, Japan); Panoramic 250 Flash (3Dhistech, Budapest, Hungary)	ImageJ software (NIH, Bethesda, MD, USA)	[106]
Breast cancer	CD3, CD8, CD20	NanoZoomer (Hamamatsu Photonics, Hamamatsu City, Japan)	ImageJ software (NIH, Bethesda, MD, USA)	[107]
Breast cancer	CD4, CD68, CD8, FOXP3, PD-L1	Vectra 3 (PerkinElmer, Waltham, MA, USA)	inForm (PerkinElmer, Waltham, MA, USA)	[108]
Breast cancer	CD4, CD8, FOXP3, CD20, CD33, PD-1	Vectra 3 (Akoya Biosciences, Marlborough, MA, USA)	inForm (Akoya, Marlborough, MA, USA)	[109]
CRC	CD3, CD8	n.s.	Developer XD (Definiens, Munich, Germany)	[101]
CRC	CD3, CD8	VENTANA iScan HT (Roche, Basel, Switzerland)	automated image analysis algorithm	[110]

Table 2. *Cont.*

Cancer Type	Markers	Scanner/Microscope	Quantification System	Reference
CRC	CD8	Aperio XT Scanner (Leica Biosystems Inc., Richmond, IL, USA)	HALO (Indica Labs, Albuquerque, NM, USA)	[105]
CRC	CD3, CD8	Zeiss Axio Scan.Z1 (Zeiss, Jena, Germany)	HALO (Indica Labs, Albuquerque, NM, USA)	[111]
CRC	CD3, CD4, CD8, CD45RO, FOXP3, Granzyme B, CD15, CD20, S100, CD68, IL17, CD57,	microscope (Leica, Wetzlar, Germany)	TMAJ software (Johns Hopkins University, Baltimore, MD, USA)	[112]
CRC	FoxP3, CD8, CD66b, CD20, CD68	Vectra 3 (PerkinElmer, Waltham, MA, USA)	inForm (PerkinElmer, Waltham, MA, USA)	[113]
CRC	SOX2, CD3, CD8 FoxP3, ALDH1, CD44v6, CD133, Lgr5, PD-L1	Aperio XT Scanner (Leica Biosystems Inc., Richmond, IL, USA)	Aperio Imagescope (Leica Biosystems Inc., Richmond, IL, USA)	[114]
CRC	CD8, CD11c, PD-L1	Pannoramic MIDI II (3Dhistech, Budapest, Hungary)	StrataQuest (TissueGnostics, Vienna, Austria)	[115]
CRC	CD8, CD4, CD20, Foxp3, CD45RO,	Vectra Polaris (PerkinElmer, Waltham, MA, USA)	inForm (PerkinElmer, Waltham, MA, USA)	[116]
CRC, CRCLM	CD20, CD3, Ki67, CD27	TissueFAXS PLUS (TissueGnostics, Vienna, Austria)	HistoQuest, TissueQuest (TissueGnostics, Vienna, Austria)	[117]
CRC, CRCLM	CD8, Foxp3, CD68, CD31	ScanScope (Aperio Technologies, Vista, CA, USA)	GENIE (Aperio Technologies, Vista, CA, USA)	[99]
CRCLM	CD45, CD20	TissueFAXS PLUS (TissueGnostics, Vienna, Austria)	HistoQuest, TissueQuest (TissueGnostics, Vienna, Austria)	[118]
CRCLM	CD3, CD4, CD8, CD20, CD68	NanoZoomer (Hamamatsu Photonics, Hamamatsu City, Japan)	Visilog 9.0 software (Noesis, Saclay, France)	[119]
CRCLM	CD3, CD8, CD45RO, Foxp3, CD20	NanoZoomer (Hamamatsu Photonics, Hamamatsu City, Japan)	Developer XD (Definiens, Munich, Germany)	[120]
Gastric cancer	PD-L1, CD8	digital slide scanner (3Dhistech, Budapest, Hungary); TissueFAXS (TissueGnostics, Vienna, Austria)	QuantCenter (3Dhistech, Budapest, Hungary); TissueQuest (TissueGnostics, Vienna, Austria)	[121]
Gastric cancer	CD68, CD163, CD3, MPO, Foxp3.	ScanScope CS (Aperio Technologies, Vista, CA, USA)	ImageScope (Aperio Technologies, Vista, CA, USA)	[122]
Gastric cancer	CD3, CD4, CD8, PD-1	ScanScope CS2 (Aperio Technologies, Vista, CA, USA)	ImageScope (Aperio Technologies, Vista, CA, USA)	[122]
Gastric cancer	CD8, FoxP3	ScanScope XT (Aperio Technologies, Vista, CA, USA)	image analysis system—ScanScope XT (Aperio Technologies, Vista, CA, USA)	[123]
Gastric cancer	CD8, Foxp3	n.s.	Aperio image analysis system (Leica Biosystems Inc., Richmond, IL, USA)	[124]
Gastric cancer	CD8, Foxp3, CD3, CD56	Vectra Multispectral Imaging System version 2 (PerkinElmer, Waltham, MA, USA)	inForm (PerkinElmer, Waltham, MA, USA)	[125]
Gastric and esophageal cancer	CD3, CD8	n.s.	HALO (Indica Labs, Albuquerque, NM, USA	[126]

Table 2. *Cont.*

Cancer Type	Markers	Scanner/Microscope	Quantification System	Reference
Gastric cancer and metastasis	PD-L1	n.s.	Aperio Imagescope IHC Membrane Image Analysis software (Aperio Technologies, Vista, CA, USA)	[127]
HCC	CD3, CD8	n.s.	ImagePro Plus (Media Cybernetics, Rockville, MD, USA)	[128]
HCC	CD3, CD8	Nikon E600 (Nikon, Tokyo, Japan);	ImageJ software (NIH, Bethesda, MD, USA)	[129]
HCC	CD3, CD15, CD20, CD23, CD68, Foxp3, LTß	Ariol SL-50 (Applied Imaging)	Image analysis system (Applied Imaging)	[4]
HCC	CD3, CD8, PD-1, TIM3	Vectra 3 (PerkinElmer, Waltham, MA, USA)	inForm (PerkinElmer, Waltham, MA, USA)	[130]
HCC	CD3, CD4, CD8, CD20, CD27, CD40, CD38, CD56, CD68, CD138, S100, Granzyme B, Ki67	Mantra (PerkinElmer, Waltham, MA, USA)	ImagePro Plus (Media Cybernetics, Rockville, MD, USA)	[131]
HCC	CD3, CD8, CD45RO,	n.s.	ImagePro Plus (Media Cybernetics, Rockville, MD, USA)	[132]
HCC	FoxP3, CD4, CD8, CD34	Olympus BX51 (Olympus, Tokyo, Japan)	ImagePro Plus (Media Cybernetics, Rockville, MD, USA)	[133]
HNSCC	FOXP3, CD8	n.s.	Visiopharm image analysis software (Visiopharm, Copenhagen, Denmark)	[134]
HNSCC	CD3, CD8	Aperio AT2 scanner (Leica Biosystems Inc., Richmond, IL, USA)	StrataQuest (TissueGnostics, Vienna, Austria)	[135]
Melanoma	PD-L1	Philips Ultra Fast Scanner 300 (Philips, Amsterdam, Netherlands)	HALO (Indica Labs, Albuquerque, NM, USA	[136]
Melanoma	CD20	TissueFAXS (TissueGnostics, Vienna, Austria)	HistoQuest (TissueGnostics, Vienna, Austria)	[137]
Melanoma	CD3, CD8, CD68, SOX10, Ki67	Mantra (PerkinElmer, Waltham, MA, USA)	inForm (PerkinElmer, Waltham, MA, USA)	[138]
Melanoma	CD19, CD20, CD27, CD38, CD138, CD5, CD8, Foxp3, CD4, CD69, CD103, CD45RO, CXCL13, CD21, CD23, Bcl6	Vectra Multispectral Imaging System version 2 (PerkinElmer, Waltham, MA, USA)	inForm (PerkinElmer, Waltham, MA, USA)	[139]
NSCLC	CD8, PD-1	Philips Ultra Fast Scanner 300 (Philips, Amsterdam, Netherlands)	HALO (Indica Labs, Albuquerque, NM, USA	[140]
NSCLC	CD8	NanoZoomer (Hamamatsu Photonics, Hamamatsu City, Japan)	Calopix software (TRIBVN Healthcare, Paris, France)	[141]
NSCLC	PD-L1, TIM, CD3, CD4, CD8, CD57, granzyme B, CD45RO, PD-1, FOXP3	Aperio AT scanner (Leica Biosystems Inc., Richmond, IL, USA)	Aperio GENIE (Leica Biosystems Inc., Richmond, IL, USA)	[142]
NSCLC	CD8, CD4, FOXP3, CD163, CCL17, IL-13	Vectra Automated Quantitative Pathology Imaging System (PerkinElmer, Waltham, MA, USA)		[143]

Table 2. *Cont.*

Cancer Type	Markers	Scanner/Microscope	Quantification System	Reference
NSCLC	CD3, CD4, CD8, CD57, granzyme B, CD45RO, PD-1, FOXP3, CD68	Aperio AT scanner (Leica Biosystems Inc., Richmond, IL, USA)	Aperio GENIE (Leica Biosystems Inc., Richmond, IL, USA)	[144]
NSCLC	CD4, CD20, CD8, Foxp3	NanoZoomer (Hamamatsu Photonics, Hamamatsu City, Japan)	Tissue Studio (Definiens, Munich, Germany)	[145]
NSCLC	CD68, CD163, PD-L1,	Mantra (PerkinElmer, Waltham, MA, USA)	inForm (PerkinElmer, Waltham, MA, USA)	[146]
NSCLC	CD8, CD4, Foxp3, CD68	Vectra Multispectral Imaging System (PerkinElmer, Waltham, MA, USA)	inForm (PerkinElmer, Waltham, MA, USA)	[147]
NSCLC	CD3, CD8, Foxp3	ScanScope CS (Aperio Technologies, Vista, CA, USA)	GENIE (Aperio Technologies, Vista, CA, USA)	[148]
NSCLC	CD8, PD-L1	Aperio AT scanner (Leica Biosystems Inc., Richmond, IL, USA)	Developer XD (Definiens, Munich, Germany)	[149]
pulmonary squamous cell carcinoma	CD8, PD-1	ScanScope (Aperio Technologies, Vista CA, USA)	ImageScope (Aperio Technologies, Vista, CA, USA)	[150]
pulmonary squamous cell carcinoma	CD20, CD21, CD23, PNAD, DC-LAMP	Vectra 3 (PerkinElmer, Waltham, MA, USA)	inForm (PerkinElmer, Waltham, MA, USA)	[151]
Oral squamous cell cancer	CD3, CD8, FoxP3, CD163, PD-L1	Vectra (PerkinElmer, Waltham, MA, USA)	inForm (PerkinElmer, Waltham, MA, USA)	[152]
Ovarian cancer	CD8, MHC I, FAP ISH	Panoramic 250 (3Dhistech, Budapest, Hungary),	Developer XD (Definiens, Munich, Germany)	[153]
Ovarian cancer	CD8	TissueFAXS (TissueGnostics, Vienna, Austria)	HistoQuest (TissueGnostics, Vienna, Austria)	[154]
Ovarian cancer	CD8, CD45RO, CD68	Panoramic Flash (3Dhistech, Budapest, Hungary)	Tissue Studio (Definiens, Munich, Germany)	[155]
Ovarian cancer	CD4, CD8, CD20	Aperio scanner (Leica Biosystems Inc., Richmond, IL, USA)	ImageScope (Aperio Technologies, Vista, CA, USA)	[156]
Ovarian cancer	CD8	Vectra (PerkinElmer, Waltham, MA, USA)	inForm (PerkinElmer, Waltham, MA, USA)	[157]
Ovarian cancer	CD8, CD103	TissueFAXS (TissueGnostics, Vienna Austria)	Fiji, Image J software (NIH, Bethesda, MD, USA)	[158]
Ovarian cancer	CD3, CD4, CD8	n.s.	CD3 Quantifier (VM Scope, Berlin, Germany)	[159]
Pancreatic cancer	CD3, CD8, CD4, Foxp3, CK8	Vectra Multispectral Imaging System version 2 (PerkinElmer, Waltham, MA, USA)	Nuance Image Analysis software; inForm (PerkinElmer, Waltham, MA, USA)	[160]
Pancreatic cancer	DC-LAMP, FoxP3, CD68, CD3, CD8, CD4, CD20	Panoramic Flash (3Dhistech, Budapest, Hungary)	ImageJ software (NIH, Bethesda, MD, USA)	[161]
Pancreatic cancer	CD20, CD8, PD1	dotSlide (Olympus, Tokyo, Japan)	ad hoc software	[162]
Pancreatic cancer	CD8	NanoZoomer (Hamamatsu Photonics, Hamamatsu City, Japan)	HALO (Indica Labs, Albuquerque, NM, USA	[163]
Pancreatic cancer	CD8, PD-L1, CD44, CD133	TissueFAXS (TissueGnostics, Vienna, Austria)	TissueQuest (TissueGnostics, Vienna, Austria)	[164]

Table 2. *Cont.*

Cancer Type	Markers	Scanner/Microscope	Quantification System	Reference
Pancreatic cancer	CD3	NanoZoomer (Hamamatsu Photonics, Hamamatsu City, Japan)	Tissue Studio (Definiens, Munich, Germany)	[165]
Pancreatic cancer	CD3, CD8, CD20, CD66b	n.s.	ImageJ software (NIH, Bethesda, MD, USA)	[166]
Pancreatic cancer	CD3, CD8	Aperio AT scanner (Leica Biosystems Inc., Richmond, IL, USA)	ImageJ software (NIH, Bethesda, MD, USA)	[167]
Prostate cancer	CD3, CD8, CD20, CD56, CD68, Foxp3	ScanScope XT(Aperio Technologies, Vista, CA, USA)	ImageScope (Aperio Technologies, Vista, CA, USA)	[168]
Prostate cancer	CD20	ScanScope XT (Aperio Technologies, Vista, CA, USA)	ImageScope (Aperio Technologies, Vista, CA, USA)	[169]
Prostate cancer	CD3, CD8, Foxp3	NanoZoomer (Hamamatsu Photonics, Hamamatsu City, Japan)	Aperio Digital Pathology software (Leica Biosystems Inc., Richmond IL, USA)	[170]
Clear cell renal cell carcinoma	CD8, PD-1, LAG-3, PD-L1, PD-L2	NanoZoomer (Hamamatsu Photonics, Hamamatsu City, Japan)	Calopix software (TRIBVN Healthcare, Paris, France)	[171]

CRC, colorectal cancer; CRCLM, colorectal cancer metastasis in the liver; HCC, hepatocellular carcinoma; HNSCC, head and neck squamous cell carcinoma; NSCLC, non-small cell lung cancer; n.s., not specified.

The examples summarized in Table 2, as well as the example shown in Figure 3 from Desbois et al. [153] show the immense power of the applications of this technique utilizing next-generation digital pathology for the assessment of the immune tumor microenvironment. In order to integrate the Immunoscore or other immune cell screening strategies also into clinical research, such fully automated next-generation digital pathology platforms should be implemented into the process of quantification of the rate of infiltration of various immune cell populations/markers. Ongoing clinical studies are aiming at the integration of such platforms in combination with the staining of a set of immune-related biomarkers including main subpopulation markers and immune checkpoint markers [14].

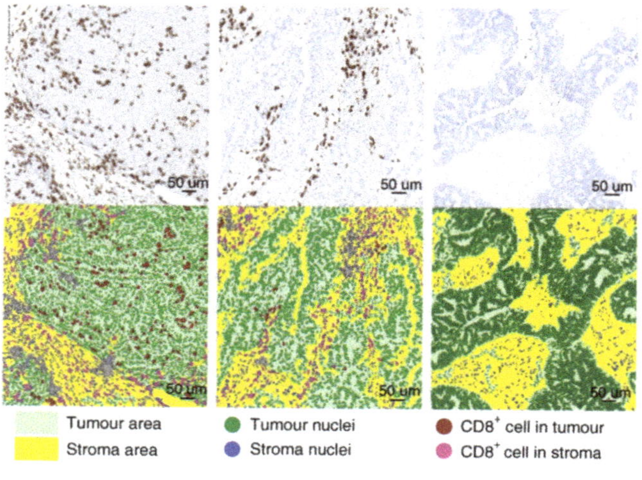

Figure 3. Analysis of the tumor immune microenvironment using next-generation digital pathology. A representative example of the automated detection of CD8+ immune cells within the tumor microenvironment of ovarian cancer by Developer XD (Definiens, Munich, Germany). Figure adapted from Desbois et al., 2020 [153].

To sum up, the need to automatically assess immune cell markers in situ, as well as analyzing spatial relationships, and thereby providing a better understanding of various immune cells populations and their interactions, is crucial for the detection of novel predictive and prognostic biomarkers as well as for clinical therapy strategy.

5.3. Detection of Blood Vessels

Neoangiogenesis and the resulting vascularization are equally required by the tumor, as in healthy tissues. In both types of tissue, normal and tumor, cell survival and proliferation depend on oxygen and nutrition supply as well as on removal of carbon dioxide and metabolic wastes. In contrast to regulated neoangiogenesis in healthy tissues, tumor angiogenesis is characterized by an uncontrolled, ineffective, often incomplete (and therefore leaky) growth of new blood vessels within the tumor tissue in order to supply the tumor mass with oxygen and nutrition [172]. However, the in situ assessment of the density of blood vessels stained by specific markers such as CD31 or CD34 was shown to correlate with the aggressiveness of the tumor in a variety of tumor types such as CRC, breast cancer, gastric cancer, and small cell and non-small cell lung cancer [173]. Furthermore, specific therapies such as neutralizing antibodies targeting anti-vascular endothelial growth factor are widely used in several cancer types [174]. However, inhibition of vessel growth has only been shown to provide limited or even no long-term improvement for cancer types including hepatocellular carcinoma and CRC [175,176]. However, the use of different non-standardized methods for detection and quantitation of blood vessel density leads to contradicting data in terms of influence on patient survival [177]. Therefore, the unbiased automated quantification of blood vessels could help to identify patient groups that would benefit from anti-angiogenic therapies.

Summarized in Table 3 are studies where next-generation digital pathology was used to detected blood vessels/blood vessel densities. Thereby we want to emphasize that the next-generation digital pathology approach is highly versatile and can be applied to various research needs and questions, not only to single cell detection or dot (RNA ISH) detection but also for the analysis of more complex structures such as blood vessels.

Table 3. Studies using next-generation digital pathology for the quantification of blood vessels.

Cancer Type	Markers	Scanner/Microscope	Quantification System	References
Breast cancer	CD34	Olympus BX41 (Olympus, Tokyo, Japan)	Cell D software (Olympus, Tokyo, Japan)	[178]
Breast cancer	CD34	NanoZoomer (Hamamatsu Photonics, Hamamatsu City, Japan)	Slidepath Image Analysis system (Leica Biosystems Inc., Richmond, IL, USA)	[179]
Breast cancer	CD34	TissueFAXS (TissueGnostics, Vienna, Austria)	HistoQuest (TissueGnostics, Vienna, Austria)	[180]
Breast cancer metastasis	CD31	Panoramic 250 (3Dhistech, Budapest, Hungary)	Visiopharm image analysis software (Visiopharm, Copenhagen, Denmark)	[181]
CRC	CD31	Mirax slide scanner system (3Dhistech, Budapest, Hungary)	Image J software (NIH, Bethesda, MD, USA)	[182]
CRC	CD31	TissueFAXS (TissueGnostics, Vienna, Austria)	StrataQuest (TissueGnostics, Vienna, Austria)	[183]

Table 3. *Cont.*

Cancer Type	Markers	Scanner/Microscope	Quantification System	References
ESCC	CD31	TissueFAXS (TissueGnostics, Vienna, Austria)	HistoQuest, TissueQuest (TissueGnostics, Vienna, Austria)	[184]
Human tumor	CD31, CD34	Aperio (Leica Biosystems Inc., Richmond, IL, USA)	Fiji, Image J software (NIH, Bethesda, MD, USA)	[185]
Melanoma	CD31	Aperio CS Scanner (Leica Biosystems Inc., Richmond, IL, USA)	Aperio image analysis system (Leica Biosystems Inc., Richmond, IL, USA)	[186]
Pancreatic cancer	CD31	n.s.	The Ariol™ image analysis system (Genetix, New Milton, England)	[187]
Renal cancer	CD34	Zeiss Axio Scan.Z1 (Zeiss, Jena, Germany)	Developer XD, Tissue Studio (Definiens, Munich, Germany)	[188]
Rectal cancer	CD34	ScanScope CS (Aperio Technologies, Vista, CA, USA)	ImageScope (Aperio Technologies, Vista, CA, USA)	[189]
Tongue cancer	PNAd	ScanScope T3 (Aperio Technologies, Vista, CA, USA)	Image J software (NIH, Bethesda, MD, USA)	[190]

CRC, colorectal cancer; n.s., not specified; ESCC, esophageal squamous cell carcinoma.

6. Conclusions

Within this review, we show several application fields that contribute to next-generation digital pathology, including the analysis of RNA ISH, conventional and/or multiplexed immunophenotyping, and blood vessel detection in the tumor microenvironment. Due to new staining technologies that allow a higher number of markers, one of the new challenges is high-dimensional data mining, which needs to be addressed by next-generation digital pathology platform providers. Several platforms are available on the market tackling different kinds of requests, including slide scanning, management of a large amount of data, follow-up image analysis with integrated AI modules, as well as high-dimensional data mining. Next-generation digital pathology has the potential to elevate research and clinics by providing automated, unbiased, fast, reproducible, and therefore reliable image cytometry.

The integration of data obtained by automated analyses, referring to the levels of DNA, RNA—including non-coding RNA—and proteins, will allow the development of tools for research and diagnostics within the scope of precision medicine, i.e., with a focus on the molecular mechanisms involved in disease formation in individual patients rather than averaged cohorts, in which individual yet decisive details get lost. This is the expanding area of tissue cytometry—and the era of next-generation digital pathology.

Author Contributions: Conception and design, F.M., R.C.E. and J.B.; Writing—original draft, F.M., A.F., B.B., B.L. and R.N.; Writing—review and editing, R.C.E. and J.B. All authors have read and agreed to the published version of the manuscript.

Funding: This research received no external funding.

Institutional Review Board Statement: Not applicable.

Informed Consent Statement: Not applicable.

Conflicts of Interest: Some of the authors (F.M., R.C.E., B.B., R.N. and B.L.) are employees of TissueGnostics, which is a for-profit company.

References

1. World Health Organization. Available online: https://www.who.int/health-topics/cancer#tab=tab_1 (accessed on 7 January 2021).
2. Macconaill, L.E.; Garraway, L.A. Clinical implications of the cancer genome. *J. Clin. Oncol.* **2010**, *28*, 5219–5228. [CrossRef]
3. Whiteside, T.L. The tumor microenvironment and its role in promoting tumor growth. *Oncogene* **2008**, *27*, 5904–5912. [CrossRef]
4. Finkin, S.; Yuan, D.; Stein, I.; Taniguchi, K.; Weber, A.; Unger, K.; Browning, J.L.; Goossens, N.; Nakagawa, S.; Gunasekaran, G.; et al. Ectopic lymphoid structures function as microniches for tumor progenitor cells in hepatocellular carcinoma. *Nat. Immunol.* **2015**, *16*, 1235–1244. [CrossRef]
5. Dikshit, A.; Phatak, J.; Kernag, S.; Pimental, H.; Zong, H.; Todorov, C.; Hernandez, L.; Kim, J.; Zhang, B.; Anderson, C.; et al. Abstract 2706: Spatially resolve RNA and protein simultaneously in FFPE tumor samples by combining RNAscope in situ hybridization and immunohistochemistry assays. *Cancer Res.* **2020**, *80*, 2706. [CrossRef]
6. Morley-Bunker, A.; Pearson, J.; Currie, M.J.; Morrin, H.; Whitehead, M.R.; Eglinton, T.; Walker, L.C. Assessment of intra-tumoural colorectal cancer prognostic biomarkers using RNA in situ hybridisation. *Oncotarget* **2019**, *10*, 1425–1439. [CrossRef]
7. Hanahan, D.; Weinberg, R.A. Hallmarks of cancer: The next generation. *Cell* **2011**, *144*, 646–674. [CrossRef]
8. Perou, C.M.; Sorlie, T.; Eisen, M.B.; van de Rijn, M.; Jeffrey, S.S.; Rees, C.A.; Pollack, J.R.; Ross, D.T.; Johnsen, H.; Akslen, L.A.; et al. Molecular portraits of human breast tumours. *Nature* **2000**, *406*, 747–752. [CrossRef]
9. Zhan, F.; Huang, Y.; Colla, S.; Stewart, J.P.; Hanamura, I.; Gupta, S.; Epstein, J.; Yaccoby, S.; Sawyer, J.; Burington, B.; et al. The molecular classification of multiple myeloma. *Blood* **2006**, *108*, 2020–2028. [CrossRef]
10. Michaelevski, I.; Chikvashvili, D.; Tsuk, S.; Singer-Lahat, D.; Kang, Y.; Linial, M.; Gaisano, H.Y.; Fili, O.; Lotan, I. Direct interaction of target SNAREs with the Kv2.1 channel. Modal regulation of channel activation and inactivation gating. *J. Biol. Chem.* **2003**, *278*, 34320–34330. [CrossRef]
11. Waylen, L.N.; Nim, H.T.; Martelotto, L.G.; Ramialison, M. From whole-mount to single-cell spatial assessment of gene expression in 3D. *Commun. Biol.* **2020**, *3*, 602. [CrossRef]
12. Voith von Voithenberg, L.; Fomitcheva Khartchenko, A.; Huber, D.; Schraml, P.; Kaigala, G.V. Spatially multiplexed RNA in situ hybridization to reveal tumor heterogeneity. *Nucleic Acids Res.* **2020**, *48*, e17. [CrossRef]
13. Matos, L.L.; Trufelli, D.C.; de Matos, M.G.; da Silva Pinhal, M.A. Immunohistochemistry as an important tool in biomarkers detection and clinical practice. *Biomark Insights* **2010**, *5*, 9–20. [CrossRef]
14. Shakya, R.; Nguyen, T.H.; Waterhouse, N.; Khanna, R. Immune contexture analysis in immuno-oncology: Applications and challenges of multiplex fluorescent immunohistochemistry. *Clin. Transl. Immunol.* **2020**, *9*, e1183. [CrossRef]
15. Ecker, R.C.; Rogojanu, R.; Streit, M.; Oesterreicher, K.; Steiner, G.E. An improved method for discrimination of cell populations in tissue sections using microscopy-based multicolor tissue cytometry. *Cytom. Part A J. Int. Soc. Anal. Cytol.* **2006**, *69*, 119–123. [CrossRef]
16. Ecker, R.C.; Steiner, G.E. Microscopy-based multicolor tissue cytometry at the single-cell level. *Cytom. Part A J. Int. Soc. Anal. Cytol.* **2004**, *59*, 182–190. [CrossRef]
17. Parwani, A.V. Next generation diagnostic pathology: Use of digital pathology and artificial intelligence tools to augment a pathological diagnosis. *Diagn. Pathol.* **2019**, *14*, 138. [CrossRef]
18. Nagpal, K.; Foote, D.; Liu, Y.; Chen, P.-H.C.; Wulczyn, E.; Tan, F.; Olson, N.; Smith, J.L.; Mohtashamian, A.; Wren, J.H.; et al. Development and validation of a deep learning algorithm for improving Gleason scoring of prostate cancer. *NPJ Digit. Med.* **2019**, *2*, 48. [CrossRef]
19. Jiao, Y.; Li, J.; Qian, C.; Fei, S. Deep learning-based tumor microenvironment analysis in colon adenocarcinoma histopathological whole-slide images. *Comput. Methods Programs Biomed.* **2021**, *204*, 106047. [CrossRef]
20. Kwak, M.S.; Lee, H.H.; Yang, J.M.; Cha, J.M.; Jeon, J.W.; Yoon, J.Y.; Kim, H.I. Deep Convolutional Neural Network-Based Lymph Node Metastasis Prediction for Colon Cancer Using Histopathological Images. *Front. Oncol.* **2021**, *10*, 3053. [CrossRef]
21. Bilal, M.; Raza, S.E.A.; Azam, A.; Graham, S.; Ilyas, M.; Cree, I.A.; Snead, D.; Minhas, F.; Rajpoot, N.M. Novel deep learning algorithm predicts the status of molecular pathways and key mutations in colorectal cancer from routine histology images. *medRxiv* **2021**. [CrossRef]
22. Mittal, S.; Yeh, K.; Leslie, L.S.; Kenkel, S.; Kajdacsy-Balla, A.; Bhargava, R. Simultaneous cancer and tumor microenvironment subtyping using confocal infrared microscopy for all-digital molecular histopathology. *Proc. Natl. Acad. Sci. USA* **2018**, *115*, E5651–E5660. [CrossRef]
23. Wang, S.; Rong, R.; Yang, D.M.; Fujimoto, J.; Yan, S.; Cai, L.; Yang, L.; Luo, D.; Behrens, C.; Parra, E.R.; et al. Computational Staining of Pathology Images to Study the Tumor Microenvironment in Lung Cancer. *Cancer Res.* **2020**, *80*, 2056–2066. [CrossRef]
24. Diao, J.A.; Wang, J.K.; Chui, W.F.; Mountain, V.; Gullapally, S.C.; Srinivasan, R.; Mitchell, R.N.; Glass, B.; Hoffman, S.; Rao, S.K.; et al. Human-interpretable image features derived from densely mapped cancer pathology slides predict diverse molecular phenotypes. *Nat. Commun.* **2021**, *12*, 1613. [CrossRef]
25. Dieu-Nosjean, M.C.; Goc, J.; Giraldo, N.A.; Sautès-Fridman, C.; Fridman, W.H. Tertiary lymphoid structures in cancer and beyond. *Trends Immunol.* **2014**, *35*, 571–580. [CrossRef]

26. Tan, W.C.C.; Nerurkar, S.N.; Cai, H.Y.; Ng, H.H.M.; Wu, D.; Wee, Y.T.F.; Lim, J.C.T.; Yeong, J.; Lim, T.K.H. Overview of multiplex immunohistochemistry/immunofluorescence techniques in the era of cancer immunotherapy. *Cancer Commun.* **2020**, *40*, 135–153. [CrossRef]
27. Remark, R.; Lupo, A.; Alifano, M.; Biton, J.; Ouakrim, H.; Stefani, A.; Cremer, I.; Goc, J.; Regnard, J.F.; Dieu-Nosjean, M.C.; et al. Immune contexture and histological response after neoadjuvant chemotherapy predict clinical outcome of lung cancer patients. *Oncoimmunology* **2016**, *5*, e1255394. [CrossRef]
28. Glass, G.; Papin, J.A.; Mandell, J.W. SIMPLE: A sequential immunoperoxidase labeling and erasing method. *J. Histochem. Cytochem.* **2009**, *57*, 899–905. [CrossRef]
29. Remark, R.; Merghoub, T.; Grabe, N.; Litjens, G.; Damotte, D.; Wolchok, J.D.; Merad, M.; Gnjatic, S. In-depth tissue profiling using multiplexed immunohistochemical consecutive staining on single slide. *Sci. Immunol.* **2016**, *1*, aaf6925. [CrossRef]
30. Sun, Z.; Nyberg, R.; Wu, Y.; Bernard, B.; Redmond, W.L. Developing an enhanced 7-color multiplex IHC protocol to dissect immune infiltration in human cancers. *PLoS ONE* **2021**, *16*, e0247238. [CrossRef]
31. Blenman, K.R.M.; Bosenberg, M.W. Immune Cell and Cell Cluster Phenotyping, Quantitation, and Visualization Using In Silico Multiplexed Images and Tissue Cytometry. *Cytom. Part A J. Int. Soc. Anal. Cytol.* **2019**, *95*, 399–410. [CrossRef]
32. Lin, J.R.; Izar, B.; Wang, S.; Yapp, C.; Mei, S.; Shah, P.M.; Santagata, S.; Sorger, P.K. Highly multiplexed immunofluorescence imaging of human tissues and tumors using t-CyCIF and conventional optical microscopes. *eLife* **2018**, *7*. [CrossRef]
33. Gerdes, M.J.; Sevinsky, C.J.; Sood, A.; Adak, S.; Bello, M.O.; Bordwell, A.; Can, A.; Corwin, A.; Dinn, S.; Filkins, R.J.; et al. Highly multiplexed single-cell analysis of formalin-fixed, paraffin-embedded cancer tissue. *Proc. Natl. Acad. Sci. USA* **2013**, *110*, 11982–11987. [CrossRef]
34. Goltsev, Y.; Samusik, N.; Kennedy-Darling, J.; Bhate, S.; Hale, M.; Vazquez, G.; Black, S.; Nolan, G.P. Deep Profiling of Mouse Splenic Architecture with CODEX Multiplexed Imaging. *Cell* **2018**, *174*, 968–981.e915. [CrossRef]
35. Parra, E.R.; Francisco-Cruz, A.; Wistuba, I.I. State-of-the-Art of Profiling Immune Contexture in the Era of Multiplexed Staining and Digital Analysis to Study Paraffin Tumor Tissues. *Cancers* **2019**, *11*, 247. [CrossRef]
36. Du, Z.; Lin, J.R.; Rashid, R.; Maliga, Z.; Wang, S.; Aster, J.C.; Izar, B.; Sorger, P.K.; Santagata, S. Qualifying antibodies for image-based immune profiling and multiplexed tissue imaging. *Nat. Protoc.* **2019**, *14*, 2900–2930. [CrossRef]
37. Bhargava, R.; Madabhushi, A. Emerging Themes in Image Informatics and Molecular Analysis for Digital Pathology. *Annu. Rev. Biomed. Eng.* **2016**, *18*, 387–412. [CrossRef]
38. Ghaznavi, F.; Evans, A.; Madabhushi, A.; Feldman, M. Digital Imaging in Pathology: Whole-Slide Imaging and Beyond. *Annu. Rev. Pathol. Mech. Dis.* **2013**, *8*, 331–359. [CrossRef]
39. Fertig, R.M.; Sangueza, O.; Gaudi, S.; Gamret, A.C.; Cervantes, J.; Jukic, D.M. Whole Slide Imaging. *Am. J. Dermatopathol.* **2018**, *40*, 938–939. [CrossRef]
40. Hamilton, P.W.; Bankhead, P.; Wang, Y.; Hutchinson, R.; Kieran, D.; McArt, D.G.; James, J.; Salto-Tellez, M. Digital pathology and image analysis in tissue biomarker research. *Methods* **2014**, *70*, 59–73. [CrossRef]
41. Meshcheryakova, A.; Mungenast, F.; Ecker, R.; Mechtcheriakova, D. Tissue Image Cytometry. In *COMULIS IoP-IPEM Ebook/Textbook: A Compendium of Imaging Modalities for Biological and Preclinicial Research*; accepted for publication.
42. Stevens, S.S. On the Theory of Scales of Measurement. *Science* **1946**, *103*, 677–680. [CrossRef]
43. Meshcheryakova, A.; Zimmermann, P.; Ecker, R.; Mungenast, F.; Heinze, G.; Mechtcheriakova, D. An Integrative MuSiCO Algorithm: From the Patient-Specific Transcriptional Profiles to Novel Checkpoints in Disease Pathobiology. In *Systems Biology*; Rajewsky, N., Jurga, S., Barciszewski, J., Eds.; Springer International Publishing: Cham, Switzerland, 2018; pp. 351–372. [CrossRef]
44. Pantanowitz, L.; Sharma, A.; Carter, A.B.; Kurc, T.; Sussman, A.; Saltz, J. Twenty Years of Digital Pathology: An Overview of the Road Travelled, What is on the Horizon, and the Emergence of Vendor-Neutral Archives. *J. Pathol. Inform.* **2018**, *9*, 40. [CrossRef]
45. Meijering, E. A bird's-eye view of deep learning in bioimage analysis. *Comput. Struct. Biotechnol. J.* **2020**, *18*, 2312–2325. [CrossRef]
46. Zhang, M.; Zhou, Y.; Zhao, J.; Man, Y.; Liu, B.; Yao, R. A survey of semi- and weakly supervised semantic segmentation of images. *Artif. Intell. Rev.* **2020**, *53*, 4259–4288. [CrossRef]
47. Wilson, G.; Cook, D. A Survey of Unsupervised Deep Domain Adaptation. *Assoc. Comput. Mach.* **2020**, *11*. [CrossRef]
48. Weiss, K.; Khoshgoftaar, T.M.; Wang, D. A survey of transfer learning. *J. Big Data* **2016**, *3*, 9. [CrossRef]
49. Yang, Q.; Liu, Y.; Chen, T.; Tong, Y. Federated Machine Learning: Concept and Applications. *ACM Trans. Intell. Syst. Technol.* **2019**, *10*. [CrossRef]
50. Wang, J.; Su, X.; Zhao, L.; Zhang, J. Deep Reinforcement Learning for Data Association in Cell Tracking. *Front. Bioeng. Biotechnol.* **2020**, *8*, 298. [CrossRef]
51. Wong, M.L.; Medrano, J.F. Real-time PCR for mRNA quantitation. *BioTechniques* **2005**, *39*, 75–85. [CrossRef]
52. Wang, F.; Flanagan, J.; Su, N.; Wang, L.C.; Bui, S.; Nielson, A.; Wu, X.; Vo, H.T.; Ma, X.J.; Luo, Y. RNAscope: A novel in situ RNA analysis platform for formalin-fixed, paraffin-embedded tissues. *J. Mol. Diagn. JMD* **2012**, *14*, 22–29. [CrossRef]
53. Grün, D.; van Oudenaarden, A. Design and Analysis of Single-Cell Sequencing Experiments. *Cell* **2015**, *163*, 799–810. [CrossRef]
54. Jin, L.; Lloyd, R.V. In situ hybridization: Methods and applications. *J. Clin. Lab. Anal.* **1997**, *11*, 2–9. [CrossRef]
55. Kenny, D.; Shen, L.P.; Kolberg, J.A. Detection of viral infection and gene expression in clinical tissue specimens using branched DNA (bDNA) in situ hybridization. *J. Histochem. Cytochem.* **2002**, *50*, 1219–1227. [CrossRef]
56. Thomsen, R.; Nielsen, P.S.; Jensen, T.H. Dramatically improved RNA in situ hybridization signals using LNA-modified probes. *RNA* **2005**, *11*, 1745–1748. [CrossRef]

57. Yin, V.P. In Situ Detection of MicroRNA Expression with RNAscope Probes. *Methods Mol. Biol.* **2018**, *1649*, 197–208. [CrossRef]
58. Cassidy, A.; Jones, J. Developments in in situ hybridisation. *Methods* **2014**, *70*, 39–45. [CrossRef]
59. Wang, H.; Su, N.; Wang, L.C.; Wu, X.; Bui, S.; Nielsen, A.; Vo, H.T.; Luo, Y.; Ma, X.J. Dual-color ultrasensitive bright-field RNA in situ hybridization with RNAscope. *Methods Mol. Biol.* **2014**, *1211*, 139–149. [CrossRef]
60. Carossino, M.; Loynachan, A.T.; James MacLachlan, N.; Drew, C.; Shuck, K.M.; Timoney, P.J.; Del Piero, F.; Balasuriya, U.B. Detection of equine arteritis virus by two chromogenic RNA in situ hybridization assays (conventional and RNAscope(®)) and assessment of their performance in tissues from aborted equine fetuses. *Arch. Virol.* **2016**, *161*, 3125–3136. [CrossRef]
61. Deleage, C.; Wietgrefe, S.W.; Del Prete, G.; Morcock, D.R.; Hao, X.P.; Piatak, M., Jr.; Bess, J.; Anderson, J.L.; Perkey, K.E.; Reilly, C.; et al. Defining HIV and SIV Reservoirs in Lymphoid Tissues. *Pathog. Immun.* **2016**, *1*, 68–106. [CrossRef]
62. Roe, C.J.; Siddiqui, M.T.; Lawson, D.; Cohen, C. RNA In Situ Hybridization for Epstein-Barr Virus and Cytomegalovirus: Comparison With In Situ Hybridization and Immunohistochemistry. *Appl. Immunohistochem. Mol. Morphol. AIMM* **2019**, *27*, 155–159. [CrossRef]
63. Wang, H.; Wang, M.X.; Su, N.; Wang, L.C.; Wu, X.; Bui, S.; Nielsen, A.; Vo, H.T.; Nguyen, N.; Luo, Y.; et al. RNAscope for in situ detection of transcriptionally active human papillomavirus in head and neck squamous cell carcinoma. *J. Vis. Exp. JoVE* **2014**. [CrossRef]
64. Zhang, W.; Svensson Akusjärvi, S.; Sönnerborg, A.; Neogi, U. Characterization of Inducible Transcription and Translation-Competent HIV-1 Using the RNAscope ISH Technology at a Single-Cell Resolution. *Front. Microbiol.* **2018**, *9*, 2358. [CrossRef]
65. Venteicher, A.S.; Tirosh, I.; Hebert, C.; Yizhak, K.; Neftel, C.; Filbin, M.G.; Hovestadt, V.; Escalante, L.E.; Shaw, M.L.; Rodman, C.; et al. Decoupling genetics, lineages, and microenvironment in IDH-mutant gliomas by single-cell RNA-seq. *Science* **2017**, *355*. [CrossRef]
66. Zeisel, A.; Hochgerner, H.; Lönnerberg, P.; Johnsson, A.; Memic, F.; van der Zwan, J.; Häring, M.; Braun, E.; Borm, L.E.; La Manno, G.; et al. Molecular Architecture of the Mouse Nervous System. *Cell* **2018**, *174*, 999–1014.e1022. [CrossRef]
67. Lee, J.C.; Jeng, Y.M.; Su, S.Y.; Wu, C.T.; Tsai, K.S.; Lee, C.H.; Lin, C.Y.; Carter, J.M.; Huang, J.W.; Chen, S.H.; et al. Identification of a novel FN1-FGFR1 genetic fusion as a frequent event in phosphaturic mesenchymal tumour. *J. Pathol.* **2015**, *235*, 539–545. [CrossRef]
68. Wu, G.; Barnhill, R.L.; Lee, S.; Li, Y.; Shao, Y.; Easton, J.; Dalton, J.; Zhang, J.; Pappo, A.; Bahrami, A. The landscape of fusion transcripts in spitzoid melanoma and biologically indeterminate spitzoid tumors by RNA sequencing. *Mod. Pathol.* **2016**, *29*, 359–369. [CrossRef]
69. Nasir, A.; Reising, L.O.; Nedderman, D.M.; Fulford, A.D.; Uhlik, M.T.; Benjamin, L.E.; Schade, A.E.; Holzer, T.R. Heterogeneity of Vascular Endothelial Growth Factor Receptors 1, 2, 3 in Primary Human Colorectal Carcinoma. *Anticancer Res.* **2016**, *36*, 2683–2696.
70. Wang, Z.; Portier, B.P.; Gruver, A.M.; Bui, S.; Wang, H.; Su, N.; Vo, H.T.; Ma, X.J.; Luo, Y.; Budd, G.T.; et al. Automated quantitative RNA in situ hybridization for resolution of equivocal and heterogeneous ERBB2 (HER2) status in invasive breast carcinoma. *J. Mol. Diagn. JMD* **2013**, *15*, 210–219. [CrossRef]
71. Barry, E.R.; Morikawa, T.; Butler, B.L.; Shrestha, K.; de la Rosa, R.; Yan, K.S.; Fuchs, C.S.; Magness, S.T.; Smits, R.; Ogino, S.; et al. Restriction of intestinal stem cell expansion and the regenerative response by YAP. *Nature* **2013**, *493*, 106–110. [CrossRef]
72. Jang, B.G.; Lee, B.L.; Kim, W.H. Distribution of LGR5+ cells and associated implications during the early stage of gastric tumorigenesis. *PLoS ONE* **2013**, *8*, e82390. [CrossRef]
73. Ziskin, J.L.; Dunlap, D.; Yaylaoglu, M.; Fodor, I.K.; Forrest, W.F.; Patel, R.; Ge, N.; Hutchins, G.G.; Pine, J.K.; Quirke, P.; et al. In situ validation of an intestinal stem cell signature in colorectal cancer. *Gut* **2013**, *62*, 1012–1023. [CrossRef]
74. Payne, R.E.; Wang, F.; Su, N.; Krell, J.; Zebrowski, A.; Yagüe, E.; Ma, X.J.; Luo, Y.; Coombes, R.C. Viable circulating tumour cell detection using multiplex RNA in situ hybridisation predicts progression-free survival in metastatic breast cancer patients. *Br. J. Cancer* **2012**, *106*, 1790–1797. [CrossRef]
75. Yu, M.; Bardia, A.; Wittner, B.S.; Stott, S.L.; Smas, M.E.; Ting, D.T.; Isakoff, S.J.; Ciciliano, J.C.; Wells, M.N.; Shah, A.M.; et al. Circulating breast tumor cells exhibit dynamic changes in epithelial and mesenchymal composition. *Science* **2013**, *339*, 580–584. [CrossRef]
76. Kim, S.Y.; Theunissen, J.W.; Balibalos, J.; Liao-Chan, S.; Babcock, M.C.; Wong, T.; Cairns, B.; Gonzalez, D.; van der Horst, E.H.; Perez, M.; et al. A novel antibody-drug conjugate targeting SAIL for the treatment of hematologic malignancies. *Blood Cancer J.* **2015**, *5*, e316. [CrossRef]
77. Naipal, K.A.; Verkaik, N.S.; Ameziane, N.; van Deurzen, C.H.; Ter Brugge, P.; Meijers, M.; Sieuwerts, A.M.; Martens, J.W.; O'Connor, M.J.; Vrieling, H.; et al. Functional ex vivo assay to select homologous recombination-deficient breast tumors for PARP inhibitor treatment. *Clin. Cancer Res.* **2014**, *20*, 4816–4826. [CrossRef]
78. Carossino, M.; Ip, H.S.; Richt, J.A.; Shultz, K.; Harper, K.; Loynachan, A.T.; Del Piero, F.; Balasuriya, U.B.R. Detection of SARS-CoV-2 by RNAscope(®) in situ hybridization and immunohistochemistry techniques. *Arch. Virol.* **2020**, *165*, 2373–2377. [CrossRef]
79. Boukerroucha, M.; Josse, C.; ElGuendi, S.; Boujemla, B.; Frères, P.; Marée, R.; Wenric, S.; Segers, K.; Collignon, J.; Jerusalem, G.; et al. Evaluation of BRCA1-related molecular features and microRNAs as prognostic factors for triple negative breast cancers. *BMC Cancer* **2015**, *15*, 755. [CrossRef]

80. Fabbri, E.; Brognara, E.; Montagner, G.; Ghimenton, C.; Eccher, A.; Cantù, C.; Khalil, S.; Bezzerri, V.; Provezza, L.; Bianchi, N.; et al. Regulation of IL-8 gene expression in gliomas by microRNA miR-93. *BMC Cancer* **2015**, *15*, 661. [CrossRef]
81. Lessard, L.; Liu, M.; Marzese, D.M.; Wang, H.; Chong, K.; Kawas, N.; Donovan, N.C.; Kiyohara, E.; Hsu, S.; Nelson, N.; et al. The CASC15 Long Intergenic Noncoding RNA Locus Is Involved in Melanoma Progression and Phenotype Switching. *J. Investig. Dermatol.* **2015**, *135*, 2464–2474. [CrossRef]
82. Xing, Z.; Park, P.K.; Lin, C.; Yang, L. LncRNA BCAR4 wires up signaling transduction in breast cancer. *RNA Biol.* **2015**, *12*, 681–689. [CrossRef]
83. Welti, J.; Rodrigues, D.N.; Sharp, A.; Sun, S.; Lorente, D.; Riisnaes, R.; Figueiredo, I.; Zafeiriou, Z.; Rescigno, P.; de Bono, J.S.; et al. Analytical Validation and Clinical Qualification of a New Immunohistochemical Assay for Androgen Receptor Splice Variant-7 Protein Expression in Metastatic Castration-resistant Prostate Cancer. *Eur. Urol.* **2016**, *70*, 599–608. [CrossRef]
84. Guedes, L.B.; Morais, C.L.; Almutairi, F.; Haffner, M.C.; Zheng, Q.; Isaacs, J.T.; Antonarakis, E.S.; Lu, C.; Tsai, H.; Luo, J.; et al. Analytic Validation of RNA In Situ Hybridization (RISH) for AR and AR-V7 Expression in Human Prostate Cancer. *Clin. Cancer Res.* **2016**, *22*, 4651–4663. [CrossRef]
85. Anderson, C.M.; Zhang, B.; Miller, M.; Butko, E.; Wu, X.; Laver, T.; Kernag, C.; Kim, J.; Luo, Y.; Lamparski, H.; et al. Fully Automated RNAscope In Situ Hybridization Assays for Formalin-Fixed Paraffin-Embedded Cells and Tissues. *J. Cell. Biochem.* **2016**, *117*, 2201–2208. [CrossRef]
86. Anderson, N.M.; Simon, M.C. The tumor microenvironment. *Curr. Biol. CB* **2020**, *30*, R921–R925. [CrossRef]
87. Du, Z.; Abedalthagafi, M.; Aizer, A.A.; McHenry, A.R.; Sun, H.H.; Bray, M.A.; Viramontes, O.; Machaidze, R.; Brastianos, P.K.; Reardon, D.A.; et al. Increased expression of the immune modulatory molecule PD-L1 (CD274) in anaplastic meningioma. *Oncotarget* **2015**, *6*, 4704–4716. [CrossRef]
88. Patel, K.R.; Liu, T.C.; Vaccharajani, N.; Chapman, W.C.; Brunt, E.M. Characterization of inflammatory (lymphoepithelioma-like) hepatocellular carcinoma: A study of 8 cases. *Arch. Pathol. Lab. Med.* **2014**, *138*, 1193–1202. [CrossRef]
89. De Monte, L.; Wörmann, S.; Brunetto, E.; Heltai, S.; Magliacane, G.; Reni, M.; Paganoni, A.M.; Recalde, H.; Mondino, A.; Falconi, M.; et al. Basophil Recruitment into Tumor-Draining Lymph Nodes Correlates with Th2 Inflammation and Reduced Survival in Pancreatic Cancer Patients. *Cancer Res.* **2016**, *76*, 1792–1803. [CrossRef]
90. Ukpo, O.C.; Thorstad, W.L.; Lewis, J.S., Jr. B7-H1 expression model for immune evasion in human papillomavirus-related oropharyngeal squamous cell carcinoma. *Head Neck Pathol.* **2013**, *7*, 113–121. [CrossRef]
91. Kim, J.; Li, N.; He, M.; Zhang, B.; Su, N.; Ma, X.-J.; Park, E. Abstract A24: Evaluation of the expression of immune functional markers in the tumor microenvironment. *Cancer Immunol. Res.* **2017**, *5*, A24. [CrossRef]
92. Staudt, N.D.; Jo, M.; Hu, J.; Bristow, J.M.; Pizzo, D.P.; Gaultier, A.; VandenBerg, S.R.; Gonias, S.L. Myeloid cell receptor LRP1/CD91 regulates monocyte recruitment and angiogenesis in tumors. *Cancer Res.* **2013**, *73*, 3902–3912. [CrossRef]
93. Shames, D.S.; Carbon, J.; Walter, K.; Jubb, A.M.; Kozlowski, C.; Januario, T.; Do, A.; Fu, L.; Xiao, Y.; Raja, R.; et al. High heregulin expression is associated with activated HER3 and may define an actionable biomarker in patients with squamous cell carcinomas of the head and neck. *PLoS ONE* **2013**, *8*, e56765. [CrossRef]
94. Lee, Y.; Shin, J.H.; Longmire, M.; Wang, H.; Kohrt, H.E.; Chang, H.Y.; Sunwoo, J.B. CD44+ Cells in Head and Neck Squamous Cell Carcinoma Suppress T-Cell-Mediated Immunity by Selective Constitutive and Inducible Expression of PD-L1. *Clin. Cancer Res.* **2016**, *22*, 3571–3581. [CrossRef]
95. bitesizebio.com. New-ISH on the Block: Introduction to RNAscope®. Available online: https://bitesizebio.com/40621/new-ish-on-the-block-introduction-to-rnascope (accessed on 7 January 2021).
96. Annese, T.; Tamma, R.; De Giorgis, M.; Ruggieri, S.; Maiorano, E.; Specchia, G.; Ribatti, D. RNAscope dual ISH-IHC technology to study angiogenesis in diffuse large B-cell lymphomas. *Histochem. Cell Biol.* **2020**, *153*, 185–192. [CrossRef]
97. Chan, S.; Filézac de L'Etang, A.; Rangell, L.; Caplazi, P.; Lowe, J.B.; Romeo, V. A method for manual and automated multiplex RNAscope in situ hybridization and immunocytochemistry on cytospin samples. *PLoS ONE* **2018**, *13*, e0207619. [CrossRef]
98. Yang, J.Q.; Wu, M.; Han, F.Y.; Sun, Y.M.; Zhang, L.; Liu, H.X. High risk HPV detection by RNAscope in situ hybridization combined with Cdc2 protein expression by immunohistochemistry for prognosis of oropharyngeal squamous cell carcinoma. *Int. J. Clin. Exp. Pathol.* **2020**, *13*, 2192–2200.
99. Angell, H.K.; Gray, N.; Womack, C.; Pritchard, D.I.; Wilkinson, R.W.; Cumberbatch, M. Digital pattern recognition-based image analysis quantifies immune infiltrates in distinct tissue regions of colorectal cancer and identifies a metastatic phenotype. *Br. J. Cancer* **2013**, *109*, 1618–1624. [CrossRef]
100. Bindea, G.; Mlecnik, B.; Fridman, W.-H.; Galon, J. The prognostic impact of anti-cancer immune response: A novel classification of cancer patients. *Semin. Immunopathol.* **2011**, *33*, 335–340. [CrossRef]
101. Pagès, F.; Mlecnik, B.; Marliot, F.; Bindea, G.; Ou, F.S.; Bifulco, C.; Lugli, A.; Zlobec, I.; Rau, T.T.; Berger, M.D.; et al. International validation of the consensus Immunoscore for the classification of colon cancer: A prognostic and accuracy study. *Lancet* **2018**, *391*, 2128–2139. [CrossRef]
102. Seo, A.N.; Lee, H.J.; Kim, E.J.; Kim, H.J.; Jang, M.H.; Lee, H.E.; Kim, Y.J.; Kim, J.H.; Park, S.Y. Tumour-infiltrating CD8+ lymphocytes as an independent predictive factor for pathological complete response to primary systemic therapy in breast cancer. *Br. J. Cancer* **2013**, *109*, 2705–2713. [CrossRef]

103. García-Martínez, E.; Gil, G.L.; Benito, A.C.; González-Billalabeitia, E.; Conesa, M.A.; García García, T.; García-Garre, E.; Vicente, V.; Ayala de la Peña, F. Tumor-infiltrating immune cell profiles and their change after neoadjuvant chemotherapy predict response and prognosis of breast cancer. *Breast Cancer Res. BCR* **2014**, *16*, 488. [CrossRef]
104. Humphries, M.P.; Hynes, S.; Bingham, V.; Cougot, D.; James, J.; Patel-Socha, F.; Parkes, E.E.; Blayney, J.K.; O'Rorke, M.A.; Irwin, G.W.; et al. Automated Tumour Recognition and Digital Pathology Scoring Unravels New Role for PD-L1 in Predicting Good Outcome in ER-/HER2+ Breast Cancer. *J. Oncol.* **2018**, *2018*, 2937012. [CrossRef]
105. Rasmusson, A.; Zilenaite, D.; Nestarenkaite, A.; Augulis, R.; Laurinaviciene, A.; Ostapenko, V.; Poskus, T.; Laurinavicius, A. Immunogradient Indicators for Antitumor Response Assessment by Automated Tumor-Stroma Interface Zone Detection. *Am. J. Pathol.* **2020**, *190*, 1309–1322. [CrossRef]
106. Martinet, L.; Filleron, T.; Le Guellec, S.; Rochaix, P.; Garrido, I.; Girard, J.P. High endothelial venule blood vessels for tumor-infiltrating lymphocytes are associated with lymphotoxin β-producing dendritic cells in human breast cancer. *J. Immunol.* **2013**, *191*, 2001–2008. [CrossRef]
107. Martinet, L.; Garrido, I.; Filleron, T.; Le Guellec, S.; Bellard, E.; Fournie, J.J.; Rochaix, P.; Girard, J.P. Human solid tumors contain high endothelial venules: Association with T- and B-lymphocyte infiltration and favorable prognosis in breast cancer. *Cancer Res.* **2011**, *71*, 5678–5687. [CrossRef]
108. Sobral-Leite, M.; Van de Vijver, K.; Michaut, M.; van der Linden, R.; Hooijer, G.K.J.; Horlings, H.M.; Severson, T.M.; Mulligan, A.M.; Weerasooriya, N.; Sanders, J.; et al. Assessment of PD-L1 expression across breast cancer molecular subtypes, in relation to mutation rate, BRCA1-like status, tumor-infiltrating immune cells and survival. *Oncoimmunology* **2018**, *7*, e1509820. [CrossRef]
109. He, T.F.; Yost, S.E.; Frankel, P.H.; Dagis, A.; Cao, Y.; Wang, R.; Rosario, A.; Tu, T.Y.; Solomon, S.; Schmolze, D.; et al. Multi-panel immunofluorescence analysis of tumor infiltrating lymphocytes in triple negative breast cancer: Evolution of tumor immune profiles and patient prognosis. *PLoS ONE* **2020**, *15*, e0229955. [CrossRef]
110. Yoon, H.H.; Shi, Q.; Heying, E.N.; Muranyi, A.; Bredno, J.; Ough, F.; Djalilvand, A.; Clements, J.; Bowermaster, R.; Liu, W.W.; et al. Intertumoral Heterogeneity of CD3(+) and CD8(+) T-Cell Densities in the Microenvironment of DNA Mismatch-Repair-Deficient Colon Cancers: Implications for Prognosis. *Clin. Cancer Res.* **2019**, *25*, 125–133. [CrossRef]
111. Nearchou, I.P.; Lillard, K.; Gavriel, C.G.; Ueno, H.; Harrison, D.J.; Caie, P.D. Automated Analysis of Lymphocytic Infiltration, Tumor Budding, and Their Spatial Relationship Improves Prognostic Accuracy in Colorectal Cancer. *Cancer Immunol. Res.* **2019**, *7*, 609–620. [CrossRef]
112. Chen, Y.; Yuan, R.; Wu, X.; He, X.; Zeng, Y.; Fan, X.; Wang, L.; Wang, J.; Lan, P.; Wu, X. A Novel Immune Marker Model Predicts Oncological Outcomes of Patients with Colorectal Cancer. *Ann. Surg. Oncol.* **2016**, *23*, 826–832. [CrossRef]
113. Edin, S.; Kaprio, T.; Hagström, J.; Larsson, P.; Mustonen, H.; Böckelman, C.; Strigård, K.; Gunnarsson, U.; Haglund, C.; Palmqvist, R. The Prognostic Importance of CD20(+) B lymphocytes in Colorectal Cancer and the Relation to Other Immune Cell subsets. *Sci. Rep.* **2019**, *9*, 19997. [CrossRef]
114. Miller, T.J.; McCoy, M.J.; Hemmings, C.; Bulsara, M.K.; Iacopetta, B.; Platell, C.F. The prognostic value of cancer stem-like cell markers SOX2 and CD133 in stage III colon cancer is modified by expression of the immune-related markers FoxP3, PD-L1 and CD3. *Pathology* **2017**, *49*, 721–730. [CrossRef]
115. Miller, T.J.; Anyaegbu, C.C.; Lee-Pullen, T.F.; Spalding, L.J.; Platell, C.F.; McCoy, M.J. PD-L1+ dendritic cells in the tumor microenvironment correlate with good prognosis and CD8+ T cell infiltration in colon cancer. *Cancer Sci.* **2020**. [CrossRef]
116. Herrera, M.; Mezheyeuski, A.; Villabona, L.; Corvigno, S.; Strell, C.; Klein, C.; Hölzlwimmer, G.; Glimelius, B.; Masucci, G.; Sjöblom, T.; et al. Prognostic Interactions between FAP+ Fibroblasts and CD8a+ T Cells in Colon Cancer. *Cancers* **2020**, *12*, 3238. [CrossRef]
117. Mungenast, F.; Meshcheryakova, A.; Beer, A.; Salzmann, M.; Tamandl, D.; Gruenberger, T.; Pietschmann, P.; Koperek, O.; Birner, P.; Kirsch, I.; et al. The Immune Phenotype of Isolated Lymphoid Structures in Non-Tumorous Colon Mucosa Encrypts the Information on Pathobiology of Metastatic Colorectal Cancer. *Cancers* **2020**, *12*, 3117. [CrossRef]
118. Meshcheryakova, A.; Tamandl, D.; Bajna, E.; Stift, J.; Mittlboeck, M.; Svoboda, M.; Heiden, D.; Stremitzer, S.; Jensen-Jarolim, E.; Grünberger, T.; et al. B cells and ectopic follicular structures: Novel players in anti-tumor programming with prognostic power for patients with metastatic colorectal cancer. *PLoS ONE* **2014**, *9*, e99008. [CrossRef]
119. Tanis, E.; Julié, C.; Emile, J.F.; Mauer, M.; Nordlinger, B.; Aust, D.; Roth, A.; Lutz, M.P.; Gruenberger, T.; Wrba, F.; et al. Prognostic impact of immune response in resectable colorectal liver metastases treated by surgery alone or surgery with perioperative FOLFOX in the randomised EORTC study 40983. *Eur. J. Cancer* **2015**, *51*, 2708–2717. [CrossRef]
120. Mlecnik, B.; Van den Eynde, M.; Bindea, G.; Church, S.E.; Vasaturo, A.; Fredriksen, T.; Lafontaine, L.; Haicheur, N.; Marliot, F.; Debetancourt, D.; et al. Comprehensive Intrametastatic Immune Quantification and Major Impact of Immunoscore on Survival. *J. Natl. Cancer Inst.* **2018**, *110*, 97–108. [CrossRef]
121. Ma, J.; Li, J.; Qian, M.; Han, W.; Tian, M.; Li, Z.; Wang, Z.; He, S.; Wu, K. PD-L1 expression and the prognostic significance in gastric cancer: A retrospective comparison of three PD-L1 antibody clones (SP142, 28–8 and E1L3N). *Diagn. Pathol.* **2018**, *13*, 91. [CrossRef]
122. Kim, J.W.; Nam, K.H.; Ahn, S.H.; Park, D.J.; Kim, H.H.; Kim, S.H.; Chang, H.; Lee, J.O.; Kim, Y.J.; Lee, H.S.; et al. Prognostic implications of immunosuppressive protein expression in tumors as well as immune cell infiltration within the tumor microenvironment in gastric cancer. *Gastric Cancer* **2016**, *19*, 42–52. [CrossRef]

123. Kim, K.J.; Lee, K.S.; Cho, H.J.; Kim, Y.H.; Yang, H.K.; Kim, W.H.; Kang, G.H. Prognostic implications of tumor-infiltrating FoxP3+ regulatory T cells and CD8+ cytotoxic T cells in microsatellite-unstable gastric cancers. *Hum. Pathol.* **2014**, *45*, 285–293. [CrossRef]
124. Koh, J.; Ock, C.Y.; Kim, J.W.; Nam, S.K.; Kwak, Y.; Yun, S.; Ahn, S.H.; Park, D.J.; Kim, H.H.; Kim, W.H.; et al. Clinicopathologic implications of immune classification by PD-L1 expression and CD8-positive tumor-infiltrating lymphocytes in stage II and III gastric cancer patients. *Oncotarget* **2017**, *8*, 26356–26367. [CrossRef]
125. Wang, M.; Huang, Y.K.; Kong, J.C.; Sun, Y.; Tantalo, D.G.; Yeang, H.X.A.; Ying, L.; Yan, F.; Xu, D.; Halse, H.; et al. High-dimensional analyses reveal a distinct role of T-cell subsets in the immune microenvironment of gastric cancer. *Clin. Transl. Immunol.* **2020**, *9*, e1127. [CrossRef]
126. Svensson, M.C.; Warfvinge, C.F.; Fristedt, R.; Hedner, C.; Borg, D.; Eberhard, J.; Micke, P.; Nodin, B.; Leandersson, K.; Jirström, K. The integrative clinical impact of tumor-infiltrating T lymphocytes and NK cells in relation to B lymphocyte and plasma cell density in esophageal and gastric adenocarcinoma. *Oncotarget* **2017**, *8*, 72108–72126. [CrossRef]
127. Kim, H.-N.; Jang, J.; Heo, Y.J.; Kim, B.; Jung, H.; Jang, Y.; Kang, S.Y.; Kim, S.T.; Lee, J.; Kang, W.K.; et al. PD-L1 expression in gastric cancer determined by digital image analyses: Pitfalls and correlation with pathologist interpretation. *Virchows Arch.* **2020**, *476*, 243–250. [CrossRef]
128. Sun, C.; Xu, J.; Song, J.; Liu, C.; Wang, J.; Weng, C.; Sun, H.; Wei, H.; Xiao, W.; Sun, R.; et al. The predictive value of centre tumour CD8^{+} T cells in patients with hepatocellular carcinoma: Comparison with Immunoscore. *Oncotarget* **2015**, *6*, 35602–35615. [CrossRef]
129. Gabrielson, A.; Wu, Y.; Wang, H.; Jiang, J.; Kallakury, B.; Gatalica, Z.; Reddy, S.; Kleiner, D.; Fishbein, T.; Johnson, L.; et al. Intratumoral CD3 and CD8 T-cell Densities Associated with Relapse-Free Survival in HCC. *Cancer Immunol. Res.* **2016**, *4*, 419–430. [CrossRef]
130. Ma, J.; Zheng, B.; Goswami, S.; Meng, L.; Zhang, D.; Cao, C.; Li, T.; Zhu, F.; Ma, L.; Zhang, Z.; et al. PD1(Hi) CD8(+) T cells correlate with exhausted signature and poor clinical outcome in hepatocellular carcinoma. *J. Immunother. Cancer* **2019**, *7*, 331. [CrossRef]
131. Garnelo, M.; Tan, A.; Her, Z.; Yeong, J.; Lim, C.J.; Chen, J.; Lim, K.H.; Weber, A.; Chow, P.; Chung, A.; et al. Interaction between tumour-infiltrating B cells and T cells controls the progression of hepatocellular carcinoma. *Gut* **2017**, *66*, 342–351. [CrossRef]
132. Yao, Q.; Bao, X.; Xue, R.; Liu, H.; Liu, H.; Li, J.; Dong, J.; Duan, Z.; Ren, M.; Zhao, J.; et al. Prognostic value of immunoscore to identify mortality outcomes in adults with HBV-related primary hepatocellular carcinoma. *Medicine* **2017**, *96*, e6735. [CrossRef]
133. Huang, Y.; Wang, F.M.; Wang, T.; Wang, Y.J.; Zhu, Z.Y.; Gao, Y.T.; Du, Z. Tumor-infiltrating FoxP3+ Tregs and CD8+ T cells affect the prognosis of hepatocellular carcinoma patients. *Digestion* **2012**, *86*, 329–337. [CrossRef]
134. Amin, D.; Richa, T.; Mollaee, M.; Zhan, T.; Tassone, P.; Johnson, J.; Luginbuhl, A.; Cognetti, D.; Martinez-Outschoorn, U.; Stapp, R.; et al. Metformin Effects on FOXP3(+) and CD8(+) T Cell Infiltrates of Head and Neck Squamous Cell Carcinoma. *Laryngoscope* **2020**, *130*, E490–E498. [CrossRef]
135. Zhang, X.M.; Song, L.J.; Shen, J.; Yue, H.; Han, Y.Q.; Yang, C.L.; Liu, S.Y.; Deng, J.W.; Jiang, Y.; Fu, G.H.; et al. Prognostic and predictive values of immune infiltrate in patients with head and neck squamous cell carcinoma. *Hum. Pathol.* **2018**, *82*, 104–112. [CrossRef]
136. Koelzer, V.H.; Gisler, A.; Hanhart, J.C.; Griss, J.; Wagner, S.N.; Willi, N.; Cathomas, G.; Sachs, M.; Kempf, W.; Thommen, D.S.; et al. Digital image analysis improves precision of PD-L1 scoring in cutaneous melanoma. *Histopathology* **2018**, *73*, 397–406. [CrossRef]
137. Garg, K.; Maurer, M.; Griss, J.; Brüggen, M.C.; Wolf, I.H.; Wagner, C.; Willi, N.; Mertz, K.D.; Wagner, S.N. Tumor-associated B cells in cutaneous primary melanoma and improved clinical outcome. *Hum. Pathol.* **2016**, *54*, 157–164. [CrossRef]
138. Gartrell, R.D.; Marks, D.K.; Hart, T.D.; Li, G.; Davari, D.R.; Wu, A.; Blake, Z.; Lu, Y.; Askin, K.N.; Monod, A.; et al. Quantitative Analysis of Immune Infiltrates in Primary Melanoma. *Cancer Immunol. Res.* **2018**, *6*, 481–493. [CrossRef]
139. Griss, J.; Bauer, W.; Wagner, C.; Simon, M.; Chen, M.; Grabmeier-Pfistershammer, K.; Maurer-Granofszky, M.; Roka, F.; Penz, T.; Bock, C.; et al. B cells sustain inflammation and predict response to immune checkpoint blockade in human melanoma. *Nat. Commun.* **2019**, *10*, 4186. [CrossRef]
140. Thommen, D.S.; Koelzer, V.H.; Herzig, P.; Roller, A.; Trefny, M.; Dimeloe, S.; Kiialainen, A.; Hanhart, J.; Schill, C.; Hess, C.; et al. A transcriptionally and functionally distinct PD-1+ CD8+ T cell pool with predictive potential in non-small-cell lung cancer treated with PD-1 blockade. *Nat. Med.* **2018**, *24*, 994–1004. [CrossRef]
141. Goc, J.; Germain, C.; Vo-Bourgais, T.K.; Lupo, A.; Klein, C.; Knockaert, S.; de Chaisemartin, L.; Ouakrim, H.; Becht, E.; Alifano, M.; et al. Dendritic cells in tumor-associated tertiary lymphoid structures signal a Th1 cytotoxic immune contexture and license the positive prognostic value of infiltrating CD8+ T cells. *Cancer Res.* **2014**, *74*, 705–715. [CrossRef]
142. Sepesi, B.; Cuentas, E.P.; Canales, J.R.; Behrens, C.; Correa, A.M.; Vaporciyan, A.; Weissferdt, A.; Kalhor, N.; Moran, C.; Swisher, S.; et al. Programmed Death Cell Ligand 1 (PD-L1) Is Associated With Survival in Stage I Non-Small Cell Lung Cancer. *Semin. Thorac. Cardiovasc. Surg.* **2017**, *29*, 408–415. [CrossRef]
143. Kinoshita, T.; Kudo-Saito, C.; Muramatsu, R.; Fujita, T.; Saito, M.; Nagumo, H.; Sakurai, T.; Noji, S.; Takahata, E.; Yaguchi, T.; et al. Determination of poor prognostic immune features of tumour microenvironment in non-smoking patients with lung adenocarcinoma. *Eur. J. Cancer* **2017**, *86*, 15–27. [CrossRef]
144. Parra, E.R.; Behrens, C.; Rodriguez-Canales, J.; Lin, H.; Mino, B.; Blando, J.; Zhang, J.; Gibbons, D.L.; Heymach, J.V.; Sepesi, B.; et al. Image Analysis-based Assessment of PD-L1 and Tumor-Associated Immune Cells Density Supports Distinct Intratumoral Microenvironment Groups in Non-small Cell Lung Carcinoma Patients. *Clin. Cancer Res.* **2016**, *22*, 6278–6289. [CrossRef]

145. Kinoshita, T.; Muramatsu, R.; Fujita, T.; Nagumo, H.; Sakurai, T.; Noji, S.; Takahata, E.; Yaguchi, T.; Tsukamoto, N.; Kudo-Saito, C.; et al. Prognostic value of tumor-infiltrating lymphocytes differs depending on histological type and smoking habit in completely resected non-small-cell lung cancer. *Ann. Oncol.* **2016**, *27*, 2117–2123. [CrossRef]
146. Cao, L.; Che, X.; Qiu, X.; Li, Z.; Yang, B.; Wang, S.; Hou, K.; Fan, Y.; Qu, X.; Liu, Y. M2 macrophage infiltration into tumor islets leads to poor prognosis in non-small-cell lung cancer. *Cancer Manag. Res.* **2019**, *11*, 6125–6138. [CrossRef]
147. Barua, S.; Fang, P.; Sharma, A.; Fujimoto, J.; Wistuba, I.; Rao, A.U.K.; Lin, S.H. Spatial interaction of tumor cells and regulatory T cells correlates with survival in non-small cell lung cancer. *Lung Cancer* **2018**, *117*, 73–79. [CrossRef]
148. O'Callaghan, D.S.; Rexhepaj, E.; Gately, K.; Coate, L.; Delaney, D.; O'Donnell, D.M.; Kay, E.; O'Connell, F.; Gallagher, W.M.; O'Byrne, K.J. Tumour islet Foxp3+ T-cell infiltration predicts poor outcome in nonsmall cell lung cancer. *Eur. Respir. J.* **2015**, *46*, 1762–1772. [CrossRef]
149. Althammer, S.; Tan, T.H.; Spitzmüller, A.; Rognoni, L.; Wiestler, T.; Herz, T.; Widmaier, M.; Rebelatto, M.C.; Kaplon, H.; Damotte, D.; et al. Automated image analysis of NSCLC biopsies to predict response to anti-PD-L1 therapy. *J. Immunother. Cancer* **2019**, *7*, 121. [CrossRef]
150. Kim, M.Y.; Koh, J.; Kim, S.; Go, H.; Jeon, Y.K.; Chung, D.H. Clinicopathological analysis of PD-L1 and PD-L2 expression in pulmonary squamous cell carcinoma: Comparison with tumor-infiltrating T cells and the status of oncogenic drivers. *Lung Cancer* **2015**, *88*, 24–33. [CrossRef]
151. Siliņa, K.; Soltermann, A.; Attar, F.M.; Casanova, R.; Uckeley, Z.M.; Thut, H.; Wandres, M.; Isajevs, S.; Cheng, P.; Curioni-Fontecedro, A.; et al. Germinal Centers Determine the Prognostic Relevance of Tertiary Lymphoid Structures and Are Impaired by Corticosteroids in Lung Squamous Cell Carcinoma. *Cancer Res.* **2018**, *78*, 1308–1320. [CrossRef]
152. Feng, Z.; Bethmann, D.; Kappler, M.; Ballesteros-Merino, C.; Eckert, A.; Bell, R.B.; Cheng, A.; Bui, T.; Leidner, R.; Urba, W.J.; et al. Multiparametric immune profiling in HPV- oral squamous cell cancer. *JCI Insight* **2017**, *2*. [CrossRef]
153. Desbois, M.; Udyavar, A.R.; Ryner, L.; Kozlowski, C.; Guan, Y.; Dürrbaum, M.; Lu, S.; Fortin, J.P.; Koeppen, H.; Ziai, J.; et al. Integrated digital pathology and transcriptome analysis identifies molecular mediators of T-cell exclusion in ovarian cancer. *Nat. Commun.* **2020**, *11*, 5583. [CrossRef]
154. Bachmayr-Heyda, A.; Aust, S.; Heinze, G.; Polterauer, S.; Grimm, C.; Braicu, E.I.; Sehouli, J.; Lambrechts, S.; Vergote, I.; Mahner, S.; et al. Prognostic impact of tumor infiltrating CD8+ T cells in association with cell proliferation in ovarian cancer patients–a study of the OVCAD consortium. *BMC Cancer* **2013**, *13*, 422. [CrossRef]
155. Montfort, A.; Owen, S.; Piskorz, A.M.; Supernat, A.; Moore, L.; Al-Khalidi, S.; Böhm, S.; Pharoah, P.; McDermott, J.; Balkwill, F.R.; et al. Combining measures of immune infiltration shows additive effect on survival prediction in high-grade serous ovarian carcinoma. *Br. J. Cancer* **2020**, *122*, 1803–1810. [CrossRef]
156. Kroeger, D.R.; Milne, K.; Nelson, B.H. Tumor-Infiltrating Plasma Cells Are Associated with Tertiary Lymphoid Structures, Cytolytic T-Cell Responses, and Superior Prognosis in Ovarian Cancer. *Clin. Cancer Res.* **2016**, *22*, 3005–3015. [CrossRef]
157. Webb, J.R.; Milne, K.; Kroeger, D.R.; Nelson, B.H. PD-L1 expression is associated with tumor-infiltrating T cells and favorable prognosis in high-grade serous ovarian cancer. *Gynecol. Oncol.* **2016**, *141*, 293–302. [CrossRef]
158. Komdeur, F.L.; Wouters, M.C.; Workel, H.H.; Tijans, A.M.; Terwindt, A.L.; Brunekreeft, K.L.; Plat, A.; Klip, H.G.; Eggink, F.A.; Leffers, N.; et al. CD103+ intraepithelial T cells in high-grade serous ovarian cancer are phenotypically diverse TCR$\alpha\beta$+ CD8$\alpha\beta$+ T cells that can be targeted for cancer immunotherapy. *Oncotarget* **2016**, *7*, 75130–75144. [CrossRef]
159. Stanske, M.; Wienert, S.; Castillo-Tong, D.C.; Kreuzinger, C.; Vergote, I.; Lambrechts, S.; Gabra, H.; Gourley, C.; Ganapathi, R.N.; Kolaschinski, I.; et al. Dynamics of the Intratumoral Immune Response during Progression of High-Grade Serous Ovarian Cancer. *Neoplasia* **2018**, *20*, 280–288. [CrossRef]
160. Carstens, J.L.; Correa de Sampaio, P.; Yang, D.; Barua, S.; Wang, H.; Rao, A.; Allison, J.P.; LeBleu, V.S.; Kalluri, R. Spatial computation of intratumoral T cells correlates with survival of patients with pancreatic cancer. *Nat. Commun.* **2017**, *8*, 15095. [CrossRef]
161. Balachandran, V.P.; Łuksza, M.; Zhao, J.N.; Makarov, V.; Moral, J.A.; Remark, R.; Herbst, B.; Askan, G.; Bhanot, U.; Senbabaoglu, Y.; et al. Identification of unique neoantigen qualities in long-term survivors of pancreatic cancer. *Nature* **2017**, *551*, 512–516. [CrossRef]
162. Castino, G.F.; Cortese, N.; Capretti, G.; Serio, S.; Di Caro, G.; Mineri, R.; Magrini, E.; Grizzi, F.; Cappello, P.; Novelli, F.; et al. Spatial distribution of B cells predicts prognosis in human pancreatic adenocarcinoma. *Oncoimmunology* **2016**, *5*, e1085147. [CrossRef]
163. Danilova, L.; Ho, W.J.; Zhu, Q.; Vithayathil, T.; De Jesus-Acosta, A.; Azad, N.S.; Laheru, D.A.; Fertig, E.J.; Anders, R.; Jaffee, E.M.; et al. Programmed Cell Death Ligand-1 (PD-L1) and CD8 Expression Profiling Identify an Immunologic Subtype of Pancreatic Ductal Adenocarcinomas with Favorable Survival. *Cancer Immunol. Res.* **2019**, *7*, 886–895. [CrossRef]
164. Hou, Y.C.; Chao, Y.J.; Hsieh, M.H.; Tung, H.L.; Wang, H.C.; Shan, Y.S. Low $CD8^+$ T Cell Infiltration and High PD-L1 Expression Are Associated with Level of $CD44^+/CD133^+$ Cancer Stem Cells and Predict an Unfavorable Prognosis in Pancreatic Cancer. *Cancers* **2019**, *11*, 541. [CrossRef]
165. Ino, Y.; Oguro, S.; Yamazaki-Itoh, R.; Hori, S.; Shimada, K.; Hiraoka, N. Reliable evaluation of tumor-infiltrating lymphocytes in pancreatic cancer tissue biopsies. *Oncotarget* **2019**, *10*, 1149–1159. [CrossRef]

166. Miksch, R.C.; Schoenberg, M.B.; Weniger, M.; Bösch, F.; Ormanns, S.; Mayer, B.; Werner, J.; Bazhin, A.V.; D'Haese, J.G. Prognostic Impact of Tumor-Infiltrating Lymphocytes and Neutrophils on Survival of Patients with Upfront Resection of Pancreatic Cancer. *Cancers* **2019**, *11*, 39. [CrossRef]
167. Tahkola, K.; Leppänen, J.; Ahtiainen, M.; Väyrynen, J.; Haapasaari, K.M.; Karttunen, T.; Kellokumpu, I.; Helminen, O.; Böhm, J. Immune cell score in pancreatic cancer-comparison of hotspot and whole-section techniques. *Virchows Arch.* **2019**, *474*, 691–699. [CrossRef]
168. Gannon, P.O.; Poisson, A.O.; Delvoye, N.; Lapointe, R.; Mes-Masson, A.M.; Saad, F. Characterization of the intra-prostatic immune cell infiltration in androgen-deprived prostate cancer patients. *J. Immunol. Methods* **2009**, *348*, 9–17. [CrossRef]
169. Woo, J.R.; Liss, M.A.; Muldong, M.T.; Palazzi, K.; Strasner, A.; Ammirante, M.; Varki, N.; Shabaik, A.; Howell, S.; Kane, C.J.; et al. Tumor infiltrating B-cells are increased in prostate cancer tissue. *J. Transl. Med.* **2014**, *12*, 30. [CrossRef]
170. Kaur, H.B.; Guedes, L.B.; Lu, J.; Maldonado, L.; Reitz, L.; Barber, J.R.; De Marzo, A.M.; Tosoian, J.J.; Tomlins, S.A.; Schaeffer, E.M.; et al. Association of tumor-infiltrating T-cell density with molecular subtype, racial ancestry and clinical outcomes in prostate cancer. *Mod. Pathol.* **2018**, *31*, 1539–1552. [CrossRef]
171. Giraldo, N.A.; Becht, E.; Pagès, F.; Skliris, G.; Verkarre, V.; Vano, Y.; Mejean, A.; Saint-Aubert, N.; Lacroix, L.; Natario, I.; et al. Orchestration and Prognostic Significance of Immune Checkpoints in the Microenvironment of Primary and Metastatic Renal Cell Cancer. *Clin. Cancer Res.* **2015**, *21*, 3031–3040. [CrossRef]
172. Wang, M.; Zhao, J.; Zhang, L.; Wei, F.; Lian, Y.; Wu, Y.; Gong, Z.; Zhang, S.; Zhou, J.; Cao, K.; et al. Role of tumor microenvironment in tumorigenesis. *J. Cancer* **2017**, *8*, 761–773. [CrossRef]
173. Winter, J.; Kneitz, H.; Bröcker, E.B. Blood vessel density in Basal cell carcinomas and benign trichogenic tumors as a marker for differential diagnosis in dermatopathology. *J. Skin Cancer* **2011**, *2011*, 241382. [CrossRef]
174. Jain, R.K. Antiangiogenesis strategies revisited: From starving tumors to alleviating hypoxia. *Cancer Cell* **2014**, *26*, 605–622. [CrossRef]
175. Lugano, R.; Ramachandran, M.; Dimberg, A. Tumor angiogenesis: Causes, consequences, challenges and opportunities. *Cell. Mol. Life Sci. CMLS* **2020**, *77*, 1745–1770. [CrossRef]
176. Tolaney, S.M.; Boucher, Y.; Duda, D.G.; Martin, J.D.; Seano, G.; Ancukiewicz, M.; Barry, W.T.; Goel, S.; Lahdenrata, J.; Isakoff, S.J.; et al. Role of vascular density and normalization in response to neoadjuvant bevacizumab and chemotherapy in breast cancer patients. *Proc. Natl. Acad. Sci. USA* **2015**, *112*, 14325–14330. [CrossRef]
177. Iakovlev, V.V.; Gabril, M.; Dubinski, W.; Scorilas, A.; Youssef, Y.M.; Faragalla, H.; Kovacs, K.; Rotondo, F.; Metias, S.; Arsanious, A.; et al. Microvascular density as an independent predictor of clinical outcome in renal cell carcinoma: An automated image analysis study. *Lab. Investig.* **2012**, *92*, 46–56. [CrossRef]
178. Niemiec, J.; Sas-Korczynska, B.; Harazin-Lechowska, A.; Martynow, D.; Adamczyk, A. Lymphatic and Blood Vessels in Male Breast Cancer. *Anticancer Res.* **2015**, *35*, 1041–1048.
179. Mohammed, Z.M.A.; Orange, C.; McMillan, D.C.; Mallon, E.; Doughty, J.C.; Edwards, J.; Going, J.J. Comparison of visual and automated assessment of microvessel density and their impact on outcome in primary operable invasive ductal breast cancer. *Hum. Pathol.* **2013**, *44*, 1688–1695. [CrossRef]
180. Haisan, A.; Rogojanu, R.; Croitoru, C.; Jitaru, D.; Tarniceriu, C.; Danciu, M.; Carasevici, E. Digital microscopy assessment of angiogenesis in different breast cancer compartments. *BioMed Res. Int.* **2013**, *2013*, 286902. [CrossRef]
181. Entenberg, D.; Oktay, M.H.; D'Alfonso, T.; Ginter, P.S.; Robinson, B.D.; Xue, X.; Rohan, T.E.; Sparano, J.A.; Jones, J.G.; Condeelis, J.S. Validation of an Automated Quantitative Digital Pathology Approach for Scoring TMEM: A Prognostic Biomarker for Metastasis. *Cancers* **2020**, *12*, 846.
182. Den Uil, S.H.; van den Broek, E.; Coupé, V.M.H.; Vellinga, T.T.; Delis-van Diemen, P.M.; Bril, H.; Belt, E.J.T.; Kranenburg, O.; Stockmann, H.B.A.C.; Belien, J.A.M.; et al. Prognostic value of microvessel density in stage II and III colon cancer patients: A retrospective cohort study. *BMC Gastroenterol.* **2019**, *19*, 146. [CrossRef]
183. Buchberger, E.; Payrhuber, D.; El Harchi, M.; Zagrapan, B.; Scheuba, K.; Zommer, A.; Bugyik, E.; Dome, B.; Kral, J.B.; Schrottmaier, W.C.; et al. Inhibition of the transcriptional repressor complex Bcl-6/BCoR induces endothelial sprouting but does not promote tumor growth. *Oncotarget* **2017**, *8*, 552–564. [CrossRef]
184. Tzeng, H.T.; Tsai, C.H.; Yen, Y.T.; Cheng, H.C.; Chen, Y.C.; Pu, S.W.; Wang, Y.S.; Shan, Y.S.; Tseng, Y.L.; Su, W.C.; et al. Dysregulation of Rab37-Mediated Cross-talk between Cancer Cells and Endothelial Cells via Thrombospondin-1 Promotes Tumor Neovasculature and Metastasis. *Clin. Cancer Res.* **2017**, *23*, 2335–2345. [CrossRef]
185. Kather, J.N.; Marx, A.; Reyes-Aldasoro, C.C.; Schad, L.R.; Zöllner, F.G.; Weis, C.A. Continuous representation of tumor microvessel density and detection of angiogenic hotspots in histological whole-slide images. *Oncotarget* **2015**, *6*, 19163–19176. [CrossRef]
186. Marinaccio, C.; Giudice, G.; Nacchiero, E.; Robusto, F.; Opinto, G.; Lastilla, G.; Maiorano, E.; Ribatti, D. Interval sentinel lymph nodes in melanoma: A digital pathology analysis of Ki67 expression and microvascular density. *Clin. Exp. Med.* **2016**, *16*, 383–389. [CrossRef]
187. Di Maggio, F.; Arumugam, P.; Delvecchio, F.R.; Batista, S.; Lechertier, T.; Hodivala-Dilke, K.; Kocher, H.M. Pancreatic stellate cells regulate blood vessel density in the stroma of pancreatic ductal adenocarcinoma. *Pancreatology* **2016**, *16*, 995–1004. [CrossRef]

188. Schraml, P.; Athelogou, M.; Hermanns, T.; Huss, R.; Moch, H. Specific immune cell and lymphatic vessel signatures identified by image analysis in renal cancer. *Mod. Pathol.* **2019**, *32*, 1042–1052. [CrossRef]
189. Arimoto, A.; Uehara, K.; Tsuzuki, T.; Aiba, T.; Ebata, T.; Nagino, M. Role of bevacizumab in neoadjuvant chemotherapy and its influence on microvessel density in rectal cancer. *Int. J. Clin. Oncol.* **2015**, *20*, 935–942. [CrossRef]
190. Lee, S.Y.; Chao-Nan, Q.; Seng, O.A.; Peiyi, C.; Bernice, W.H.M.; Swe, M.S.; Chii, W.J.; Jacqueline, H.S.G.; Chee, S.K. Changes in specialized blood vessels in lymph nodes and their role in cancer metastasis. *J. Transl. Med.* **2012**, *10*, 206. [CrossRef]

genes

MDPI

Review

The Genetic Complexity of Prostate Cancer

Eva Compérat [1,2,3,*], Gabriel Wasinger [3], André Oszwald [3], Renate Kain [3], Geraldine Cancel-Tassin [1] and Olivier Cussenot [1,4]

1. CeRePP/GRC5 Predictive Onco-Urology, Sorbonne University, 75020 Paris, France; g.cancel@cerepp.org (G.C.-T.); olivier.cussenot@aphp.fr (O.C.)
2. Department of Pathology, Hôpital Tenon, Sorbonne University, 75020 Paris, France
3. Department of Pathology, Medical University of Vienna, 1090 Vienna, Austria; gabriel.wasinger@meduniwien.ac.at (G.W.); andre.oszwald@meduniwien.ac.at (A.O.); renate.kain@meduniwien.ac.at (R.K.)
4. Department of Urology, Hôpital Tenon, Sorbonne University, 75020 Paris, France

* Correspondence: eva.comperat@meduniwien.ac.at; Tel.: +33-658246024

Received: 28 September 2020; Accepted: 23 November 2020; Published: 25 November 2020

Abstract: Prostate cancer (PCa) is a major concern in public health, with many genetically distinct subsets. Genomic alterations in PCa are extraordinarily complex, and both germline and somatic mutations are of great importance in the development of this tumor. The aim of this review is to provide an overview of genetic changes that can occur in the development of PCa and their role in potential therapeutic approaches. Various pathways and mechanisms proposed to play major roles in PCa are described in detail to provide an overview of current knowledge.

Keywords: prostate cancer; germline mutations; somatic mutations; PTEN; TMPRSS2; ERG; androgen receptors

---

1. Introduction

Prostate cancer (PCa) is a major concern in public health, with more than 1.1 million cases worldwide detected every year [1]. Several risk factors for developing PCa are known, e.g., older age, family history and African ethnicity. Despite the refinement of existing treatments and emergence of new management strategies, such as active surveillance and focal therapy, metastatic disease is frequent and mortality is still relatively high, with 26,730 estimated deaths in 2017 [2].

Based on the severity of disease at diagnosis according to differentiation, extension and stage, PCa may be treated in different ways; of particular importance is its initial hormone dependency which allows specific treatments, especially during early-stage disease [1]. Further differences exist between primary, metastatic (mPCa) and castration resistant PCa (CRPCa). Therefore, it is important to know the most important actors, especially the important role of genetics in this hormone-sensitive cancer.

Several studies showed many different genetically distinct subsets of PCa. Various drivers are known, such as androgen-related fusions of ETS-related gene (ERG) and ETS family members, speckle-type pox virus and zinc finger protein (SPOP) mutations, DNA hypermethylation, PIK3/RAS/RAF pathway alterations and DNA damage repair (DDR) pathways.

For better understanding, it is necessary to mention genetic screening, which was explored in the early 2000s and abandoned in 2012 after it was demonstrated that many of the detected PCa were clinically insignificant and did not affect patient life expectancy [3]. In recent years, powerful genetic tests were developed that provide polygenic risk scores for individual patients [4,5]. However, a remaining challenge is the recrudescence of clinically relevant PCa, which may also benefit from personalized approaches for risk assessment or therapy.

Despite recent advances, standard pathology remains a fundamental tool in managing PCa. The Gleason score (GS), reflecting tumor differentiation, is a staple of clinical decision-making, and recent meetings could refine the consensus and diminish interobserver variability [6]. Nevertheless, GS alone will not give all the necessary information; molecular profiling of PCa could provide further information. For instance, a study by Haffner et al. [7] showed that metastasis was not strictly and may result from a tumor region with lower grade, alongside observing PTEN loss. Clinicians need to be increasingly aware that although classical histopathology is a firmly established basis for clinical decisions, it is not the single determinant of PCa behavior.

In recent studies taking a scrutinous approach to Gleason grading of PCa, pathologists showed that cribriform and intraductal PCa, which should be considered Gleason grade 4, were probably more aggressive than the classical Gleason grade 4 pattern. Moreover, several studies underlined a more aggressive behavior of cribriform PCa, partly explained by the underlying molecular aberrations in these tumor patterns. A recent study tested genomic instability by determining the portion of the genome altered and somatic copy number alterations (CNA). Patients with cribriform and/or intraductal PCa and ≥ GS 7 had significantly higher percentages of the genome altered than men without this pattern in both cohorts of The Cancer Genome Atlas (TCGA) (2.2 fold; $p = 0.0003$) and the Canadian Prostate Cancer Genome Network (CPC-GENE) (1.7 fold; $p = 0.004$) [8]. These patterns were associated with deletions of different chromosomes, such as 8p, 16q, 10q23, 13q22, 17p13 and 1q22, and amplification of 8q24, which plays a major role in PCa evolution and is specifically addressed in this review. CNAs comprised a total of 1299 gene deletions and 369 amplifications in the TCGA dataset. Several of the affected genes were known to be associated with aggressive prostate cancer, such as loss of PTEN, CDH1 and BCAR1 and gain of MYC. Point mutations in TP53, SPOP and FOXA1 were also associated with these PCa patterns but occurred less frequently than CNAs. This study clearly shows that cribriform/intraductal patterns are associated with increased genomic instability, clustering to genetic regions involved in aggressive PCa.

The very complex genomic situation of PCa can be broken down into two major aspects, which need to be considered, namely, the germline genetic background and the somatic changes in PCa (Figure 1).

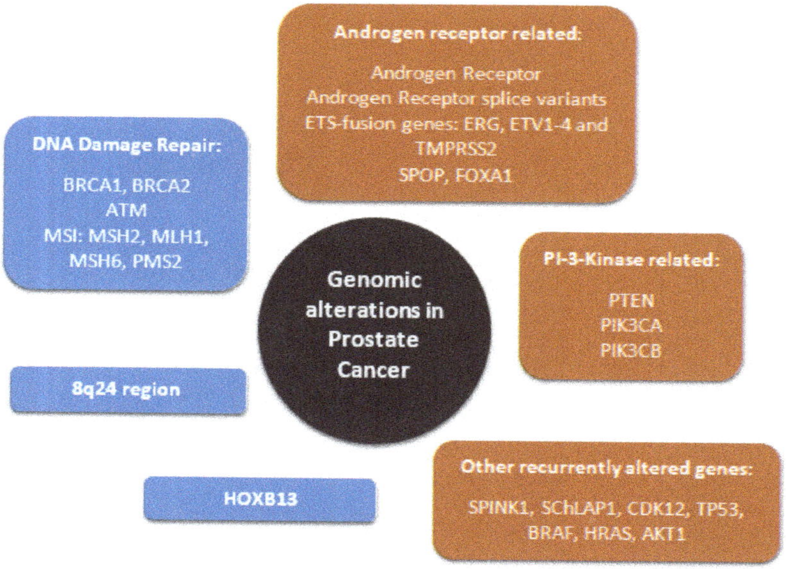

Figure 1. Common genomic alterations in prostate cancer. Germline mutations are highlighted in blue, while typical somatic mutations are highlighted in orange. For abbreviations see full text.

2. Germline Mutations Driving PCa

Hereditary prostate cancer (HPC) is defined by strict clinical criteria and represents 5% of all newly diagnosed PCa [9]. Inherited predisposition to acquire PCa is genetically determined by the presence of a deleterious mutation of DNA repair genes also related to breast/ovarian cancers (i.e., BRCA1 and BRCA2, ATM, etc.) or PCa-specific risk genes (HOXB13 and 8q24 region) [10]. A recent study performed germline sequencing and analysis of DNA repair genes [11] in 5545 men of European ancestry, including 2775 nonaggressive (localized disease, stage T1/T2 and GS $\geq$ 6 tumors) and 2770 aggressive (lethal or metastatic disease, stage T4 or both T3 and GS $\geq$ 8) PCa cases. The authors found that BRCA2 and PALB2 showed the most statistically significant gene-based disease associations, with 2.5% of aggressive and 0.8% of nonaggressive cases carrying deleterious BRCA2 alleles, and 0.65% of aggressive and 0.11% of nonaggressive cases carrying deleterious PALB2 alleles. ATM had a nominal association, with 1.6% of aggressive and 0.8% of nonaggressive cases carrying ATM alleles. According to typology of genetic risk, predisposition exposes the individual to an earlier age of onset or a more aggressive form of the disease, increasing the risk of death from this cancer. Germline mutations of DNA repair factors are found in only up to 8% of these patients due to the rarity of the mutations [12,13].

DNA damage response (DDR) pathways are extremely promising targets in PCa treatment. Two prominent factors are breast cancer 1 and 2 genes (BRCA1 and BRCA2). The TCGA research network, testing 333 primary PCas, reported mutations in DNA repair genes in 19% of primary PCas. Among these, 3% were affected by BRCA2, including germline as well as somatic truncating mutations. Only one case displayed BRCA1 as a frameshift mutation [14]. If these tumor suppressor genes are mutated heterozygously in the germ line, they can generate aggressive forms, especially mPCa [15]. In a normal setting, BRCA1 and BRCA2 repair double-strand breaks by homologous recombination. In case of a BRCA germline mutation, a somatic loss of function in the wild-type BRCA allele is consequently frequently observed together with defective homologous recombination. Although deleterious BRCA germline variants are rare, these patients were shown to more frequently develop PCa and mPCa, and also exhibit high Gleason scores (grade group 3–5) and worse outcomes [16]. Nevertheless, these tumors did not seem to have specific histological aspects allowing them to be recognized on a standard slide. Interestingly, it was observed that ATM and BRCA1/2 germline mutations were associated with Gleason grade reclassification during active surveillance of carrier patients, with an upgrading of GS from 6(3 + 3) to GS 7(3 + 4) or 7(4 + 3) [17].

Another gene implicated in DNA repair mechanisms is ATM, which also plays a role in DNA damage repair and mediates downstream checkpoint signaling. Its prevalence in mPCa is 1.6% and therefore it must be considered, especially in these forms of PCa [15]. ATM also integrates the concept of the so-called homologous repair deficiency profile [18]. These mutations were reported in nearly 2.5% of PCa patients, some detected in lethal PCa (death due to metastatic PCa), with fewer in localized PCa (low-risk disease, GS $\leq$ 6, organ confined) [19]. Interestingly, in Chinese patients, ATM mutations were found in 4.5% of lethal PCa. In the mentioned American–African–Asian study, no co-mutations BRCA1/2–ATM were detected, indicating that these mutations are probably simultaneously exclusive. This study also demonstrated that mutation carriers displayed higher overall lethality, higher mPCa rates and lower PCa-specific survival in patients with diagnosed mPCa, especially in young patients under 60 years. On the other hand, the lowest carrier rate was among patients who died of PCa > 75 years or > 10 years after their initial PCa diagnosis.

The potential of detecting patients with the above-mentioned germline mutations opens the door to specific treatment approaches, one of the most promising drugs being inhibitors of anti-poly ADP-ribose polymerase (PARP). PARP inhibitors induce cell death because they interfere with a cellular mechanism of single-strand DNA break repair. These occur normally during the cell cycle, but also during oncogenesis. In case of mutation of other abovementioned repair genes, PARP is required to repair both strands, which means that the cell is entirely dependent on PARP for single-strand repair. In context of BRCA mutation and concomitant PARP inhibition, for example, a tumor cell would not be able to perform these measures, resulting in chromosomal instability, cell-cycle arrest and apoptosis [20].

A recent study by Hussain et al. [21] showed that patients with PCa harboring at least one mutation in BRCA1, BRCA2 or ATM who received the PARP inhibitor olaparib had significantly longer survival than those who received enzalutamide or abiraterone plus prednisone as control therapy. Other studies also demonstrated that metastatic, castration-resistant PCa (mCRPCa) with BRCA2 germline mutations or deleterious variants showed greater response to platinum-based chemotherapy [22]. This relationship between genotype and responsiveness to platinum chemotherapy was also observed among BRCA2 patients with breast and ovarian cancers. Another study showed concordant results, underlining that biallelic BRCA2 inactivation in mCRPCa could serve as a biomarker for predicting sensitivity to platinum chemotherapy, which showed a clear benefit in biallelic, BRCA-mutated patients [23].

These types of studies underline the necessity of germline testing in special patient groups. Considerations exist regarding who to test for genetic counseling, such as a known mutation in a cancer-susceptibility gene within the family, mPCa, high-risk, localized PCa (grade group 4/5, PSA $\geq$ 20, WHO group $\geq$ 3), young age at PCa detection, family history suggestive of Lynch syndrome and hereditary breast ovarian or prostate cancer. Prostate cancers with deficiency of the DNA mismatch repair (MMR) system are rare, with 1% at the localized stage and 5% in the advanced stages [24]. PCa rarely occurs as an unconventional malignancy of the Lynch syndrome spectrum (HNPCC) [25]. These forms of PCa have aggressive anatomopathological criteria, with intraductal forms (25%), a high incidence of Gleason score 8 and high rates (40–50%) of metastatic de novo disease which are frequently visceral (30% of metastatic patients). At the molecular level, they are characterized by a high rate of mutation (tumor mutational burden). Considering their advanced stage, they respond relatively well to androgenic deprivation, but less so to taxanes once castration-resistant. Afflicted patients are a subgroup currently considered candidates and evaluated for anti-PD1 or anti-CTL4 immunotherapy [26,27]. On the germinal level, Pritchard et al. [15] reported that, of 692 patients with mPCa, 12% showed deleterious germline mutations (1% involving MSH2, 1% MSH6 and 2% PMS2). Abida et al. [26] reported that, of 1346 PCa patients, 3% showed high microsatellite instability, of which 22% possessed a germ mutation in a gene associated with Lynch syndrome. Most patients (46%) had MSH6 mutations.

3. 8q24 Region

8q24 is a hotspot of susceptibility loci for PCa. These risk loci, identified by genome-wide association studies (GWAS), do not affect the coding DNA and are frequently associated with single-nucleotide polymorphisms (SNPs). It was shown that amplification of 8q24 (harboring MYC) is frequent. Often, enhancers such as rs-6983267 interact with MYC and alter sensitivity to certain crucial signaling pathways, e.g., WNT [28,29]. With ongoing research, it is evident that 8q24 variants play a role in PCa carcinogenesis. A recent meta-analysis determined significant associations between PCa risk and 15 variants in 8q24. The 8q24 region is dense with SNPs; some of these variants might enhance genes implicated in PCa carcinogenesis [29].

Although the inherited component for PCa was previously acknowledged, the identification of genetic variation on 8q24 conferring cancer-specific susceptibility may help to improve screening strategies. Genome-wide studies reported low penetrance of signals influencing the risk. Nevertheless, since risk alleles are relatively common in the population, their cumulative impact is potentially substantial. One recent study underlined several independent signals in different regions and yielded a biological annotation enriched with different elements, such as promoters, enhancers and transcription factor-binding sites, such as AR, ERG and FOXA1 [30]. In another study by the same group, 12 independent risk signals with different variants were described, some of them for the first time (rs1914295, rs190257175 and rs12549761). They were weakly correlated with already known PCa risk markers. On the other hand, this study showed that men with a cumulative risk score had a greater risk of developing PCa than the average population. The described 12 variants accounted for around 25% of what could be explained as familial genetic risk factor, highlighting the contribution of germline variation on 8q24. However, although the 8q24 region is now established as a major region for PCa susceptibility, the underlying biological mechanisms still require further elucidation [10].

4. Somatic Mutations Driving PCa

Outlier androgen-regulated genes play an important role in PCa development. These genes are generally expressed at low levels, but can show variation in expression of different genomic subsets [31,32]. Especially in light of emerging personalized medicine, it is increasingly important to take into consideration the individual mechanisms driving aggressive PCa.

The gene fusions in PCa are mostly controlled by androgen, fusing to members of the ETS (E26 transformation-specific) family of transcription factors [33]. One of the most cited gene fusions in PCa is the overexpression of TMPRSS2–ERG (T2E) gene fusion. A consequence of this fusion is an overexpression of oncogenic factors; this is frequently present globally, in about 50% of PCa [33]. TMPRSS2 also fuses to other ETS family genes, such as ETV1, ETV4 and, rarely, ETV5. From a morphological point of view, these tumors frequently display particular aspects, such as macronuclei, signet cell rings, cribriform aspects and intraductal carcinoma. The prognostic impact of T2E is still a matter of controversy. T2E acts as an aberrant transcription factor with oncogenic properties. Several papers indicated that its presence was an indicator of aggressive phenotype and poor prognosis [34–36]. Some authors claimed that it was the most important prognostic factor in patients treated with prostatectomy [37]. Furthermore, it was proposed to be a risk marker for lymph-node metastasis and poorly differentiated disease, as well as biochemical recurrence at five years [37,38]. Nevertheless, the data are still controversial, and T2E is not taken into consideration for decision-making at the moment. Several contrasting studies found no clinical significance of the TMPRSS2-ERG fusion [38–40]. Interestingly, recent studies showed that T2E-positive and -negative PCa are two different molecular groups of PCa, suggesting that the T2E status determines the nature of metastasis-related gene signatures [40,41]. In T2E-positive cases overexpression in mPCa was seen for several genes (GMNN, TROAP and WEE1). In T2E-negative cases, completely different metastasis-associated genes were expressed, such as ASPN, BGN and TYMS. Therefore, the authors concluded that, according to the T2E status, different genes are linked to the development of mPCa. In patients with overexpression of ASPN, BGN and TYMS, shorter event-free survivals were exhibited. TYMS (thymidylate synthetase), for instance, plays a role in DNA replication and repair. Interestingly, neither PTEN nor TP53 mutations biased the results. Finally, the study showed that the T2E status is not a strong prognostic biomarker per se, but determines the prognostic value of other biomarkers [41].

Another interesting outlier gene is SPINK1, which seems to be, in most cases, mutually exclusive with ERG overexpression [41]. SPINK1, located on 5q32, encodes a protein which functions as a trypsin inhibitor. SPINK1 positivity was shown to be an independent predictor of shorter biochemical recurrence and progression-free survival. Recombinant SPINK1 protein is able to stimulate cell proliferation in benign prostate tissue. A recent study showed that SPINK1 overexpression is linked to higher PTEN expression and lower AR (androgen receptor) expression, with the authors further suggesting that SPINK1 protein expression may not be a predictor of recurrence or lethal PCa amongst men treated by radical prostatectomy. SPINK1 and ERG proteins were not entirely mutually exclusive in this study, as some previous studies suggested [42].

Another important outlier to mention is SChLAP1, a noncoding RNA gene, located on 2q31. Apparently, SChLAP1 has no coding potential, is located in the nucleus and is associated with ETS gene fusion as well as with mPCa. SChLAP1 seems to coordinate cell proliferation and metastatic spread. High expression is associated with lethal PCa, independently of tumor differentiation (Gleason score), tumor stage, PTEN status and age. In a multivariate analysis, SChLAP1 predicted mPCa within 10 years (odds ratio = 2.45) [43].

Obviously, single gene outliers alone are limited in their results and application. These genes might become the basis of patient risk stratification and adaptations in treatment, e.g., more intense earlier treatment or targeted therapies, and will probably become more available in the upcoming years to facilitate genetic and individual subclassification [44].

5. Recurrently Altered Genes

PCa has a lower mutational burden than many other epithelial tumors. The recent TCGA paper showed several significantly mutated genes, such as SPOP, TP53, FOXA1, PTEN and others [14]. Clinically relevant genes, such as BRAF, HRAS and AKT1, as well as the β-catenin pathway and the DNA repair pathway (see above), also showed importance.

Loss of PTEN (located on 10q23) is frequently observed and is closely related to MYC overexpression. The latter is found on 8q24. Both together seem to play a role in high-risk PCa.

PTEN loss is associated with adverse findings in early PCa and occurs in approximatively 15% as homozygous deletions. PTEN loss or mRNA-based genomic signatures can be useful to help determine whether definitive therapy is required, and its loss seems to be more frequent in patients with African ancestry [45]. Early investigations already showed that loss of PTEN, even when detected by immunohistochemistry, was a predictor of aggressive metastatic disease. A paper by Haffner et al. [7] demonstrated that PTEN mutation was not present in morphological, higher grade lesions surrounding a tumor with PTEN loss when the Gleason score was lower. Associated TP53 and SPOP mutations were reported in the same patient. Interestingly, lymph node metastasis did not harbor the same mutations, suggesting an independent clonal/subclonal origin of these lesions. Genetically, there is strong evidence to suggest that poorly differentiated PCa (Gleason score 9/10) has a higher level of genomic instability with an increased rate of copy-number alterations and alterations in key signaling pathways (TP53, PTEN and RB1) associated with resistance to androgen deprivation therapy (ADT) [14]. The function of PTEN is closely linked to the PIK3 pathway, in which PTEN is generally considered a negative regulator. PIK3CA was shown to be frequently mutated, either via activation of mutational hotspots, coincidently activating mutations or amplification. According to the literature, PTEN-deleted tumors are likely to be PIK3CB-dependent; a coexistent loss and mutation of PTEN and PIK3CB might increase PIK3 pathway output and indicate PCa with AR signaling inhibition. PIK3 pathway and DNA repair alterations seem to be more frequent in metastatic specimens [14].

MYC overexpression is an early event in prostate cancer development. Apparently, concomitant with PTEN loss via characterized HOXB13 transcription control, genomic instability and aggressive disease with a high risk of metastases are initiated. These tumors typically show a high Gleason score and disease progression. On the other hand, isolated MYC activation and PTEN were reported to be insufficient to induce invasive PCa, with cells remaining in a precancerous stage (high-grade prostate intraepithelial neoplasia (PIN)) [46]. A recent study from Liu et al. [47] showed that gain of MYC and loss of PTEN resulted in elevated PCa mortality associated even with single copy-number changes. Regarding the interplay of these two factors, MYC overexpression can induce genetic instability, while PTEN might repress this process in PCa cells. Recent results suggested that PTEN might play a role in DNA repair, and if PTEN is lost, high levels of DNA damage that normally repress apoptosis due to increased PIK3 signaling are introduced [48].

One of the genes recurrently mutated in PCa is SPOP, an E3 ubiquitin ligase adaptor protein of the ubiquitin–proteasome system. Mutations may affect the degradation of developmental regulators of PCa, including AR. SPOP plays a role in several further important cellular functions, and intervenes in transcription, cell-cycle regulation and apoptosis.

SPOP mutations were previously described in precancerous lesions and primary tumors of the prostate, suggesting that SPOP mutations are early and recurrent events in the development of PCa [7]. SPOP mutations were identified in 6–13% of primary prostate adenocarcinomas and 14.5% of metastatic prostate cancers, but data regarding expression in distant metastasis are sparse [49]. Recent studies showed that SPOP mutations are less frequent in metastatic than in primary tumors (8% versus 11%) [14]. In the abovementioned article by Haffner et al. [7] describing the clonal evolution of a PCa in metastasis, SPOP was shown to be mutated in the lethal metastatic cell clone. In this context, it is intriguing that SPOP is also a component of the DNA damage response (DDR).

One PCa-relevant direct substrate of SPOP is the androgen receptor (AR) [50], which harbors a SPOP-binding motif. When binding to SPOP, AR undergoes ubiquitin-mediated degradation. AR splice variants that lack the SPOP-binding motif escape this degradation. Interestingly, PCa-associated SPOP mutants do not bind to AR or promote its degradation. SPOP-mediated degradation of AR is driven by antiandrogens and blocked by androgens. It is not clear whether SPOP can interact with other nuclear receptors [51].

It is unclear whether the same SPOP mutations are present in different ethnicities, since some authors described differences according to the patients' ancestry. We [52] and others observed different mutation frequencies between African and European patients, with SPOP mutations differing significantly between both groups. In our study, SPOP mutations were found in more than 20% of patients with African origin compared to 10% of European origin. In contrast, a recent study analyzing 720 PCa samples from six international cohorts, including Caucasian, African–American and Asian PCa patients, showed that SPOP mutations were variably frequent (4.6–14.4%), but found no association with ethnicity. Hence, the authors concluded that SPOP mutations were not associated with ethnicity, biochemical recurrence, clinical parameters or pathological parameters [53]. In light of inconsistent data, more studies are needed to make a conclusive statement.

CDK12 is a gene also implicated in DNA repair by regulation of the expression levels of different DNA repair damage response genes. CDK12 is recurrently mutated in aggressive localized PCa [52] and in mPCa [54]. CDK12 biallelic loss is associated with focal tandem duplications. Moreover, CDK12 mutations are related to increased gene fusion, which can yield neoantigens and induce strong immune infiltration, suggesting that patients with these mutations could benefit from immune checkpoint immunotherapy [54].

6. Androgen Receptors

PCa is a hormone-sensitive cancer, and androgen receptors (AR) play a major role in the treatment and development of PCa. For advanced PCa, ADT is the standard of care, but is decisively ineffective in castration-resistant PCa (CRPCa) [1]. Under ADT treatment, circulating testosterone indicates the androgen suppression level. In CRPCa, the activity of AR remains elevated, despite reduced androgen levels. In recent years, second-generation, AR-targeting therapies were developed in order to treat CRPCa with agents to antagonize AR and to suppress extragonadal androgen (coming from the adrenal gland, for instance) [55].

The AR gene, located on Xq11-12, is a major transcriptional regulator in the normal prostate, but also in PCa cells. The AR, a steroid hormone receptor, forms a complex with heat-shock protein 90 (HSP-90). When binding with androgen, it undergoes a change, allowing nuclear translocation, DNA binding and regulation of gene transcription [56]. Different structural variants exist, and some AR were previously implicated in aggressive tumor activity (see below). The AR gene may undergo genomic alterations such as point mutations, which can induce structural changes. These alterations, specifically seen in CRPCa, can help to understand the dependence of CRPCa on androgen-independent AR signaling.

After ligand binding, the AR is translocated into the nucleolus, forms a dimer and binds to the androgen-response element of the promoter or enhancer of target genes [57]. Furthermore, the AR dimer forms a complex with coactivator- and coregulatory proteins in different regions and regulates gene expression with diverse functions. These are located downstream of the androgen response element, including fusion genes (TMPRESS2–ERG), transcription factors (FOXP1, NKX3.1) and others. Many coactivators interacting with different AR domains are implicated in AR activation in therapy-resistant PCa. While normal AR activity on transcription is ligand-driven, AR transcript variants can encode truncated AR proteins lacking a ligand-binding domain, which can activate AR-target genes in the absence of androgens [58].

The spectrum of genes regulated by AR deserves special attention. AR has transcriptional activity and structural variants exist which play roles in the outcome of the patient. However, it is unknown to what extent individual primary PCa tumors differ in androgen sensitivity or dependence. Androgen activity is a central axis in PCa evolution, and drives most ETS fusion genes [26]. ETS fusion genes are under AR control, but the ETS fusion-positive groups have different AR transcriptional activity. More frequent events, such as androgen-regulated fusions of ERG or ETS family members, form distinct PCa groups (T2E, see above). The most frequent drivers are ERG, ETV1-4, SPOP and FOXA1 mutations [14]. TMPRSS2 is the most frequent fusion partner of all ETS fusions and is androgen-related. Tumors with SPOP or FOXA1 mutations have the highest AR transcriptional activity, as SPOP mutations deregulate AR and AR co-activators [59]. In PCa. two different changes affect the AR pathway. The output is controlled by AR mRNA and protein expression, but also by the expression and mutations of AR cofactors [60]. FOXA1, which can be mutated, is a transcription factor that targets AR and plays an obvious role in PCa oncogenesis. A subset of these mutations also harbor SPOP mutations and consequently produce high AR levels.

It must be mentioned that the AR is the most frequently aberrant gene in mCRPCa, with around 63% aberrant expression [12]. Point mutations are found in 15–30% of CRPCa, most of the mutations residing in the ligand-binding domain. These point mutations can activate AR with a specific point mutation (T878A), which also activates resistance to second generation AR agonists [61]. These mutations were detected in 13% of CRPCa patients with abiraterone resistance. A second observed mechanism was the amplification of the AR receptor, seen in up to 50% of patients with CRPCa [62]. In case of ADT, low androgen levels still exist. In case of AR amplifications, PCa cells can survive under ADT, leading to progression of CRPCa. Change in androgen biosynthesis also plays a role. For instance, during ADT, the adrenal gland still produces androgens, while CRPCa overexpresses converting enzymes, converting weak androstenedione levels into DHT to activate AR [63].

The AR splice variants (AR-Vs) have different mRNA sequences than the full-length AR (AR-FL). Around 22 AR-Vs are currently known, most of them lacking the ligand-binding domain (LBD), which is the target of existing AR therapy. Most of these AR-Vs mediate active AR signaling, which means that they can act without the presence of androgens or AR-FL [64]. One of the best-described variants is AR-V7, which also lacks the LBD. AR-V7, known to be an important factor in the treatment of CRPCa, can already be detected in primary tumors and surrounding normal prostate tissue. This is surprising, since truncated AR splice variants were proposed to be expressed predominantly in mCRPCa, and their presence was associated with hormone-therapy resistance, at least for AR-V7. However, the TCGA study showed that these splice variants are already expressed in therapy-naive primary PCa [14]. Structural alterations of AR-Vs may occur in the same allele, leading to a generation of AR-Vs displaying deletions and duplications within the AR LBD. AR-V expression can be regulated by both splicing enhancer sequences and also protein kinase pathways [65].

It could be beneficial to obtain AR-V7 data before treating patients with mCRPCa, as a recent study suggested [66]. The authors showed that patients with mCRPCa exhibiting nuclear localized AR-V7 in circulating tumor cells had better outcomes when treated by taxanes than by AR signaling inhibitors. In a follow-up study, Graf et al. [67] showed that AR-V7-positive patients who were treated with taxanes exhibited better survival, while those who were AR-V7-negative exhibited better survival when treated with AR signaling inhibitors. Therefore, in the era of emerging personalized medicine, the status of the most important splice variants could become an important clinical criterion in the future.

7. Conclusions

As this short overview shows, genomic changes in PCa development are extremely complex and many different factors need to be considered. One major challenge of advanced disease is that many proposed underlying mechanisms are still insufficiently understood or unclear due to contrasting results, and thus cannot yet be leveraged as refined therapeutic approaches. However, science is progressing rapidly. This review provides a snapshot of the current knowledge, but in the upcoming years, more data will be available to treat this frequent cancer more effectively.

Author Contributions: Conceptualization, E.C. and O.C.; methodology, E.C., O.C. and G.C.-T.; software, G.W., A.O.; validation, E.C., O.C. and G.C.-T.; formal analysis, G.W., A.O., R.K.; investigation, E.C.; writing—original draft preparation, E.C., O.C., G.W., A.O.; writing—review and editing, R.K.; visualization, G.W.; supervision, E.C. and O.C.; project administration, R.K.; All authors have read and agreed to the published version of the manuscript.

Funding: This research received no external funding.

Conflicts of Interest: The authors declare no conflict of interest.

References

1. Lam, T.B.L.; MacLennan, S.; Willemse, P.-P.M.; Mason, M.D.; Plass, K.; Shepherd, R.; Baanders, R.; Bangma, C.H.; Bjartell, A.; Bossi, A.; et al. EAU-EANM-ESTRO-ESUR-SIOG Prostate Cancer Guideline Panel Consensus Statements for Deferred Treatment with Curative Intent for Localised Prostate Cancer from an International Collaborative Study (DETECTIVE Study). *Eur. Urol.* **2019**, *76*, 790–813. [CrossRef] [PubMed]
2. Siegel, R.L.; Miller, K.D.; Jemal, A. Cancer Statistics, 2017. *CA Cancer J. Clin.* **2017**, *67*, 7–30. [CrossRef] [PubMed]
3. Giri, V.N.; Knudsen, K.E.; Kelly, W.K.; Abida, W.; Andriole, G.L.; Bangma, C.H.; Bekelman, J.E.; Benson, M.C.; Blanco, A.; Burnett, A.; et al. Role of Genetic Testing for Inherited Prostate Cancer Risk: Philadelphia Prostate Cancer Consensus Conference 2017. *J. Clin. Oncol.* **2018**, *36*, 414–424. [CrossRef] [PubMed]
4. Huynh-Le, M.-P.; Fan, C.C.; Karunamuni, R.; Walsh, E.I.; Turner, E.L.; Lane, J.A.; Martin, R.M.; Neal, D.E.; Donovan, J.L.; Hamdy, F.C.; et al. A Genetic Risk Score to Personalize Prostate Cancer Screening, Applied to Population Data. *Cancer Epidemiol. Biomark. Prev.* **2020**, *29*, 1731–1738. [CrossRef] [PubMed]
5. Viste, E.; Vinje, C.A.; Lid, T.G.; Skeie, S.; Evjen-Olsen, Ø.; Nordström, T.; Thorsen, O.; Gilje, B.; Janssen, E.A.M.; Kjosavik, S.R. Effects of replacing PSA with Stockholm3 for diagnosis of clinically significant prostate cancer in a healthcare system-the Stavanger experience. *Scand. J. Prim. Health Care* **2020**, *38*, 315–322. [CrossRef]
6. Epstein, J.I.; Egevad, L.; Amin, M.B.; Delahunt, B.; Srigley, J.R.; Humphrey, P.A. Grading Committee The 2014 International Society of Urological Pathology (ISUP) Consensus Conference on Gleason Grading of Prostatic Carcinoma: Definition of Grading Patterns and Proposal for a New Grading System. *Am. J. Surg. Pathol.* **2016**, *40*, 244–252.
7. Haffner, M.C.; Mosbruger, T.; Esopi, D.M.; Fedor, H.; Heaphy, C.M.; Walker, D.A.; Adejola, N.; Gürel, M.; Hicks, J.; Meeker, A.K.; et al. Tracking the clonal origin of lethal prostate cancer. *J. Clin. Investig.* **2013**, *123*, 4918–4922. [CrossRef]
8. Böttcher, R.; Kweldam, C.F.; Livingstone, J.; Lalonde, E.; Yamaguchi, T.N.; Huang, V.; Yousif, F.; Fraser, M.; Bristow, R.G.; van der Kwast, T.; et al. Cribriform and intraductal prostate cancer are associated with increased genomic instability and distinct genomic alterations. *BMC Cancer* **2018**, *18*, 8. [CrossRef]
9. Cancel-Tassin, G.; Cussenot, O. Genetic susceptibility to prostate cancer. *BJU Int.* **2005**, *96*, 1380–1385. [CrossRef]
10. Matejcic, M.; Saunders, E.J.; Dadaev, T.; Brook, M.N.; Wang, K.; Sheng, X.; Olama, A.A.A.; Schumacher, F.R.; Ingles, S.A.; Govindasami, K.; et al. Germline variation at 8q24 and prostate cancer risk in men of European ancestry. *Nat. Commun.* **2018**, *9*, 4616. [CrossRef]
11. Darst, B.F.; Dadaev, T.; Saunders, E.; Sheng, X.; Wan, P.; Pooler, L.; Xia, L.Y.; Chanock, S.; Berndt, S.I.; Gapstur, S.M.; et al. Germline sequencing DNA repair genes in 5545 men with aggressive and non-aggressive prostate cancer. *J. Natl. Cancer Inst.* **2020**. [CrossRef] [PubMed]
12. Robinson, D.; Van Allen, E.M.; Wu, Y.-M.; Schultz, N.; Lonigro, R.J.; Mosquera, J.-M.; Montgomery, B.; Taplin, M.-E.; Pritchard, C.C.; Attard, G.; et al. Integrative Clinical Genomics of Advanced Prostate Cancer. *Cell* **2015**, *162*, 454. [CrossRef] [PubMed]
13. Daly, M.B.; Pilarski, R.; Berry, M.; Buys, S.S.; Farmer, M.; Friedman, S.; Garber, J.E.; Kauff, N.D.; Khan, S.; Klein, C.; et al. NCCN Guidelines Insights: Genetic/Familial High-Risk Assessment: Breast and Ovarian, Version 2.2017. *J. Natl. Compr. Cancer Netw.* **2017**, *15*, 9–20. [CrossRef] [PubMed]
14. Cancer Genome Atlas Research Network The Molecular Taxonomy of Primary Prostate Cancer. *Cell* **2015**, *163*, 1011–1025. [CrossRef]
15. Pritchard, C.C.; Mateo, J.; Walsh, M.F.; De Sarkar, N.; Abida, W.; Beltran, H.; Garofalo, A.; Gulati, R.; Carreira, S.; Eeles, R.; et al. Inherited DNA-Repair Gene Mutations in Men with Metastatic Prostate Cancer. *N. Engl. J. Med.* **2016**, *375*, 443–453. [CrossRef]

16. Mitra, A.; Fisher, C.; Foster, C.S.; Jameson, C.; Barbachanno, Y.; Bartlett, J.; Bancroft, E.; Doherty, R.; Kote-Jarai, Z.; Peock, S.; et al. Prostate cancer in male BRCA1 and BRCA2 mutation carriers has a more aggressive phenotype. *Br. J. Cancer* **2008**, *98*, 502–507. [CrossRef]
17. Carter, H.B.; Helfand, B.; Mamawala, M.; Wu, Y.; Landis, P.; Yu, H.; Wiley, K.; Na, R.; Shi, Z.; Petkewicz, J.; et al. Germline Mutations in ATM and BRCA1/2 Are Associated with Grade Reclassification in Men on Active Surveillance for Prostate Cancer. *Eur. Urol.* **2019**, *75*, 743–749. [CrossRef]
18. Pennington, K.P.; Walsh, T.; Harrell, M.I.; Lee, M.K.; Pennil, C.C.; Rendi, M.H.; Thornton, A.; Norquist, B.M.; Casadei, S.; Nord, A.S.; et al. Germline and somatic mutations in homologous recombination genes predict platinum response and survival in ovarian, fallopian tube, and peritoneal carcinomas. *Clin. Cancer Res.* **2014**, *20*, 764–775. [CrossRef]
19. Na, R.; Zheng, S.L.; Han, M.; Yu, H.; Jiang, D.; Shah, S.; Ewing, C.M.; Zhang, L.; Novakovic, K.; Petkewicz, J.; et al. Germline Mutations in ATM and BRCA1/2 Distinguish Risk for Lethal and Indolent Prostate Cancer and are Associated with Early Age at Death. *Eur. Urol.* **2017**, *71*, 740–747. [CrossRef]
20. Ashworth, A.; Lord, C.J. Synthetic lethal therapies for cancer: What's next after PARP inhibitors? *Nat. Rev. Clin. Oncol.* **2018**, *15*, 564–576. [CrossRef]
21. Hussain, M.; Mateo, J.; Fizazi, K.; Saad, F.; Shore, N.; Sandhu, S.; Chi, K.N.; Sartor, O.; Agarwal, N.; Olmos, D.; et al. Survival with Olaparib in Metastatic Castration-Resistant Prostate Cancer. *N. Engl. J. Med.* **2020**. [CrossRef] [PubMed]
22. Pomerantz, M.M.; Spisák, S.; Jia, L.; Cronin, A.M.; Csabai, I.; Ledet, E.; Sartor, A.O.; Rainville, I.; O'Connor, E.P.; Herbert, Z.T.; et al. The association between germline BRCA2 variants and sensitivity to platinum-based chemotherapy among men with metastatic prostate cancer. *Cancer* **2017**, *123*, 3532–3539. [CrossRef] [PubMed]
23. Lord, C.J.; Ashworth, A. The DNA damage response and cancer therapy. *Nature* **2012**, *481*, 287–294. [CrossRef] [PubMed]
24. Sharma, M.; Yang, Z.; Miyamoto, H. Loss of DNA mismatch repair proteins in prostate cancer. *Medicine (Baltimore)* **2020**, *99*, e20124. [CrossRef] [PubMed]
25. Beebe-Dimmer, J.L.; Kapron, A.L.; Fraser, A.M.; Smith, K.R.; Cooney, K.A. Risk of Prostate Cancer Associated With Familial and Hereditary Cancer Syndromes. *J. Clin. Oncol.* **2020**, *38*, 1807–1813. [CrossRef]
26. Abida, W.; Cheng, M.L.; Armenia, J.; Middha, S.; Autio, K.A.; Vargas, H.A.; Rathkopf, D.; Morris, M.J.; Danila, D.C.; Slovin, S.F.; et al. Analysis of the Prevalence of Microsatellite Instability in Prostate Cancer and Response to Immune Checkpoint Blockade. *JAMA Oncol.* **2019**, *5*, 471–478. [CrossRef]
27. Graham, L.S.; Montgomery, B.; Cheng, H.H.; Yu, E.Y.; Nelson, P.S.; Pritchard, C.; Erickson, S.; Alva, A.; Schweizer, M.T. Mismatch repair deficiency in metastatic prostate cancer: Response to PD-1 blockade and standard therapies. *PLoS ONE* **2020**, *15*, e0233260. [CrossRef]
28. Tuupanen, S.; Turunen, M.; Lehtonen, R.; Hallikas, O.; Vanharanta, S.; Kivioja, T.; Björklund, M.; Wei, G.; Yan, J.; Niittymäki, I.; et al. The common colorectal cancer predisposition SNP rs6983267 at chromosome 8q24 confers potential to enhanced Wnt signaling. *Nat. Genet.* **2009**, *41*, 885–890. [CrossRef]
29. Tong, Y.; Yu, T.; Li, S.; Zhao, F.; Ying, J.; Qu, Y.; Mu, D. Cumulative Evidence for Relationships Between 8q24 Variants and Prostate Cancer. *Front. Physiol.* **2018**, *9*, 915. [CrossRef]
30. Dadaev, T.; Saunders, E.J.; Newcombe, P.J.; Anokian, E.; Leongamornlert, D.A.; Brook, M.N.; Cieza-Borrella, C.; Mijuskovic, M.; Wakerell, S.; Olama, A.A.A.; et al. Fine-mapping of prostate cancer susceptibility loci in a large meta-analysis identifies candidate causal variants. *Nat. Commun.* **2018**, *9*, 2256. [CrossRef]
31. Hellwig, B.; Hengstler, J.G.; Schmidt, M.; Gehrmann, M.C.; Schormann, W.; Rahnenführer, J. Comparison of scores for bimodality of gene expression distributions and genome-wide evaluation of the prognostic relevance of high-scoring genes. *BMC Bioinform.* **2010**, *11*, 276. [CrossRef] [PubMed]
32. Kamoun, A.; Cancel-Tassin, G.; Fromont, G.; Elarouci, N.; Armenoult, L.; Ayadi, M.; Irani, J.; Leroy, X.; Villers, A.; Fournier, G.; et al. Comprehensive molecular classification of localized prostate adenocarcinoma reveals a tumour subtype predictive of non-aggressive disease. *Ann. Oncol.* **2018**, *29*, 1814–1821. [CrossRef] [PubMed]
33. Kumar-Sinha, C.; Tomlins, S.A.; Chinnaiyan, A.M. Recurrent gene fusions in prostate cancer. *Nat. Rev. Cancer* **2008**, *8*, 497–511. [CrossRef] [PubMed]
34. Hemminki, K.; Ji, J.; Försti, A.; Sundquist, J.; Lenner, P. Concordance of survival in family members with prostate cancer. *J. Clin. Oncol.* **2008**, *26*, 1705–1709. [CrossRef] [PubMed]

35. Brandt, A.; Sundquist, J.; Hemminki, K. Risk for incident and fatal prostate cancer in men with a family history of any incident and fatal cancer. *Ann. Oncol.* **2012**, *23*, 251–256. [CrossRef]
36. Jansson, K.F.; Akre, O.; Garmo, H.; Bill-Axelson, A.; Adolfsson, J.; Stattin, P.; Bratt, O. Concordance of tumor differentiation among brothers with prostate cancer. *Eur. Urol.* **2012**, *62*, 656–661. [CrossRef]
37. Nam, R.K.; Sugar, L.; Yang, W.; Srivastava, S.; Klotz, L.H.; Yang, L.-Y.; Stanimirovic, A.; Encioiu, E.; Neill, M.; Loblaw, D.A.; et al. Expression of the TMPRSS2:ERG fusion gene predicts cancer recurrence after surgery for localised prostate cancer. *Br. J. Cancer* **2007**, *97*, 1690–1695. [CrossRef]
38. Attard, G.; Clark, J.; Ambroisine, L.; Fisher, G.; Kovacs, G.; Flohr, P.; Berney, D.; Foster, C.S.; Fletcher, A.; Gerald, W.L.; et al. Duplication of the fusion of TMPRSS2 to ERG sequences identifies fatal human prostate cancer. *Oncogene* **2008**, *27*, 253–263. [CrossRef]
39. Yoshimoto, M.; Joshua, A.M.; Chilton-Macneill, S.; Bayani, J.; Selvarajah, S.; Evans, A.J.; Zielenska, M.; Squire, J.A. Three-color FISH analysis of TMPRSS2/ERG fusions in prostate cancer indicates that genomic microdeletion of chromosome 21 is associated with rearrangement. *Neoplasia* **2006**, *8*, 465–469. [CrossRef]
40. Saramäki, O.R.; Harjula, A.E.; Martikainen, P.M.; Vessella, R.L.; Tammela, T.L.J.; Visakorpi, T. TMPRSS2:ERG fusion identifies a subgroup of prostate cancers with a favorable prognosis. *Clin. Cancer Res.* **2008**, *14*, 3395–3400. [CrossRef]
41. Gerke, J.S.; Orth, M.F.; Tolkach, Y.; Romero-Pérez, L.; Wehweck, F.S.; Stein, S.; Musa, J.; Knott, M.M.L.; Hölting, T.L.B.; Li, J.; et al. Integrative clinical transcriptome analysis reveals TMPRSS2-ERG dependency of prognostic biomarkers in prostate adenocarcinoma. *Int. J. Cancer* **2020**, *146*, 2036–2046. [CrossRef]
42. Flavin, R.; Pettersson, A.; Hendrickson, W.K.; Fiorentino, M.; Finn, S.; Kunz, L.; Judson, G.L.; Lis, R.; Bailey, D.; Fiore, C.; et al. SPINK1 protein expression and prostate cancer progression. *Clin. Cancer Res.* **2014**, *20*, 4904–4911. [CrossRef] [PubMed]
43. Prensner, J.R.; Zhao, S.; Erho, N.; Schipper, M.; Iyer, M.K.; Dhanasekaran, S.M.; Magi-Galluzzi, C.; Mehra, R.; Sahu, A.; Siddiqui, J.; et al. RNA biomarkers associated with metastatic progression in prostate cancer: A multi-institutional high-throughput analysis of SChLAP1. *Lancet Oncol.* **2014**, *15*, 1469–1480. [CrossRef]
44. Kim, H.; Skowronski, J.; Den, R.B. Prognostic outlier genes for enhanced prostate cancer treatment. *Future Oncol.* **2017**, *13*, 249–261. [CrossRef] [PubMed]
45. Yuan, J.; Kensler, K.H.; Hu, Z.; Zhang, Y.; Zhang, T.; Jiang, J.; Xu, M.; Pan, Y.; Long, M.; Montone, K.T.; et al. Integrative comparison of the genomic and transcriptomic landscape between prostate cancer patients of predominantly African or European genetic ancestry. *PLoS Genet.* **2020**, *16*, e1008641. [CrossRef] [PubMed]
46. Hubbard, G.K.; Mutton, L.N.; Khalili, M.; McMullin, R.P.; Hicks, J.L.; Bianchi-Frias, D.; Horn, L.A.; Kulac, I.; Moubarek, M.S.; Nelson, P.S.; et al. Combined MYC Activation and Pten Loss Are Sufficient to Create Genomic Instability and Lethal Metastatic Prostate Cancer. *Cancer Res.* **2016**, *76*, 283–292. [CrossRef]
47. Liu, W.; Xie, C.C.; Thomas, C.Y.; Kim, S.-T.; Lindberg, J.; Egevad, L.; Wang, Z.; Zhang, Z.; Sun, J.; Sun, J.; et al. Genetic markers associated with early cancer-specific mortality following prostatectomy. *Cancer* **2013**, *119*, 2405–2412. [CrossRef]
48. Bassi, C.; Ho, J.; Srikumar, T.; Dowling, R.J.O.; Gorrini, C.; Miller, S.J.; Mak, T.W.; Neel, B.G.; Raught, B.; Stambolic, V. Nuclear PTEN controls DNA repair and sensitivity to genotoxic stress. *Science* **2013**, *341*, 395–399. [CrossRef]
49. Barbieri, C.E.; Baca, S.C.; Lawrence, M.S.; Demichelis, F.; Blattner, M.; Theurillat, J.-P.; White, T.A.; Stojanov, P.; Van Allen, E.; Stransky, N.; et al. Exome sequencing identifies recurrent SPOP, FOXA1 and MED12 mutations in prostate cancer. *Nat. Genet.* **2012**, *44*, 685–689. [CrossRef]
50. An, J.; Wang, C.; Deng, Y.; Yu, L.; Huang, H. Destruction of full-length androgen receptor by wild-type SPOP, but not prostate-cancer-associated mutants. *Cell Rep.* **2014**, *6*, 657–669. [CrossRef]
51. Mani, R.-S. The emerging role of speckle-type POZ protein (SPOP) in cancer development. *Drug Discov. Today* **2014**, *19*, 1498–1502. [CrossRef]
52. Tonon, L.; Fromont, G.; Boyault, S.; Thomas, E.; Ferrari, A.; Sertier, A.-S.; Kielbassa, J.; Le Texier, V.; Kamoun, A.; Elarouci, N.; et al. Mutational Profile of Aggressive, Localised Prostate Cancer from African Caribbean Men Versus European Ancestry Men. *Eur. Urol.* **2019**, *75*, 11–15. [CrossRef]
53. Blattner, M.; Lee, D.J.; O'Reilly, C.; Park, K.; MacDonald, T.Y.; Khani, F.; Turner, K.R.; Chiu, Y.-L.; Wild, P.J.; Dolgalev, I.; et al. SPOP mutations in prostate cancer across demographically diverse patient cohorts. *Neoplasia* **2014**, *16*, 14–20. [CrossRef] [PubMed]

54. Wu, Y.-M.; Cieślik, M.; Lonigro, R.J.; Vats, P.; Reimers, M.A.; Cao, X.; Ning, Y.; Wang, L.; Kunju, L.P.; de Sarkar, N.; et al. Inactivation of CDK12 Delineates a Distinct Immunogenic Class of Advanced Prostate Cancer. *Cell* **2018**, *173*, 1770–1782.e14. [CrossRef] [PubMed]
55. Scher, H.I.; Fizazi, K.; Saad, F.; Taplin, M.-E.; Sternberg, C.N.; Miller, K.; de Wit, R.; Mulders, P.; Chi, K.N.; Shore, N.D.; et al. Increased survival with enzalutamide in prostate cancer after chemotherapy. *N. Engl. J. Med.* **2012**, *367*, 1187–1197. [CrossRef] [PubMed]
56. Chen, Y.; Clegg, N.J.; Scher, H.I. Anti-androgens and androgen-depleting therapies in prostate cancer: New agents for an established target. *Lancet Oncol.* **2009**, *10*, 981–991. [CrossRef]
57. Mizokami, A.; Koh, E.; Fujita, H.; Maeda, Y.; Egawa, M.; Koshida, K.; Honma, S.; Keller, E.T.; Namiki, M. The adrenal androgen androstenediol is present in prostate cancer tissue after androgen deprivation therapy and activates mutated androgen receptor. *Cancer Res.* **2004**, *64*, 765–771. [CrossRef]
58. Watson, P.A.; Chen, Y.F.; Balbas, M.D.; Wongvipat, J.; Socci, N.D.; Viale, A.; Kim, K.; Sawyers, C.L. Constitutively active androgen receptor splice variants expressed in castration-resistant prostate cancer require full-length androgen receptor. *Proc. Natl. Acad. Sci. USA* **2010**, *107*, 16759–16765. [CrossRef]
59. Hernández-Llodrà, S.; Segalés, L.; Safont, A.; Juanpere, N.; Lorenzo, M.; Fumadó, L.; Rodríguez-Vida, A.; Cecchini, L.; Bellmunt, J.; Lloreta-Trull, J. SPOP and FOXA1 mutations are associated with PSA recurrence in ERG wt tumors, and SPOP downregulation with ERG-rearranged prostate cancer. *Prostate* **2019**, *79*, 1156–1165. [CrossRef]
60. Heemers, H.V.; Tindall, D.J. Androgen receptor (AR) coregulators: A diversity of functions converging on and regulating the AR transcriptional complex. *Endocr. Rev.* **2007**, *28*, 778–808. [CrossRef]
61. Azad, A.A.; Volik, S.V.; Wyatt, A.W.; Haegert, A.; Le Bihan, S.; Bell, R.H.; Anderson, S.A.; McConeghy, B.; Shukin, R.; Bazov, J.; et al. Androgen Receptor Gene Aberrations in Circulating Cell-Free DNA: Biomarkers of Therapeutic Resistance in Castration-Resistant Prostate Cancer. *Clin. Cancer Res.* **2015**, *21*, 2315–2324. [CrossRef] [PubMed]
62. LaTulippe, E.; Satagopan, J.; Smith, A.; Scher, H.; Scardino, P.; Reuter, V.; Gerald, W.L. Comprehensive gene expression analysis of prostate cancer reveals distinct transcriptional programs associated with metastatic disease. *Cancer Res.* **2002**, *62*, 4499–4506. [PubMed]
63. Chang, K.-H.; Li, R.; Kuri, B.; Lotan, Y.; Roehrborn, C.G.; Liu, J.; Vessella, R.; Nelson, P.S.; Kapur, P.; Guo, X.; et al. A gain-of-function mutation in DHT synthesis in castration-resistant prostate cancer. *Cell* **2013**, *154*, 1074–1084. [CrossRef] [PubMed]
64. Lu, C.; Luo, J. Decoding the androgen receptor splice variants. *Transl. Androl. Urol.* **2013**, *2*, 178–186.
65. Lu, C.; Brown, L.C.; Antonarakis, E.S.; Armstrong, A.J.; Luo, J. Androgen receptor variant-driven prostate cancer II: Advances in laboratory investigations. *Prostate Cancer Prostatic Dis.* **2020**, *23*, 381–397. [CrossRef]
66. Bastos, D.A.; Antonarakis, E.S. CTC-derived AR-V7 detection as a prognostic and predictive biomarker in advanced prostate cancer. *Expert Rev. Mol. Diagn.* **2018**, *18*, 155–163. [CrossRef]
67. Graf, R.P.; Hullings, M.; Barnett, E.S.; Carbone, E.; Dittamore, R.; Scher, H.I. Clinical Utility of the Nuclear-localized AR-V7 Biomarker in Circulating Tumor Cells in Improving Physician Treatment Choice in Castration-resistant Prostate Cancer. *Eur. Urol.* **2020**, *77*, 170–177. [CrossRef]

Publisher's Note: MDPI stays neutral with regard to jurisdictional claims in published maps and institutional affiliations.

genes

MDPI

Review

Long Non-Coding RNAs at the Chromosomal Risk Loci Identified by Prostate and Breast Cancer GWAS

Panchadsaram Janaththani [1,2], Sri Lakshmi Srinivasan [1,2] and Jyotsna Batra [1,2,*]

1 School of Biomedical Sciences, Faculty of Health, Queensland University of Technology (QUT), Brisbane 4000, Australia; janaththani@gmail.com (P.J.); sri.srinivasan2016@gmail.com (S.L.S.)
2 Translational Research Institute, Queensland University of Technology (QUT), Brisbane 4102, Australia
* Correspondence: jyotsna.batra@qut.edu.au

Abstract: Long non-coding RNAs (lncRNAs) are emerging as key players in a variety of cellular processes. Deregulation of the lncRNAs has been implicated in prostate and breast cancers. Recently, germline genetic variations associated with cancer risk have been correlated with lncRNA expression and/or function. In addition, single nucleotide polymorphisms (SNPs) at well-characterized cancer-associated lncRNAs have been analyzed for their association with cancer risk. These SNPs may occur within the lncRNA transcripts or spanning regions that may alter the structure, function, and expression of these lncRNA molecules and contribute to cancer progression and may have potential as therapeutic targets for cancer treatment. Additionally, some of these lncRNA have a tissue-specific expression profile, suggesting them as biomarkers for specific cancers. In this review, we highlight some of the cancer risk-associated SNPs that modulated lncRNAs with a potential role in prostate and breast cancers and speculate on how these lncRNAs may contribute to cancer development.

Keywords: long non-coding RNA; prostate cancer; breast cancer; single nucleotide polymorphisms; genome-wide association studies

Citation: Janaththani, P.; Srinivasan, S.L.; Batra, J. Long Non-Coding RNAs at the Chromosomal Risk Loci Identified by Prostate and Breast Cancer GWAS. *Genes* **2021**, *12*, 2028. https://doi.org/10.3390/genes12122028

Academic Editor: Mikko P. Turunen

Received: 4 November 2021
Accepted: 17 December 2021
Published: 20 December 2021

1. Introduction

Hormone-related cancers, prostate and breast cancer, accounted for more than 3.6 million newly diagnosed cancer cases worldwide in 2020 [1]. Genetic predisposition has been identified as one of the factors contributing to the risk of these cancers. Genome-wide association studies (GWASs), analyzing common low-penetrance variants, have identified specific risk loci for these cancers [2,3]. Most of the risk-associated single nucleotide polymorphisms (SNPs) identified through GWAS are present in non-protein-coding DNA [4–7]. This non-coding DNA can regulate the expression of protein-coding genes and maintain the 3D structure of the genome by serving as a scaffold for transcription factors. Alternatively, some non-coding DNA is now found to be transcribed as non-protein-coding RNA (ncRNA) using high-throughput next-generation sequencing platforms [8]. Even though ncRNAs are not translated into proteins, they play a vital part in human complexity from maintaining normal cellular function to playing a broader role in human diseases including cancer [9–12]. Both ncRNAs and protein machinery involved in the development of diseases have become targets of novel therapeutic approaches [13–15]. Based on transcript size, these ncRNAs are grouped into two major classes: small non-coding RNAs (<200 bp) and long non-coding RNAs (lncRNAs) (>200 bp). The small ncRNA class comprises miRNAs, tRNAs, snRNAs, siRNAs, and piRNAs [3,16,17]., LncRNAs have recently been identified as important mediators in many diseases, including cancer [16,18,19]. Long non-coding RNAs (lncRNAs) are RNA transcripts that lack translational potential into functional proteins. The biogenesis of lncRNAs is similar to mRNAs. Most lncRNAs are transcribed by RNA polymerase II while some are also transcribed by RNA polymerase III. Most of the lncRNAs undergo post-transcriptional modifications, such as splicing, polyadenylation, and 5′ capping-like protein-coding RNAs [20]. However, these molecules have several short open reading frames (sORFs) and have very little protein coding potential, which discriminates them from mRNA [21]. Based on their origin, these lncRNAs can be classified as intronic, exonic, intergenic, intragenic, antisense, 3′ and 5′ UTR, promoter-associated (paRNA), and enhancer-associated (eRNA) [22]. Most of the lncRNAs are localized in the nucleus while some are found in both the nucleus and cytoplasm, and some are specifically distributed in the cytoplasm. These lncRNAs play a functional role in gene expression regulation by either cis (targeting genomically local genes) or trans (targeting distant genes) action [22]. Interestingly, recent advanced research has identified several putative coding sORFs suggesting that lncRNAs may be translated into micro peptides with a functional role [23–25].

Even though lncRNAs constitute a major part of the human transcriptome, the functional characterization and identification of molecular pathways in which these lncRNAs are involved remains a challenge. Nevertheless, considerable variabilities in the function of lncRNAs can be observed through well-characterized lncRNAs to date. It is evidenced that many lncRNAs are deregulated in prostate and breast cancer and some of their expression has been significantly associated with different stages of cancer. These lncRNAs are proposed to be involved in cancer development by playing functional roles in chromatin remodeling, transcriptional regulation, or post-transcriptional regulation. They show tumor-suppressive or oncogenic potential, emphasizing their potential in targeted therapeutics for prostate and breast cancer [26,27]. In addition, lncRNAs show tissue- and cancer-specific expression patterns, enabling them to be better diagnostic and prognostic tools for cancer therapies [26,28]. Moreover, SNPs could affect the expression and molecular function of lncRNAs, for instance, by disrupting their secondary structure and playing critical roles in tumorigenesis [29]. Recently, there has been increasing evidence in studies linking genetic variants modulating lncRNA expression and prostate or breast cancer risks [6,30]. In this review, we summarize lncRNAs regulated by risk-associated genetic variants (Table 1) in these two hereditary cancers to gain insights into the contribution of lncRNAs to cancer etiology, oncogenic function, and treatment resistance.

2. Prostate Cancer Risk-Associated SNPs Modulating lncRNAs

As a multifactorial disease, prostate cancer has several aspects contributing to its etiology, comprising both modifiable and non-modifiable factors [31]. Diet and environmental exposure disruptors, such as bisphenol A, chlordecone, and pesticides [31], are reported as modifiable prostate cancer risk factors. Age is a well-known non-modifiable risk factor for prostate cancer, where the risk of developing cancer increases with age [32,33]. Ethnicity is another non-modifiable contributing factor to the development of prostate cancer, where Asians have been reported to have lower prostate cancer incidence rates than European and American populations [31]. Furthermore, family history and/or heredity is also a known non-modifiable prostate cancer risk factor [33]. There is a considerable amount of evidence for a genetic basis (up to ~57%) contributing to the risk of prostate cancer [34,35]. Recently, a large prostate cancer GWAS identified novel risk loci making it to a total of 269 risk loci to date [36] and the study led to the identification of a genetic risk score of prostate cancer predisposition. Nevertheless, identification of the causal genes has been a major challenge, given the location of a large proportion of these variants are in the non-coding regions. Functional studies are known to complement GWAS results to identify specific genes whose expressions are associated with disease phenotype. One such approach is by expression quantitative trait locus (eQTL), which can identify the association between risk genotype and gene expression, and transcriptome-wide association studies (TWASs), which can assess the association with disease risk throughout the transcriptome.

One of the few studies to explore prostate cancer GWAS SNPs-associated lncRNAs identified that the prostate cancer-associated SNPs are less polymorphic in the flanking regions, but the SNP density was similar in protein-coding and lncRNA gene regions, indicating the sequences of lncRNA are evolutionarily conserved [37]. This study reported that 52 loci were located within the lncRNA genes, including a new prostate cancer risk-related SNP rs3787016 in a predicted lncRNA *AC1127096.1* [37]. This locus has been initially identified to be associated with prostate cancer risk in members with families having multiple cases of prostate cancer [38]. An independent case-control study in Chinese men reported that the rs3787016 SNP 'A' allele was associated with a higher risk of developing prostate cancer in younger individuals as well as individuals with a smoking history or Gleason score ≥7 (4 + 3) or aggressive disease [39].

By using a systemic approach for lncRNA based on their position in the promoter region, intercellular functional correlation, eQTL with one or more risk SNPs (cis-eQTL), and differential expression between tumor and normal prostate tissues, Guo et al. shortlisted 45 potential lncRNAs with 50% of prostate cancer risk loci from 122 loci [30]. This included already well-known lncRNAs, such as *KCNQ1OT1*, *H19*, and prostate cancer-associated transcript 1 (*PCAT1*). LncRNA *KCNQ1OT1* was known to act as an miRNA sponge and compete with either miR-211-5p or miR-15 to promote prostate cancer progression [40,41]. Polymorphisms in *H19* were also associated with risk and some of the clinical parameters of bladder cancer [42], hepatocellular carcinoma [43], cervical cancer [42], and urothelial cell carcinoma [44]. Two additional SNPs in *H19*, rs3024270 or rs3741219, were shown to be associated with the risk of perineural invasion of prostate cancer [45]. Increased expression of *H19* was observed in prostate cancer patients with a high Gleason score compared to a low Gleason score and benign prostatic hyperplasia (BPH) [46]. The function of *H19* remains controversial in multiple cancers, including prostate cancer. While knockdown of *H19* in PC3 and DU145 prostate cancer cells reduced cell proliferation and glucose and lactate levels [46], the H19-derived miR-675 axis is described as a suppressor of prostate cancer metastasis, regulating extracellular matrix protein and transforming growth factor β-induced protein (TGFB1) [47]. This suggests that the tumor microenvironment and cell types should be accounted for when determining the functional role of *H19*. In addition, *H19* overexpression increased the expression of stem cell markers *Oct4* and *Sox2* and increased colony formation in the RWPE-1 prostate epithelial cell line [48]. *H19*-dependent transcriptional regulation by estrogen and hypoxia redirected the cells from epithelial to mesenchymal transition (EMT) to β integrin-mediated invasion [49].

Guo et al. also reported that of the 45 lncRNAs regulated by non-coding SNPs, 18 lncRNAs' expression was significantly correlated with 15 prostate cancer risk loci [30]. Moreover, some of these risk SNPs are enriched in the promoter regions of five lncRNAs –*PCAT1*, *RP11-400F19.18*, *RP11-24D8.1*, *RP11-552F3.10*, and *RP11-328M4.2* [30]. This study further identified that the risk SNP rs7463708 in the enhancer region of the *PCAT1* increased the binding of a novel AR interacting partner, ONECUT2, which then looped to the *PCAT1* promoter. Moreover, *PCAT1* was identified as an androgen late-response gene and interacted with AR and lysine-specific demethylase 1 (LSD1) upon prolonged androgen treatment to promote prostate cancer growth [30]. *PCAT1*, which has been identified as the top-ranked lncRNA based on its overexpression in prostate tumor [50], is also known to promote prostate cell proliferation through upregulation of the cMyc protein [51]. Apart from this distal enhancer locus at rs7463708 SNP, another independent risk locus tagged by rs10086908 SNP was associated with *PCAT1* expression, with nine SNPs located across the promoter and exons of the *PCAT1* gene [30]. Moreover, rs1902432 SNP in *PCAT1* was also identified to be associated with an increased risk of PCa [52]. Interestingly, a meta-analysis of five lncRNA polymorphisms in prostate cancer-associated non-coding RNA1 (*PRNCR1*, also known as *PCAT8*) and multiple cancer susceptibility reported that four of the SNPs (rs16901946, rs13252298, rs1016343, and rs1456315) were associated with overall cancer risk while no association was found with rs7007694 SNP [53]. Another small case-control study using 178 prostate cancer patients and 180 BPH cases in the Iranian population identified rs13252298, rs1456315, and rs7841060 SNPs in *PRNCR1* to be associated with prostate cancer risk [54]. There was no significant association between rs7007694 SNP and prostate cancer risk [54], as also reported with overall cancer risk previously [53] and after adjusting for clinicopathological characteristics, such as age, tumor stage, prostate-specific antigen (PSA) levels, Gleason score, perineural invasion, and surgical margin [54]. However, it is important to validate these associations in a larger cohort. *PRNCR1* was reported to have high expression in aggressive prostate cancer and was reported to enhance both ligand-dependent and ligand-independent AR-mediated transcriptional activity by directly binding to the region of 549-623 amino acids of AR and therefore promotes prostate cancer growth [55]. On the contrary, another study by Parolia et al. reported significantly lower expression levels of *PRNCR1* in the prostate cancer models they tested, raising questions about its involvement in AR activation in prostate cancer [56]. Similarly, *PRNCR1* was excluded from the study by Guo et al., due to its undetectable expression in the LNCaP prostate cancer cell line and TCGA prostate adenocarcinoma RNA-sequencing data [30]. In addition to prostate cancer, polymorphisms in *PRNCR1* at the *8q24* locus are also associated with gastric [57], colorectal [58], and lung cancer risk [59], indicating its functional role in multiple cancers.

A multiethnic meta-analysis study of prostate cancer GWAS in >10 million SNPs in ~80,000 individuals identified a novel prostate cancer risk locus at 9p21 [60]. The prostate cancer risk-associated variant at this locus, rs17694493, is predicted to disrupt the binding motifs of transcription factors STAT1 and RUNX1 and positioned in the intronic region of a novel lncRNA gene *CDKN2B-AS1* (also known as *ANRIL*). Moreover, SNPs rs4977574, rs1333048, and rs10757278 in the *ANRIL* gene were also associated with BPH and prostate cancer risk in the Iranian population [61]. Overexpression of *ANRIL* in prostate cancer cells increased cell proliferation and migration by regulating the let-7a/TGFB1/Smad signaling pathway [62], demonstrating the potential molecular mechanism by which this lncRNA mediates cancer progression.

Prostate cancer risk-associated SNPs, rs11672691 and rs887391, were identified to regulate two *PCAT19* lncRNA isoforms with two distinct transcription start sites, *PCAT19*-short and *PCAT19*-long, through a promoter-to-enhancer switching mechanism [63]. The rs11672691 SNP on chromosome 19 was identified to be associated with both non-aggressive and aggressive prostate cancer risk [64], prostate cancer-specific mortality [65], and poor prognosis after diagnosis [63]. *PCAT19*-long promoted prostate cancer progression by interacting with a nuclear riboprotein, Heterogeneous Nuclear Ribonucleoprotein A/B (HNRNPAB), to upregulate a subset of cell-cycle genes [63], suggesting a novel mechanism for the HNRNPAB role in prostate cancer progression.

Some SNPs are also found to regulate distant lncRNAs by chromosome looping. For instance, prostate cancer risk SNP rs378854 at the 8q24 locus was found to regulate the expression of lncRNA *PVT1*, which is located 0.5 Mb away from this variant by long-range chromatin looping [66]. Exon 9 of the *PVT1* gene was overexpressed in aggressive prostate cancer cases with African ancestry, suggesting this could be used as a biomarker for metastatic disease [67]. Knockdown of *PVT1* was shown to reduce prostate cancer growth in vitro and in vivo and increase cell apoptosis in prostate cancer cells [68]. Some of these studies are summarized in Table 1.

3. Breast Cancer Risk-Associated SNPs Modulating lncRNAs

Breast cancer is the commonly diagnosed cancer in females worldwide. It is a heterogeneous disease on a molecular and clinical level, and has four distinct subtypes: Luminal A, Luminal B, human epidermal growth factor receptor 2 (HER2) overexpression, and triple-negative, based on the status of estrogen receptor (ER), progesterone receptor (PR), and HER2 [69,70]. Breast cancer GWASs have identified more than 200 risk loci, including differential associations with ER+, ER−, or triple-negative breast cancer [7,71,72].

A transcriptome-wide association study by Wu et al. identified 26 lncRNAs through eQTL analysis of breast cancer risk loci [73]. The functional role of three of these lncRNAs: *RP11–218M22.1*, *RP11–467J12.4*, and *CTD-3032H12.1*, was confirmed by the significant reduction in cell proliferation on lncRNA knockdown in three breast cancer cell lines, 184A1, MCF7, and T47D, and reduced colony-forming efficiency in MCF7 cells. *RP11–467J12.4*, also known as *PR-lncRNA-1*, is mainly localized in the nucleus, and regulated by P53 in human and mouse cells [74]. LncRNA *CTD-3032H12.1* is predicted to interact with another lncRNA *RP11-20F24.2* and mRNA of *ANKRD30A*, a transcription factor implicated in breast cancer progression, using a tissue-specific co-expression regulatory network model [75].

A recent study by Marjaneh et al. exploring multi-exonic non-coding RNA (mencRNA) genes at 139 breast cancer GWAS loci identified more than 4000 mencRNAs using RNA-capture sequencing [76]. Interestingly, the breast cancer risk variants were enriched in the exonic regions of these RNAs, suggesting that these risk variants may impact RNA stability, structure, or function. One example reported in this study for enriched risk variants in exons is the 2q14.2 locus, where three of the four independent risk signals were in the exonic regions of mencRNAs. Furthermore, eQTL analysis shortlisted 800 mencRNAs, including seven signals: XLOC_022678, XLOC_093918, XLOC_112072, XLOC_142280, XLOC_169717, XLOC_195543, and XLOC_209276, overlapping with breast cancer risk signals [76]. Four of these eQTLs were identified to regulate mencRNAs through distal interactions as confirmed by Capture Hi-Seq. This includes the potential causal variants at the estrogen-regulated enhancer of two lncRNAs: *CUPID1* and *CUPID2* at the 11q13 locus, which were previously known to promote homologous-based DNA repair [77].

A study carried out by Suvanto et al., analyzing 84 lncRNA and 44 transcribed-ultra conserved RNA (T-UCR, a subtype of lncRNAs) regions, identified SNPs in seven lncRNAs and eight T-UCRs associated with breast cancer risk, which were not previously reported by GWAS studies [6]. This includes risk SNPs, rs71124350, and rs28489579 at the 15q21.1 locus, which correlates with the expression of GA-binding protein transcription factor-β subunit 1 antisense RNA 1 (*GABPB1-AS1*). This lncRNA was predicted to be associated with two miRNA networks: hsa-miR-3613-3p and hsa-miR-7106-5p, which were differentially regulated in breast cancer compared to adjacent normal breast tissues [78]. Although the functional role of *GABPB1-AS1* is not known in breast cancer, it was reported to have a function in other cancers. For instance, *GABPB1-AS1* is known to regulate oxidative stress by regulating translation of its sense protein GABPB1 when exposed to a small molecule compound, Erastin, that induces non-apoptotic iron-dependent oxidative cell death (ferroptosis) in hepatocellular carcinoma (HCC) cells [79]. Moreover, high expression of *GABPB1-AS1* was correlated with better overall survival of HCC patients [79]. Similarly, high *GABPB1-AS1* expression was correlated with better prognosis and inversely correlated with tumor size, TNM stage, and Furhman stage of clear cell renal cell carcinoma patients [80]. This was further validated using in vitro and in vivo assays with *GABPB1-AS1* overexpression models, resulting in reduced proliferation, migration, and invasion in 786-o and caki-1 renal cell cancer cells and reduced tumor growth in xenograft models [80].

In addition to these transcriptome-wide lncRNA findings of risk loci, some studies have focused on the risk association of genetic variants in well-known breast cancer-related genes. For instance, rs1899663 and rs7958904 SNPs at lncRNA *HOTAIR*, *HOX* transcript antisense intergenic RNA, were associated with an increased risk of breast cancer in the Southeast Chinese Han population of 969 breast cancer cases and 970 healthy controls [81]. Moreover, rs1899663 SNP was also associated with both disease-free and overall survival in younger cases. On the contrary, Yan et al. reported that rs1899663 and rs4759314 SNPs were associated with reduced breast cancer risk among women with age at menarche >14 while rs920778 SNP was associated with an increased risk in the Chinese population of 502 cases and 504 matched healthy controls [82]. rs1899663 and rs12826786 SNPs were associated with a reduced breast cancer risk in the southeast Iranian population while rs920778 SNP was associated with an increased risk similar to the association reported by Yan et al. [83]. A recent meta-analysis study showed that rs12826786 and rs920778 SNPs at *HOTAIR* were correlated with an increased overall cancer risk [84]. *HOTAIR* belongs to the conserved genomic region of several *HOX* family coding and non-coding genes, and known to play a functional role in embryonic development [85]. Overexpression of *HOTAIR* was correlated with metastasis and poor prognosis of various cancers, including breast cancer [85,86]. Overexpression of *HOTAIR* in breast cancer cells increased invasiveness of these cells in a polycomb repressive complex 2 (PRC2)-

dependent manner by reprogramming the polycomb binding profile similar to embryonic fibroblast [87]. Interestingly, *HOTAIR* induction has been shown to also be important for the invasive growth of Claudin-low breast cancer cells, which are triple-negative cancer subtype with low expression of claudin-3, claudinin-4, and claudinin-7 [88].

Another well-known cancer-associated lncRNA, *H19*, a maternally inherited imprinted gene, is reported to be overexpressed in breast cancer, and associated with poor prognosis in breast cancer patients, especially in the triple-negative molecular subtype [89,90]. A genetic association study in the Chinese Han population of 464 breast cancer cases and 467 healthy controls did not observe any significant association with breast cancer risk for two SNPs, rs3741219 and rs217727, in the *H19* gene in the overall analysis, but on stratified analysis, rs217727 SNP was found to correlate with breast cancer risk in patients with ER+ or HER2+ or women who had more than two pregnancies [89]. Overexpression of *H19* in breast cancer cells promoted cell proliferation and migration [91] while *H19* knockdown reduced estrogen-induced cell growth of breast cancer cells [92]. However, a meta-analysis study by Mathias et al. analyzing 31 SNPs in 12 lncRNAs could not find any association for these SNPs with breast cancer susceptibility, including rs920778, rs1899663, rs12826786, and rs4759314 SNPs on the *HOTAIR* locus and rs217727, rs3741219, rs2107425, and rs2839698 SNPs on the *H19* locus [93] likely due to the smaller number of studies included for this analysis. This emphasizes the need for comprehensive functional analysis with experimental evidence to improve our understanding of how these genetic variants contribute to breast cancer pathology (Table 1).

Table 1. LncRNAs associated with risk SNPs in prostate and breast cancer.

	lncRNA	SNP	Function of lncRNA	References
Prostate cancer	*PCAT1*	rs7463708 rs10086908 rs1902432	Promote prostate cancer growth by interacting with AR and LSD1 upon prolonged androgen treatment, Promote prostate cancer cell proliferation through c-Myc upregulation	[30,50,51]
	PRNCR1/PCAT8	rs16901946 rs13252298 rs1016343 rs1456315	Enhance ligand-dependent and -independent AR-mediated transcriptional activity	[53–55]
	CDKN2B-AS1/ANRIL	rs4977574 rs1333048 rs10757278	Increase prostate cancer cell proliferation and migration by modulating the let-7a/TGFB1/Smad signaling pathway	[60–62]
	PCAT19	rs11672691 rs887391	*PCAT19*-long isoform promotes prostate cancer progression by upregulating a subset of cell-cycle genes via interaction with HNRNPAB	[63–65]
	PVT1	rs378854	*PVT1* knockdown inhibits prostate cancer growth in vitro and in vivo and increased cell apoptosis	[66,68]
Breast cancer	*RP11–218M22.1*	rs12422552	Knockdown reduced breast cancer cell proliferation and colony formation	[73]
	RP11–467J12.4/ PR-lncRNA1	rs3112612	Knockdown reduced breast cancer cell proliferation and colony formation	[73]
	CTD-3032H12.1	rs28539243	Knockdown reduced breast cancer cell proliferation and colony formation, predicted to interact with lncRNA RP11-20F24.2 and mRNA of ANKRD30A	[73,74]
	GABPB1-AS1	rs71124350 rs28489579	Predicted association with two miRNA networks, which are differentially regulated in breast cancer	[6,78]
	HOTAIR	rs1899663 rs7958904 rs4759314 rs920778 rs12826786	Overexpression of *HOTAIR* correlates with metastasis of breast cancer, increase breast cancer cell invasiveness via reprogramming PRC2 binding	[81–87]
	H19	rs3741219 rs217727	Overexpression of *H19* promoted breast cancer cell proliferation and migration and knockdown reduced estrogen-induced breast cancer cell growth	[89–91]

4. Conclusions

Recently, there has been remarkable progress in our understanding of the multifaceted role of lncRNAs and the genetic variants impacting lncRNA expression and function, recognizing them as critical players in prostate and breast cancer progression. Although a majority of these cancer risk-associated genetic variants are found in non-coding RNA loci, only a few studies have focused on uncovering the role of these SNPs in modulating the structure and function of lncRNAs in cancer progression. Emerging sequencing techniques and bioinformatic analysis are helpful in predicting the putative function of the lncRNA. Databases, such as lncRNASNP2 [94] and LincSNP 3.0 [95], provide information on how these SNPs modulate the lncRNA structure and function. Some of these lncRNAs are differentially expressed in disease progression models and cancer subtypes, highlighting their potential to be used as a diagnostic and prognostic biomarker. For example, PCA3 (also known as DD3) is the only FDA-approved lncRNA used for prostate cancer diagnosis [96], which is overexpressed in prostate tumors compared to non-malignant tissues. Nevertheless, using lncRNA-SNPs to predict the disease progression and/or therapeutic options is still in an early stage, since how these SNPs regulate the expression or function of lncRNAs remains uncertain. Most of these studies have identified cancer-associated lncRNAs through their expression correlation with the SNP genotype. Moreover, whether these SNPs are true causal SNPs or correlated variants in high linkage disequilibrium with the causal SNPs still needs to be clarified. Advancing techniques including CRISPR genome editing may provide comprehensive insights in this field to help identify the functional role of the cancer-associated risk variants through lncRNAs in disease progression and identify their applicability as novel therapeutic targets or biomarkers for multiple cancers.

Author Contributions: Conceptualization, P.J. and J.B.; writing—original draft preparation, P.J.; writing—review and editing, S.L.S. and J.B.; supervision, J.B. All authors have read and agreed to the published version of the manuscript.

Funding: This work was supported by the DoD Prostate Cancer Idea Development Grant under Award No. (W81XWH-19-1-0343), Opinions, interpretations, conclusions, and recommendations are those of the author and are not necessarily endorsed by the Department of Defense, J.B. and S.S. are supported by the Advance Queensland Industry Fellowship.

Institutional Review Board Statement: Not applicable.

Informed Consent Statement: Not applicable.

Data Availability Statement: Not applicable.

Conflicts of Interest: The authors declare no conflict of interest.

References

1. Sung, H.; Ferlay, J.; Siegel, R.L.; Laversanne, M.; Soerjomataram, I.; Jemal, A.; Bray, F. Global cancer statistics 2020: GLOBOCAN estimates of incidence and mortality worldwide for 36 cancers in 185 countries. *CA Cancer J. Clin.* **2021**, *71*, 209–249. [CrossRef]
2. Kar, S.P.; Beesley, J.; Amin Al Olama, A.; Michailidou, K.; Tyrer, J.; Kote-Jarai, Z.; Lawrenson, K.; Lindstrom, S.; Ramus, S.J.; Thompson, D.J.; et al. Genome-wide meta-analyses of breast, ovarian, and prostate cancer association studies identify multiple new susceptibility loci shared by at least two cancer types. *Cancer Discov.* **2016**, *6*, 1052–1067. [CrossRef]
3. Burgos-Aceves, M.A.; Smith, Y. microRNAs Downregulation in Cancer is Associated with Guanine Enrichment in the Terminal Loop Sequences of their Precursors. *MicroRNA* **2018**, *7*, 20–27. [CrossRef]
4. Edwards, S.L.; Beesley, J.; French, J.D.; Dunning, A.M. Beyond GWASs: Illuminating the Dark Road from Association to Function. *Am. J. Hum. Genet.* **2013**, *93*, 779–797. [CrossRef] [PubMed]
5. Srinivasan, S.; Clements, J.A.; Batra, J. Single nucleotide polymorphisms in clinics: Fantasy or reality for cancer? *Crit. Rev. Clin. Lab. Sci.* **2016**, *53*, 29–39. [CrossRef]
6. Suvanto, M.; Beesley, J.; Blomqvist, C.; Chenevix-Trench, G.; Khan, S.; Nevanlinna, H. SNPs in lncRNA Regions and Breast Cancer Risk. *Front. Genet.* **2020**, *11*, 550. [CrossRef] [PubMed]
7. Milne, R.L.; Kuchenbaecker, K.B.; Michailidou, K.; Beesley, J.; Kar, S.; Lindström, S.; Hui, S.; Lemaçon, A.; Soucy, P.; Lemaçon, A.; et al. Identification of ten variants associated with risk of estrogen-receptor-negative breast cancer. *Nat. Genet.* **2017**, *49*, 1767–1778. [CrossRef] [PubMed]

8. The FANTOM Consortium; Carninci, P.; Kasukawa, T.; Katayama, S.; Gough, J.; Frith, M.C.; Maeda, N.; Oyama, R.; Ravasi, T.; Lenhard, B.; et al. The Transcriptional Landscape of the Mammalian Genome. *Science* **2005**, *309*, 1559–1563. [CrossRef] [PubMed]
9. Alexander, R.P.; Fang, G.; Rozowsky, J.; Snyder, M.; Gerstein, M.B. Annotating non-coding regions of the genome. *Nat. Rev. Genet.* **2010**, *11*, 559–571. [CrossRef]
10. Li, J.; Xuan, Z.; Liu, C. Long Non-Coding RNAs and Complex Human Diseases. *Int. J. Mol. Sci.* **2013**, *14*, 18790–18808. [CrossRef] [PubMed]
11. Sparber, P.; Filatova, A.; Khantemirova, M.; Skoblov, M. The role of long non-coding RNAs in the pathogenesis of hereditary diseases. *BMC Med. Genom.* **2019**, *12*, 63–78. [CrossRef]
12. DiStefano, J.K. The Emerging Role of Long Noncoding RNAs in Human Disease. *Methods Mol. Biol.* **2018**, *1706*, 91–110. [CrossRef] [PubMed]
13. Esteller, M. Non-coding RNAs in human disease. *Nat. Rev. Genet.* **2011**, *12*, 861–874. [CrossRef] [PubMed]
14. Singh, D.; Khan, M.A.; Siddique, H.R. Emerging role of long non-coding RNAs in cancer chemoresistance: Unravelling the multifaceted role and prospective therapeutic targeting. *Mol. Biol. Rep.* **2020**, *47*, 5569–5585. [CrossRef]
15. Arun, G.; Diermeier, S.D.; Spector, D.L. Therapeutic Targeting of Long Non-Coding RNAs in Cancer. *Trends Mol. Med.* **2018**, *24*, 257–277. [CrossRef]
16. Cao, J. The functional role of long non-coding RNAs and epigenetics. *Biol. Proced. Online* **2014**, *16*, 11. [CrossRef]
17. Cohen, A.; Burgos-Aceves, M.A.; Smith, Y. Estrogen repression of microRNA as a potential cause of cancer. *Biomed. Pharmacother.* **2016**, *78*, 234–238. [CrossRef] [PubMed]
18. Clark, M.B.; Mattick, J.S. Long noncoding RNAs in cell biology. *Semin. Cell Dev. Biol.* **2011**, *22*, 366–376. [CrossRef]
19. Zhang, Y.; Pitchiaya, S.; Cieślik, M.; Niknafs, Y.S.; Tien, J.C.-Y.; Hosono, Y.; Iyer, M.K.; Yazdani, S.; Subramaniam, S.; Shukla, S.; et al. Analysis of the androgen receptor–regulated lncRNA landscape identifies a role for ARLNC1 in prostate cancer progression. *Nat. Genet.* **2018**, *50*, 814–824. [CrossRef]
20. Chen, L.-L. Linking Long Noncoding RNA Localization and Function. *Trends Biochem. Sci.* **2016**, *41*, 761–772. [CrossRef]
21. Rinn, J.L.; Chang, H.Y. Genome Regulation by Long Noncoding RNAs. *Annu. Rev. Biochem.* **2012**, *81*, 145–166. [CrossRef]
22. Nie, L.; Wu, H.-J.; Hsu, J.-M.; Chang, S.-S.; Labaff, A.M.; Li, C.-W.; Wang, Y.; Hsu, J.L.; Hung, M.-C. Long non-coding RNAs: Versatile master regulators of gene expression and crucial players in cancer. *Am. J. Transl. Res.* **2012**, *4*, 127–150. [PubMed]
23. Nelson, B.R.; Makarewich, C.A.; Anderson, D.M.; Winders, B.R.; Troupes, C.D.; Wu, F.; Reese, A.L.; McAnally, J.R.; Chen, X.; Kavalali, E.T.; et al. A peptide encoded by a transcript annotated as long noncoding RNA enhances SERCA activity in muscle. *Science* **2016**, *351*, 271–275. [CrossRef]
24. Matsumoto, A.; Pasut, A.; Matsumoto, M.; Yamashita, R.; Fung, J.; Monteleone, E.; Saghatelian, A.; Nakayama, K.I.; Clohessy, J.G.; Pandolfi, P.P. mTORC1 and muscle regeneration are regulated by the LINC00961-encoded SPAR polypeptide. *Nature* **2017**, *541*, 228–232. [CrossRef]
25. Andrews, S.J.; Rothnagel, J.A. Emerging evidence for functional peptides encoded by short open reading frames. *Nat. Rev. Genet.* **2014**, *15*, 193–204. [CrossRef] [PubMed]
26. Mitobe, Y.; Takayama, K.-I.; Horie-Inoue, K.; Inoue, S. Prostate cancer-associated lncRNAs. *Cancer Lett.* **2018**, *418*, 159–166. [CrossRef] [PubMed]
27. Zhang, T.; Hu, H.; Yan, G.; Wu, T.; Liu, S.; Chen, W.; Ning, Y.; Lu, Z. Long Non-Coding RNA and Breast Cancer. *Technol. Cancer Res. Treat.* **2019**, *18*, 1–10. [CrossRef] [PubMed]
28. Lee, B.; Mazar, J.; Aftab, M.N.; Qi, F.; Shelley, J.; Li, J.-L.; Govindarajan, S.; Valerio, F.; Rivera, I.; Thurn, T.; et al. Long Noncoding RNAs as Putative Biomarkers for Prostate Cancer Detection. *J. Mol. Diagn.* **2014**, *16*, 615–626. [CrossRef]
29. Minotti, L.; Agnoletto, C.; Baldassari, F.; Corrà, F.; Volinia, S. SNPs and Somatic Mutation on Long Non-Coding RNA: New Frontier in the Cancer Studies? *High-Throughput* **2018**, *7*, 34. [CrossRef] [PubMed]
30. Guo, H.; Ahmed, M.; Zhang, F.; Yao, C.Q.; Li, S.; Liang, Y.; Hua, J.; Soares, F.; Sun, Y.; Langstein, J.; et al. Modulation of long noncoding RNAs by risk SNPs underlying genetic predispositions to prostate cancer. *Nat. Genet.* **2016**, *48*, 1142–1150. [CrossRef] [PubMed]
31. Adjakly, M.; Ngollo, M.; Dagdemir, A.; Judes, G.; Pajon, A.; Karsli-Ceppioglu, S.; Penault-Llorca, F.; Boiteux, J.-P.; Bignon, Y.-J.; Guy, L.; et al. Prostate cancer: The main risk and protective factors—Epigenetic modifications. *Annales d'Endocrinologie* **2015**, *76*, 25–41. [CrossRef]
32. Brawley, O.W. Prostate cancer epidemiology in the United States. *World J. Urol.* **2012**, *30*, 195–200. [CrossRef]
33. Crawford, E.D. Epidemiology of prostate cancer. *Urology* **2003**, *62*, 3–12. [CrossRef]
34. Mucci, L.A.; Hjelmborg, J.B.; Harris, J.R.; Czene, K.; Havelick, D.J.; Scheike, T.; Graff, R.E.; Holst, K.; Möller, S.; Unger, R.H.; et al. Familial Risk and Heritability of Cancer Among Twins in Nordic Countries. *JAMA* **2016**, *315*, 68–76. [CrossRef]
35. Farashi, S.; Kryza, T.; Clements, J.; Batra, J. Post-GWAS in prostate cancer: From genetic association to biological contribution. *Nat. Rev. Cancer* **2019**, *19*, 46–59. [CrossRef] [PubMed]
36. Conti, D.V.; Darst, B.F.; Moss, L.C.; Saunders, E.J.; Sheng, X.; Chou, A.; Schumacher, F.R.; Al Olama, A.A.; Benlloch, S.; Dadaev, T.; et al. Trans-ancestry genome-wide association meta-analysis of prostate cancer identifies new susceptibility loci and informs genetic risk prediction. *Nat. Genet.* **2021**, *53*, 65–75. [CrossRef]

37. Jin, G.; Sun, J.; Isaacs, S.D.; Wiley, K.E.; Kim, S.-T.; Chu, L.W.; Zhang, Z.; Zhao, H.; Zheng, S.L.; Isaacs, W.B.; et al. Human polymorphisms at long non-coding RNAs (lncRNAs) and association with prostate cancer risk. *Carcinogenesis* **2011**, *32*, 1655–1659. [CrossRef] [PubMed]
38. Hsieh, C.-L.; Oakley-Girvan, I.; Balise, R.R.; Halpern, J.; Gallagher, R.P.; Wu, A.H.; Kolonel, L.N.; O'Brien, L.E.; Lin, I.G.; Berg, D.J.V.D.; et al. A Genome Screen of Families with Multiple Cases of Prostate Cancer: Evidence of Genetic Heterogeneity. *Am. J. Hum. Genet.* **2001**, *69*, 148–158. [CrossRef] [PubMed]
39. Cao, D.-L.; Gu, C.-Y.; Zhu, Y.-P.; Dai, B.; Zhang, H.-L.; Shi, G.-H.; Shen, Y.-J.; Ma, C.-G.; Xiao, W.-J.; Qin, X.-J.; et al. Polymorphisms at long non-coding RNAs and prostate cancer risk in an eastern Chinese population. *Prostate Cancer Prostatic Dis.* **2014**, *17*, 315–319. [CrossRef]
40. Chen, Q.-H.; Li, B.; Liu, D.-G.; Zhang, B.; Yang, X.; Tu, Y.-L. LncRNA KCNQ1OT1 sponges miR-15a to promote immune evasion and malignant progression of prostate cancer via up-regulating PD-L1. *Cancer Cell Int.* **2020**, *20*, 394. [CrossRef]
41. Hao, H.; Chen, H.; Xie, L.; Liu, H.; Wang, D. LncRNA KCNQ1OT1 Promotes Proliferation, Invasion and Metastasis of Prostate Cancer by Regulating miR-211-5p/CHI3L1 Pathway. *OncoTargets Ther.* **2021**, *14*, 1659–1671. [CrossRef]
42. Huang, M.-C.; Chou, Y.-H.; Shen, H.-P.; Ng, S.-C.; Lee, Y.; Sun, Y.-H.; Hsu, C.-F.; Yang, S.-F.; Wang, P.-H. The clinicopathological characteristic associations of long non-coding RNA gene H19 polymorphisms with uterine cervical cancer. *J. Cancer* **2019**, *10*, 6191–6198. [CrossRef] [PubMed]
43. Wu, E.-R.; Chou, Y.-E.; Liu, Y.-F.; Hsueh, K.-C.; Lee, H.-L.; Yang, S.-F.; Su, S.-C. Association of lncRNA H19 Gene Polymorphisms with the Occurrence of Hepatocellular Carcinoma. *Genes* **2019**, *10*, 506. [CrossRef] [PubMed]
44. Yang, P.-J.; Hsieh, M.-J.; Hung, T.-W.; Wang, S.-S.; Chen, S.-C.; Lee, M.-C.; Yang, S.-F.; Chou, Y.-E. Effects of Long Noncoding RNA H19 Polymorphisms on Urothelial Cell Carcinoma Development. *Int. J. Environ. Res. Public Health* **2019**, *16*, 1322. [CrossRef]
45. Hu, J.-C.; Lin, C.-Y.; Wang, S.-S.; Chiu, K.-Y.; Li, J.-R.; Chen, C.-S.; Hung, S.-C.; Yang, C.-K.; Ou, Y.-C.; Cheng, C.-L.; et al. Impact of H19 Polymorphisms on Prostate Cancer Clinicopathologic Characteristics. *Diagnostics* **2020**, *10*, 656. [CrossRef]
46. Sun, S.-C.; Zhao, H.; Liu, R.; Wang, B.-L.; Liu, Y.-Q.; Zhao, Y.; Shi, Z.-D. Expression of long non-coding RNA H19 in prostate cancer and its effect on the proliferation and glycometabolism of human prostate cancer cells. *Zhonghua Nan Ke Xue* **2017**, *23*, 120–124. [PubMed]
47. Zhu, M.; Chen, Q.; Liu, X.; Sun, Q.; Zhao, X.; Deng, R.; Wang, Y.; Huang, J.; Xu, M.; Yan, J.; et al. Lncrna h19/mir-675 axis represses prostate cancer metastasis by targeting tgfbi. *FEBS J.* **2014**, *281*, 3766–3775. [CrossRef] [PubMed]
48. Roy, H.B.-L.; Vennin, C.; Brocqueville, G.; Spruyt, N.; Adriaenssens, E.; Bourette, R.P. Enrichment of Human Stem-Like Prostate Cells with s-SHIP Promoter Activity Uncovers a Role in Stemness for the Long Noncoding RNA H19. *Stem Cells Dev.* **2015**, *24*, 1252–1262. [CrossRef]
49. Bacci, L.; Aiello, A.; Ripoli, C.; Loria, R.; Pugliese, D.; Pierconti, F.; Rotili, D.; Strigari, L.; Pinto, F.; Bassi, P.F.; et al. H19-dependent transcriptional regulation of beta3 and beta4 integrins upon estrogen and hypoxia favors metastatic potential in prostate cancer. *Int. J. Mol. Sci.* **2019**, *20*, 4012. [CrossRef]
50. Prensner, J.; Iyer, M.K.; Balbin, O.A.; Dhanasekaran, S.M.; Cao, Q.; Brenner, J.C.; Laxman, B.; Asangani, I.; Grasso, C.S.; Kominsky, H.D.; et al. Transcriptome sequencing across a prostate cancer cohort identifies PCAT-1, an unannotated lincRNA implicated in disease progression. *Nat. Biotechnol.* **2011**, *29*, 742–749. [CrossRef] [PubMed]
51. Prensner, J.; Chen, W.; Han, S.; Iyer, M.K.; Cao, Q.; Kothari, V.; Evans, J.R.; Knudsen, K.; Paulsen, M.T.; Ljungman, M.; et al. The Long Non-Coding RNA PCAT-1 Promotes Prostate Cancer Cell Proliferation through cMyc. *Neoplasia* **2014**, *16*, 900–908. [CrossRef]
52. Yuan, Q.; Chu, H.; Ge, Y.; Ma, G.; Du, M.; Wang, M.; Zhang, Z.; Zhang, W. LncRNA PCAT1 and its genetic variant rs1902432 are associated with prostate cancer risk. *J. Cancer* **2018**, *9*, 1414–1420. [CrossRef]
53. Huang, X.; Zhang, W.; Shao, Z. Association between long non-coding RNA polymorphisms and cancer risk: A meta-analysis. *Biosci. Rep.* **2018**, *38*, 38. [CrossRef] [PubMed]
54. Sattarifard, H.; Hashemi, M.; Hassanzarei, S.; Narouie, B.; Bahari, G. Association between genetic polymorphisms of long non-coding RNA PRNCR1 and prostate cancer risk in a sample of the Iranian population. *Mol. Clin. Oncol.* **2017**, *7*, 1152–1158. [CrossRef]
55. Yang, L.; Lin, C.; Jin, C.; Yang, J.C.; Tanasa, B.; Li, W.; Merkurjev, D.; Ohgi, K.A.; Meng, D.; Zhang, J.; et al. Lncrna-dependent mechanisms of androgen-receptor-regulated gene activation programs. *Nature* **2013**, *500*, 598–602. [CrossRef] [PubMed]
56. Parolia, A.; Crea, F.; Xue, H.; Wang, Y.; Mo, F.; Ramnarine, V.R.; Liu, H.H.; Lin, D.; Saidy, N.R.N.; Clermont, P.-L.; et al. The long non-coding RNA PCGEM1 is regulated by androgen receptor activity in vivo. *Mol. Cancer* **2015**, *14*, 46. [CrossRef] [PubMed]
57. Li, L.; Jia, F.; Bai, P.; Liang, Y.; Sun, R.; Yuan, F.; Zhang, L.; Gao, L. Association between polymorphisms in long non-coding RNA PRNCR1 in 8q24 and risk of gastric cancer. *Tumour Biol.* **2016**, *37*, 299–303. [CrossRef]
58. Li, L.; Sun, R.; Liang, Y.; Pan, X.; Li, Z.; Bai, P.; Zeng, X.; Zhang, D.; Zhang, L.; Gao, L. Association between polymorphisms in long non-coding RNA PRNCR1 in 8q24 and risk of colorectal cancer. *J. Exp. Clin. Cancer Res.* **2013**, *32*, 104. [CrossRef]
59. Li, N.; Cui, Z.; Gao, M.; Li, S.; Song, M.; Wang, Y.; Tong, L.; Bi, Y.; Zhang, Z.; Wang, S.; et al. Genetic Polymorphisms of PRNCR1 and Lung Cancer Risk in Chinese Northeast Population: A Case—Control Study and Meta-Analysis. *DNA Cell Biol.* **2021**, *40*, 132–144. [CrossRef] [PubMed]

60. Al Olama, A.A.; Kote-Jarai, Z.; Berndt, S.I.; Conti, D.V.; Schumacher, F.; Han, Y.; Benlloch, S.; Hazelett, D.; Wang, Z.; Saunders, E.; et al. A meta-analysis of 87,040 individuals identifies 23 new susceptibility loci for prostate cancer. *Nat. Genet.* **2014**, *46*, 1103–1109. [CrossRef]
61. Taheri, M.; Pouresmaeili, F.; Omrani, M.D.; Habibi, M.; Sarrafzadeh, S.; Noroozi, R.; Rakhshan, A.; Sayad, A.; Ghafouri-Fard, S. Association of ANRIL gene polymorphisms with prostate cancer and benign prostatic hyperplasia in an Iranian population. *Biomark. Med.* **2017**, *11*, 413–422. [CrossRef]
62. Zhao, B.; Lu, Y.L.; Yang, Y.; Hu, L.B.; Bai, Y.; Li, R.Q.; Zhang, G.Y.; Li, J.; Bi, C.W.; Yang, L.B.; et al. Overexpression of lncrna anril promoted the proliferation and migration of prostate cancer cells via regulating let-7a/tgf-beta1/smad signaling pathway. *Cancer Biomark.* **2018**, *21*, 613–620. [CrossRef]
63. Hua, J.T.; Ahmed, M.; Guo, H.; Zhang, Y.; Chen, S.; Soares, F.; Lu, J.; Zhou, S.; Wang, M.; Li, H.; et al. Risk SNP-Mediated Promoter-Enhancer Switching Drives Prostate Cancer through lncRNA PCAT19. *Cell* **2018**, *174*, 564–575.e18. [CrossRef]
64. Al Olama, A.A.; Kote-Jarai, Z.; Schumacher, F.; Wiklund, F.; Berndt, S.I.; Benlloch, S.; Giles, G.; Severi, G.; Neal, D.; Hamdy, F.C.; et al. A meta-analysis of genome-wide association studies to identify prostate cancer susceptibility loci associated with aggressive and non-aggressive disease. *Hum. Mol. Genet.* **2013**, *22*, 408–415. [CrossRef] [PubMed]
65. Shui, I.M.; Lindström, S.; Kibel, A.S.; Berndt, S.I.; Campa, D.; Gerke, T.; Penney, K.L.; Albanes, D.; Berg, C.; Bueno-De-Mesquita, H.B.; et al. Prostate Cancer (PCa) Risk Variants and Risk of Fatal PCa in the National Cancer Institute Breast and Prostate Cancer Cohort Consortium. *Eur. Urol.* **2014**, *65*, 1069–1075. [CrossRef] [PubMed]
66. Meyer, K.B.; Maia, A.-T.; O'Reilly, M.; Ghoussaini, M.; Prathalingam, R.; Porter-Gill, P.; Ambs, S.; Prokunina-Olsson, L.; Carroll, J.; Ponder, B.A.J. A Functional Variant at a Prostate Cancer Predisposition Locus at 8q24 Is Associated with PVT1 Expression. *PLoS Genet.* **2011**, *7*, e1002165. [CrossRef]
67. Ilboudo, A.; Chouhan, J.; McNeil, B.K.; Osborne, J.R.; Ogunwobi, O.O. PVT1 Exon 9: A Potential Biomarker of Aggressive Prostate Cancer? *Int. J. Environ. Res. Public Health* **2015**, *13*, 12. [CrossRef] [PubMed]
68. Yang, J.; Li, C.; Mudd, A.; Gu, X. LncRNA PVT1 predicts prognosis and regulates tumor growth in prostate cancer. *Biosci. Biotechnol. Biochem.* **2017**, *81*, 2301–2306. [CrossRef]
69. Al-Thoubaity, F.K. Molecular classification of breast cancer: A retrospective cohort study. *Ann. Med. Surg.* **2020**, *49*, 44–48. [CrossRef] [PubMed]
70. Harbeck, N.; Penault-Llorca, F.; Cortes, J.; Gnant, M.; Houssami, N.; Poortmans, P.; Ruddy, K.; Tsang, J.; Cardoso, F. Breast cancer. *Nat. Rev. Dis. Prim.* **2019**, *5*, 66. [CrossRef]
71. Michailidou, K.; Collaborators, N.; Lindström, S.; Dennis, J.; Beesley, J.; Hui, S.; Kar, S.; Lemaçon, A.; Soucy, P.; Glubb, D.; et al. Association analysis identifies 65 new breast cancer risk loci. *Nature* **2017**, *551*, 92–94. [CrossRef]
72. Zhang, H.; Ahearn, T.U.; Lecarpentier, J.; Barnes, D.; Beesley, J.; Qi, G.; Jiang, X.; O'Mara, T.A.; Zhao, N.; Bolla, M.K.; et al. Genome-wide association study identifies 32 novel breast cancer susceptibility loci from overall and subtype-specific analyses. *Nat. Genet.* **2020**, *52*, 572–581. [CrossRef] [PubMed]
73. Wu, L.; Shi, W.; Long, J.; Guo, X.; Michailidou, K.; Beesley, J.; Bolla, M.K.; Shu, X.-O.; Lu, Y.; Cai, Q.; et al. A transcriptome-wide association study of 229,000 women identifies new candidate susceptibility genes for breast cancer. *Nat. Genet.* **2018**, *50*, 968–978. [CrossRef] [PubMed]
74. Sánchez, Y.; Segura, V.; Marín-Béjar, O.; Athie, A.; Marchese, F.; González, J.; Bujanda, L.; Guo, S.; Matheu, A.; Huarte, M. Genome-wide analysis of the human p53 transcriptional network unveils a lncRNA tumour suppressor signature. *Nat. Commun.* **2014**, *5*, 5812. [CrossRef]
75. Cava, C.; Bertoli, G.; Castiglioni, I. Portrait of Tissue-Specific Coexpression Networks of Noncoding RNAs (miRNA and lncRNA) and mRNAs in Normal Tissues. *Comput. Math. Methods Med.* **2019**, *2019*, 9029351. [CrossRef]
76. Marjaneh, M.M.; Beesley, J.; O'Mara, T.A.; Mukhopadhyay, P.; Koufariotis, L.T.; Kazakoff, S.; Hussein, N.; Fachal, L.; Bartonicek, N.; Hillman, K.M.; et al. Non-coding RNAs underlie genetic predisposition to breast cancer. *Genome Biol.* **2020**, *21*, 7. [CrossRef] [PubMed]
77. Betts, J.A.; Marjaneh, M.M.; Al-Ejeh, F.; Lim, Y.C.; Shi, W.; Sivakumaran, H.; Tropee, R.; Patch, A.-M.; Clark, M.; Bartonicek, N.; et al. Long Noncoding RNAs CUPID1 and CUPID2 Mediate Breast Cancer Risk at 11q13 by Modulating the Response to DNA Damage. *Am. J. Hum. Genet.* **2017**, *101*, 255–266. [CrossRef]
78. Xiong, H.; Chen, Z.; Chen, W.; Li, Q.; Lin, B.; Jia, Y. Fkbp-related ncrnamrna axis in breast cancer. *Genomics* **2020**, *112*, 4595–4607. [CrossRef]
79. Qi, W.; Li, Z.; Xia, L.; Dai, J.; Zhang, Q.; Wu, C.; Xu, S. LncRNA GABPB1-AS1 and GABPB1 regulate oxidative stress during erastin-induced ferroptosis in HepG2 hepatocellular carcinoma cells. *Sci. Rep.* **2019**, *9*, 16185. [CrossRef]
80. Gao, S.; Zhang, F.; Sun, H.; Yang, X. Lncrna ga-binding protein transcription factor subunit β-1 antisense RNA 1 inhibits renal carcinoma growth through an mir-1246/phosphoenolpyruvate carboxykinase 1 pathway. *OncoTargets Ther.* **2020**, *13*, 6827–6836. [CrossRef] [PubMed]
81. Lin, Y.; Guo, W.; Li, N.; Fu, F.; Lin, S.; Wang, C. Polymorphisms of long non-coding RNA HOTAIR with breast cancer susceptibility and clinical outcomes for a southeast Chinese Han population. *Oncotarget* **2017**, *9*, 3677–3689. [CrossRef] [PubMed]
82. Yan, R.; Cao, J.; Song, C.; Chen, Y.; Wu, Z.; Wang, K.; Dai, L. Polymorphisms in lncRNA HOTAIR and susceptibility to breast cancer in a Chinese population. *Cancer Epidemiol.* **2015**, *39*, 978–985. [CrossRef] [PubMed]

83. Hassanzarei, S.; Hashemi, M.; Sattarifard, H.; Hashemi, S.M.; Bahari, G.; Ghavami, S. Genetic polymorphisms of HOTAIR gene are associated with the risk of breast cancer in a sample of southeast Iranian population. *Tumor Biol.* **2017**, *39*, 1010428317727539. [CrossRef] [PubMed]
84. Min, L.; Mu, X.; Tong, A.; Qian, Y.; Ling, C.; Yi, T.; Zhao, X. The association between HOTAIR polymorphisms and cancer susceptibility: An updated systemic review and meta-analysis. *OncoTargets Ther.* **2018**, *11*, 791–800. [CrossRef] [PubMed]
85. Mozdarani, H.; Ezzatizadeh, V.; Parvaneh, R.R. The emerging role of the long non-coding RNA HOTAIR in breast cancer development and treatment. *J. Transl. Med.* **2020**, *18*, 152. [CrossRef] [PubMed]
86. Cai, B.; Wu, Z.; Liao, K.; Zhang, S. Long noncoding RNA HOTAIR can serve as a common molecular marker for lymph node metastasis: A meta-analysis. *Tumor Biol.* **2014**, *35*, 8445–8450. [CrossRef]
87. Gupta, R.A.; Shah, N.; Wang, K.C.; Kim, J.; Horlings, H.M.; Wong, D.J.; Tsai, M.-C.; Hung, T.; Argani, P.; Rinn, J.L.; et al. Long non-coding RNA HOTAIR reprograms chromatin state to promote cancer metastasis. *Nature* **2010**, *464*, 1071–1076. [CrossRef] [PubMed]
88. Li, M.; Li, X.; Zhuang, Y.; Flemington, E.K.; Lin, Z.; Shan, B. Induction of a novel isoform of the lncrna hotair in claudin-low breast cancer cells attached to extracellular matrix. *Mol. Oncol.* **2017**, *11*, 1698–1710. [CrossRef]
89. Xia, Z.; Yan, R.; Duan, F.; Song, C.; Wang, P.; Wang, K. Genetic Polymorphisms in Long Noncoding RNA H19 Are Associated with Susceptibility to Breast Cancer in Chinese Population. *Medicine* **2016**, *95*, e2771. [CrossRef] [PubMed]
90. Shima, H.; Kida, K.; Adachi, S.; Yamada, A.; Sugae, S.; Narui, K.; Miyagi, Y.; Nishi, M.; Ryo, A.; Murata, S.; et al. Lnc RNA H19 is associated with poor prognosis in breast cancer patients and promotes cancer stemness. *Breast Cancer Res. Treat.* **2018**, *170*, 507–516. [CrossRef]
91. Vennin, C.; Spruyt, N.; Dahmani, F.; Julien, S.; Bertucci, F.; Finetti, P.; Chassat, T.; Bourette, R.; Le Bourhis, X.; Adriaenssens, E. H19 non coding RNA-derived miR-675 enhances tumorigenesis and metastasis of breast cancer cells by downregulating c-Cbl and Cbl-b. *Oncotarget* **2015**, *6*, 29209–29223. [CrossRef] [PubMed]
92. Sun, H.; Wang, G.; Peng, Y.; Zeng, Y.; Zhu, Q.N.; Li, T.L.; Cai, J.Q.; Zhou, H.H.; Zhu, Y.S. H19 lncrna mediates 17beta-estradiol-induced cell proliferation in mcf-7 breast cancer cells. *Oncol. Rep.* **2015**, *33*, 3045–3052. [CrossRef] [PubMed]
93. Mathias, C.; Garcia, L.E.; Teixeira, M.D.; Kohler, A.F.; Marchi, R.D.; Barazetti, J.F.; Gradia, D.F.; de Oliveira, J.C. Polymorphism of lncRNAs in breast cancer: Meta-analysis shows no association with susceptibility. *J. Gene Med.* **2020**, *22*, 3271. [CrossRef] [PubMed]
94. Miao, Y.-R.; Liu, W.; Zhang, Q.; Guo, A.-Y. lncRNASNP2: An updated database of functional SNPs and mutations in human and mouse lncRNAs. *Nucleic Acids Res.* **2018**, *46*, D276–D280. [CrossRef] [PubMed]
95. Gao, Y.; Li, X.; Shang, S.; Guo, S.; Wang, P.; Sun, D.; Gan, J.; Sun, J.; Zhang, Y.; Wang, J.; et al. LincSNP 3.0: An updated database for linking functional variants to human long non-coding RNAs, circular RNAs and their regulatory elements. *Nucleic Acids Res.* **2021**, *49*, D1244–D1250. [CrossRef]
96. Sartori, D.A.; Chan, D.W. Biomarkers in prostate cancer: What's new? *Curr. Opin. Oncol.* **2014**, *26*, 259–264. [CrossRef] [PubMed]

genes

MDPI

Article

Convergence of Prognostic Gene Signatures Suggests Underlying Mechanisms of Human Prostate Cancer Progression

Bogdan-Alexandru Luca [1,2], Vincent Moulton [1,†], Christopher Ellis [1], Shea P. Connell [2], Daniel S. Brewer [2,3,†] and Colin S. Cooper [2,*,†]

1 School of Computing Sciences, University of East Anglia, Norwich Research Park, Norwich NR4 7TJ, UK; b.luca@stanford.edu (B.-A.L.); v.moulton@uea.ac.uk (V.M.); c.ellis1@uea.ac.uk (C.E.)
2 Norwich Medical School, University of East Anglia, Norwich Research Park, Norwich NR4 7TJ, UK; s.connell@uea.ac.uk (S.P.C.); d.brewer@uea.ac.uk (D.S.B.)
3 The Earlham Institute, Norwich Research Park, Norwich NR4 7UZ, UK
* Correspondence: colin.cooper@uea.ac.uk
† These authors jointly supervised this work.

Received: 8 June 2020; Accepted: 10 July 2020; Published: 16 July 2020

Abstract: The highly heterogeneous clinical course of human prostate cancer has prompted the development of multiple RNA biomarkers and diagnostic tools to predict outcome for individual patients. Biomarker discovery is often unstable with, for example, small changes in discovery dataset configuration resulting in large alterations in biomarker composition. Our hypothesis, which forms the basis of this current study, is that highly significant overlaps occurring between gene signatures obtained using entirely different approaches indicate genes fundamental for controlling cancer progression. For prostate cancer, we found two sets of signatures that had significant overlaps suggesting important genes ($p < 10^{-34}$ for paired overlaps, hypergeometrical test). These overlapping signatures defined a core set of genes linking hormone signalling (HES6-AR), cell cycle progression (Prolaris) and a molecular subgroup of patients (PCS1) derived by Non Negative Matrix Factorization (NNMF) of control pathways, together designated as SIG-HES6. The second set (designated SIG-DESNT) consisted of the DESNT diagnostic signature and a second NNMF signature PCS3. Stratifications using SIG-HES6 (HES6, PCS1, Prolaris) and SIG-DESNT (DESNT) classifiers frequently detected the same individual high-risk cancers, indicating that the underlying mechanisms associated with SIG-HES6 and SIG-DESNT may act together to promote aggressive cancer development. We show that the use of combinations of a SIG-HES6 signature together with DESNT substantially increases the ability to predict poor outcome, and we propose a model for prostate cancer development involving co-operation between the SIG-HES6 and SIG-DESNT pathways that has implication for therapeutic design.

Keywords: prostate cancer; prognostic signature; diagnostic signature; biomarkers; cancer progression; aggressive cancer

1. Introduction

A major problem in management of human prostate cancer is the high variability in its clinical course making prediction of outcome at the time of diagnosis or following radical therapy extremely difficult [1,2]. A critical challenge is to improve prediction of outcome beyond the use of standard clinical predictors including D'Amico stratification and CAPRA score [3]. For prostate cancer, the development of expression-based prognostic biomarkers has proven very fruitful with over 20 predictive signatures and classifications reported. Many signatures were derived using supervised approaches involving comparisons of aggressive and nonaggressive disease [4–17]. Several biomarkers represent particular biological functions [18–22]. For example, the Prolaris biomarker [19] contains genes known to be involved in controlling transition through the cell cycle.

Unsupervised approaches may also be used for classification and biomarker identification [23–26]. We have used an unsupervised mathematical approach called Latent Process Decomposition (LPD) that takes into account the issue of heterogeneity within individual prostate samples to identify a new poor prognosis category of prostate cancer called DESNT [23,27]. In an alternative unsupervised approach, the status of control pathways deduced from expression datasets was analysed using Non-Negative Matrix Factorisation (NNMF), leading to the identification of a poor prognosis category called PSC1 [26]. The presence of somatic copy number alterations, sometimes linked to the expression of genes within regions of alteration, has also be utilised for biomarker identification [28].

An interesting feature of biomarkers discovery involving comparisons of expression linked to different clinical states are the small overlaps between different predictive gene lists for the same biological endpoint. This observation, and its underlying causes, are well documented for human breast cancer [29–31]. A series of studies demonstrated that the lack of overlap cannot be simply be attributed to trivial reasons such as the use of different patient cohorts, different detection technologies, and different analytical methods. This is illustrated by the work of Ein-Dor et al. [29], who repeated the analysis performed by van't Veer et al. [32] during their derivation of the Mammaprint 70-gene predictive signature for breast cancer outcome. Using transcriptome data from many subsets of training samples that were selected from the complete van't Veer et al. dataset, they demonstrated that multiple different but equally predictive 70-gene signatures could be derived. They noted that for many hundreds of individual genes, the correlation with survival had intermediate predictive values and that the differences between values were very small. The relative ranking of genes changed dramatically when slightly different training sets were used, leading to the selection of poorly overlapping predictive signatures.

In the current study, we wished to examine the relationship between biomarker signatures that were derived using a variety different approaches. Our hypothesis is that the progression of prostate cancer occurs via one or a small number of underlying biological process and that the significant overlaps between prognostic signatures obtained by independent methods of discovery and using different datasets may indicate sampling from genes fundamental for controlling cancer progression.

2. Materials and Methods

2.1. The You et al. Discovery Cohort (DISC)

To repeat the work of You et al., we complied and normalised expression profiles from the same set of 38 public datasets as in You et al. [26] (DISC cohort) except that we did not include the ArrayExpress dataset E-SMDB-2486, which contains the same samples as the GEO dataset GSE3933. Where available, the raw data has been retrieved from GEO and ArrayExpress repositories, otherwise the provided normalised data from the datasets were used. For the TCGA dataset, the RNA-seq level 3 raw expression data have been downloaded from the TCGA data portal. Only the 217 TCGA samples uploaded before 24/04/2013 have been included in the DISC cohort. Two-channel microarray datasets have been internally normalised using the loess method [33] and across arrays using the quantile method [34]—both implemented in the limma R package [35]. One-channel microarrays have been normalised across arrays using the RMA algorithm [36] implemented in either the *affy* [37] or *oligo* [38] R packages, depending on the microarray platform. RNA-seq raw read counts have been processed using the variance stabilizing transformation implemented in DESeq2 [39]. For datasets that contained samples from more than one platform, samples from each platform have been normalised separately. The probes from the three platforms used in the GSE6919 dataset have been merged into a single sample, for each patient id. Probes from each platform have been annotated to Entrez gene identifiers using the corresponding Bioconductor annotation packages, if available;

otherwise the probe identifiers have been converted to entrez ids using the SOURCE interface (http://source-search.princeton.edu/help/SOURCE/resultsBatchHelp.html), Agilent annotation lists or biomaRt package [40]. The Multi-Dimensional-Scaling (MDS) decomposition of the expression profiles of the DISC cohort are shown in Figure S1.

For each dataset, duplicate probes for each Entrez Id have been removed, keeping only the probe with the highest mean expression across samples. The gene intensities have been centred by subtracting the median across samples. The DISC cohort has been then assembled by matching the Entrez Ids, resulting in a cohort of 2707 samples and 32,832 genes. Subsequently, the primary tissue samples without an associated Gleason score, have been removed, resulting in a set of 1381 samples. To remove dataset and platform-specific effects, the data was median-centered and the quantile normalised (MCQ) as described in You et al. [26]. Potential differences compared to the original protocol of You et al. may arise because of the removal of sample duplication and the use of the most probable approach when the published protocol was not completely clear.

2.2. Validation Datasets

Four publicly available transcriptome microarray datasets derived from prostatectomy samples from men with prostate cancer were used as a validation dataset and are referred to as: Memorial Sloan Kettering Cancer Centre (MSKCC) [41], CancerMap [23], CamCap [24], and SWD [42]. From the MSKCC dataset, only prostatectomy specimens were used, both in the derivation of the original DESNT classification and for validation analyses in the current studies. The CamCap dataset was produced by combining two Illumina HumanHT-12 V4.0 expression beadchip datasets (GEO: GSE70768 and GSE70769) obtained from two prostatectomy series (Cambridge and Stockholm). The original CamCap [24] and CancerMap [23] datasets have 40 patients in common and thus 20 of the common samples were excluded at random from each dataset. Each Affymetrix Exon microarray dataset was normalised using the RMA algorithm [36] implemented in the Affymetrix Expression Console software. For CamCap and Stephenson, previous normalised values were used. The ComBat algorithm from the sva R package and quantile transformation, was used to mitigate series-specific effects.

2.3. Replicating You et al. Analysis

2.3.1. Pathway Activation Z-Score

For a given pathway and a given sample, the pathway activation score has been calculated as indicated in Levine et al. [43], namely:

$$Z_{tS} = \frac{\overline{X}_{tS} - \overline{X}_t}{\sigma_t} \sqrt{|S|} \quad (1)$$

where $\overline{X}_{tS}$ is the mean expression level of the genes in pathway S and sample t, $\overline{X}_t$ is the mean expression level of all genes in sample t, σ_t is the standard deviation of all genes in sample t, and $|S|$ is the number of genes in the set S.

2.3.2. Non-Negative Matrix Factorization

Non-negative matrix factorization (NNMF) algorithm implemented in the NNMF R package [44] was used with default parameters.

2.3.3. NNMF Random Forest Classifier

A random forest classifier was trained on the DISC cohort to discriminate between the three NNMF clusters, using as features the 14-pathway z-scores calculated as described above. The model has been built using the implementation from the randomForest R package. The number of trees has been set to 5001, and samples within each class are down-sampled to the frequency of the smallest class; otherwise, the default settings have been used. The model obtained an out-of-bag (OOB) overall accuracy of 92.5%, and a per-class AUC of 0.98–0.99 (Figure S2a–c).

2.4. Replicating the Ramos-Montoya Classifier

To reproduce the Ramos-Montoya classification, a random forest model has been trained on the MSKCC dataset. It uses as training labels the assignment of the MSKCC samples into two classes available in Figure 4a of Ramos-Montoya et al. [22] and as features, the 222 genes in the Ramos-Montoya signature. The model has been built using the *randomForest* R package. The number of tress has been set to 5001, and samples within each class are down-sampled to the frequency of the smallest class; otherwise, the default settings have been used. The model obtained an out-of-bag (OOB) accuracy of 92.67%, and a per-class AUC of 0.99 (Figure S2d).

2.5. Replicating the Prolaris Classifier

For the Prolaris classification on a given sample, a score is calculated by averaging the within-sample z-score normalised expression of the 31 CCP genes [19]. For a given dataset, the top 25% of patients with the highest score are considered high-risk.

2.6. LPD (Latent Process Decomposition) DESNT

LPD [45,46] is an unsupervised Bayesian approach which breaks down (decomposes) each sample into component sub-elements (signatures). Each signature is a representative gene expression pattern. LPD is able to classify complex data based on the relative representation of these signatures in each sample and can objectively assess the most likely number of signatures. The approach can take into account the heterogeneous composition of individual prostate cancer samples. The LPD procedure was carried out exactly as described previously [23,27]. The OAS-LPD algorithm is a modified version of the LPD algorithm in which new sample(s) are decomposed into LPD signatures, without retraining the model (i.e., without re-estimating the model parameters μ_{gk}, $\sigma^2{}_{gk}$, and α). OAS-LPD was carried out exactly as previously described [27].

2.7. Statistical Analysis

The statistical analyses have been carried out in R version 3.3.2. For determining the statistical significance of intersection between two sets of genes, the hypergeometrical test has been used. Genes were defined as differentially expressed if the FDR-adjusted p-value < 0.001 and a fold change >1.4 and identified for each comparison using a moderated t-test implemented in the limma R package. Gene set enrichment analysis was performed using the *Fast Gene Set Enrichment Analysis* Bioconductor package [47] using 10,000 permutations. Survival analyses were performed using the log-rank test and Kaplan–Meier estimator, as implemented in the *survival* R package with biochemical recurrence after prostatectomy as the end point. All survival analyses were performed on the combined CancerMap, CamCap and MSKCC cancer datasets (n = 482) unless otherwise stated.

3. Results

3.1. Relationships between Prostate Cancer Signatures

As a starting point for this study, we compared 25 published mRNA expression signatures derived to predict aggressive human prostate cancer (Table 1, Data S1). The majority of the gene signatures were determined by comparisons of expression patterns with clinical endpoints, and are predicted to have small overlaps [29]. The pattern of overlaps observed in general fitted this model (Figure S3). We noted two highly significant sets of overlaps (Figure 1a,b, Figure S3, $p < 10^{-34}$ for paired overlaps, hypergeometrical test) involving signatures that were derived using unsupervised approaches or that involved investigations of particular biological pathways.

Table 1. Prognostic and Classification gene signatures. Abbreviations are as follows: A, signature discovered by association with clinically distinct states, B, signature representing a biological function; U, signature identified by unsupervised approach; LPD, Latent Process Decomposition; OAS-LPD, One Added Sample-LPD; HCA, Hierarchical Cluster Analysis; ADT, Androgen Deprivation Therapy; NNMF, Non-Negative Matrix Factorisation; RP, Radical Prostatectomy; PSA Prostate Specific Antigen.

Citation	Year	Genes	Type	Discovery Method	Name
Agell et al.	2012	12	A	Association to Gleason	-
Bibkova et al.	2007	16	A	Association to Gleason	-
Bismar et al.	2006	12	A	Benign vs. Cancer vs. Metastases	-
Cuzick et al.	2011	31	B	Cell Cycle Genes	Prolaris
Erho et al.	2013	22	A	Cancers with Different Progressions	DECIPHER
Glinksy et al.	2004	11	A	PSA Failure vs. No-failure	-
Irshad et al.	2013	19	B	Aging Genes Altered in Indolent Cancer	-
Klein et al.	2014	17	A	Association with Outcome	OncotypeDX
Lalonde et al.	2014	276	U	Genes within Copy Number Changes	-
Long et al.	2011	13	A	PSA failure vs. No failure	-
Luca et al.	2017	45	U	LPD	DESNT
Luca et al.	2020	49	U	OAS-LPD	OAS-DESNT
Mo et al.	2018	93	B + A	Stroma association to metastasis	-
Planche et al.	2011	48	A	Normal vs. Tumour differential gene expression in stroma	-
Rajan et al.	2014	7	A	Before and After ADT	-
Ramos-Montoya et al.	2014	222	B	Genes Controlled by HES6	-
Ramaswamy et al.	2003	17	A	Metastases vs. Primary	-
Ross-Adams et al.	2014	100	U	Clustering of Variable Genes	-
Sharma et al.	2013	16	B	Androgen Receptor Regulated	-
Singh et al.	2002	29	A	Associated with Gleason	-
Varambally et al.	2005	44	A	Metastases vs. Primary	-
Walker et al.	2017	70	U + A	HCA and PLS Regression *	-
Wu et al.	2013	32	A	Associated with Outcome	-
You et al.	2016	428	U	NNMF of Control Pathways	PCS1, PCS2, PCS3
Yu et al.	2007	7	B	Polycomb Repression Signature	-

* Applied HCA for subgroup identification and partial-least-squares regression for signature development. All studies cited are listed in the reference section.

First, there was an overlap between the DESNT genes detected as important by two different LPD procedures, LPD-DESNT [23] and OAS-LPD DESNT [27], and the gene differentially over-expressed in the PCS3 subgroup detected by You et al. [26] ($p = 2.6 \times 10^{-35}$ and 2.1×10^{-41}; hypergeometrical test)(Figure 1a, Data S2). In the work of You et al., three groups designated PCS1 (86 genes), PSC2 (123 genes) and PCS3 (219 genes) were detected by non-negative matrix factorisation (NNMF) of the control pathway status calculated from the observed cancer expression profiles. This match could be considered as a match to PSC1 because DESNT genes are under-expressed and the genes overexpressed in PCS3 are also under-expressed in PCS1 [26] (Figure S4). This pathway is referred to as SIG-DESNT and the set of genes within PCS3 that match the DESNT genes is referred to as PCS3-U (U = underexpressed).

Secondly, there was a three-way overlap between the genes associated with PCS1 of You et al. [26], the Prolaris test genes [19], and the signature of Ramos-Montoya et al. [22] (Figure 1b, Data S3). The Prolaris genes were chosen based on their role in controlling cell cycle. Ramos-Montoya et al. selected genes that were controlled by HES6, a transcription factor that has a critical role in driving the androgen receptor (AR) program. This pathway is referred to as SIG-HES6. The 20 genes within PCS1 matching Prolaris and the HES6 signature are all overexpressed. These genes are referred to as PCS1-O (overexpressed) and are distinct from the genes overlapping with the DESNT signature (Data S2).

We were interested to compare cancers detected by these two groups of biomarkers (SIG-DESNT and SIG-HES6) to test whether they are sampling from the same or different cancers at high risk of PSA failure. To examine high risk cancers detected by NNMF of control pathways, we needed first to repeat the analyses carried out by You et al.

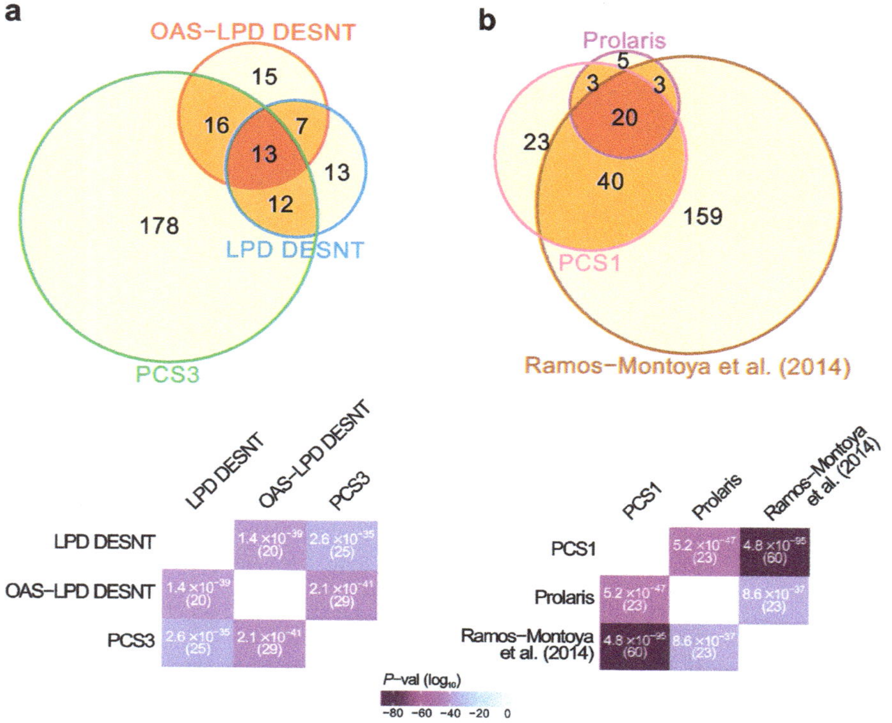

Figure 1. Highly significant signature overlaps. (**a**) Overlaps between LPD DESNT, OAS-LPD DESNT and PCS3. (**b**) Overlap between Prolaris, Ramos-Montoya and PCS1 gene signatures. For each pair of signatures, the probability of the observed overlap occurring by chance was calculated as described in the materials and methods.

3.2. Cancer Subgroups Identified by Non-Negative Matrix Factorisation of Control Pathways

To repeat the work of You et al. we initially complied and normalised expression profiles from 37 public datasets as outlined by the authors (DISC cohort). This resulted in a combine dataset with linked clinical data consisting of primary prostate cancer (n = 1059), non-malignant prostate tissue (n = 746), and metastatic samples from men with castration-resistant prostate cancer (n = 254) (Materials and Methods, Figure S1). We separately compiled a validation cohort consisting of prostatectomy specimens from the MSKCC, CancerMap, CamCap and SWD datasets. For analysis, we used the 14 pathways that were selected by You at al. on the basis of their likely involvement in prostate cancer development. Scores representing the activation status of each pathway in each sample were aggregated into a z-score. Computation of the cophenetic coefficient using a putative number of subgroups between two and six indicated three as the most likely number of subgroups (Figure 2a,b).

Assignment of samples to the three subgroups was carried out by NNMF using a 14xN matrix populated with z-scores as a starting point. The results showed that the three subgroups were detected each with a distinct pattern of pathway activation (Figure 2c). The three subgroups were designated NMF1, NMF2 and NMF3. The NMF1 subgroup exhibited overexpression of the AV, AR-V, PRF, PTEN, ES control pathways, while NMF2 had overexpression of the ERG, AR and FOXA1 pathways (Figure 2d). NMF3 was characterised by overexpression of the PRC, PN, MES and RAS pathways. These correspond to the patterns of pathway activation to the three subgroups PCS1, PCS2 and PSC3, respectively, identified by You et.al. The assignment of the majority of the samples (94%) was the same (Table S1). However, because of the small differences, a distinct nomenclature is used in our study (e.g., NMF1 instead of PCS1). The differences may reflect small deviations in the datasets and in the methodology used (Materials and Methods).

We identified 262 genes (NMF1 n = 74; NMF2 n = 85; NMF3 n = 103; FDR < 0.0001; fold-change > 1.4) that were differentially expressed between the three subgroups NMF1, NMF2 and NMF3: 155 of these overlapped with the 428 differentially expressed genes that distinguished PCS1, PCS2 and PCS3 (Figure S5, Data S4). The overlap between the 262 differentially expressed genes and DESNT genes remained highly significant (Data S5): OAS-LPD DESNT and NMF3 (13 genes, $p = 2.8 \times 10^{-17}$; hypergeometrical test); and DESNT and NMF3 (11 genes, $p = 2.2 \times 10^{-14}$).

Finally, a random forest classifier trained on the division into NMF1, NMF2, and NMF3 in the original dataset was used to interrogate four test datasets (Figure 2e). In each case, there was a significantly worse outcome for patients assigned to the NMF1 subgroup compared to the NMF2 and NMF3 datasets, consistent with the poor outcome observed for the PCS1 dataset of You et al. We conclude that we have achieved a similar although not identical stratification of cancer samples to that achieved by You et al. and that this may be used as a comparator with DESNT cancer and other stratifications.

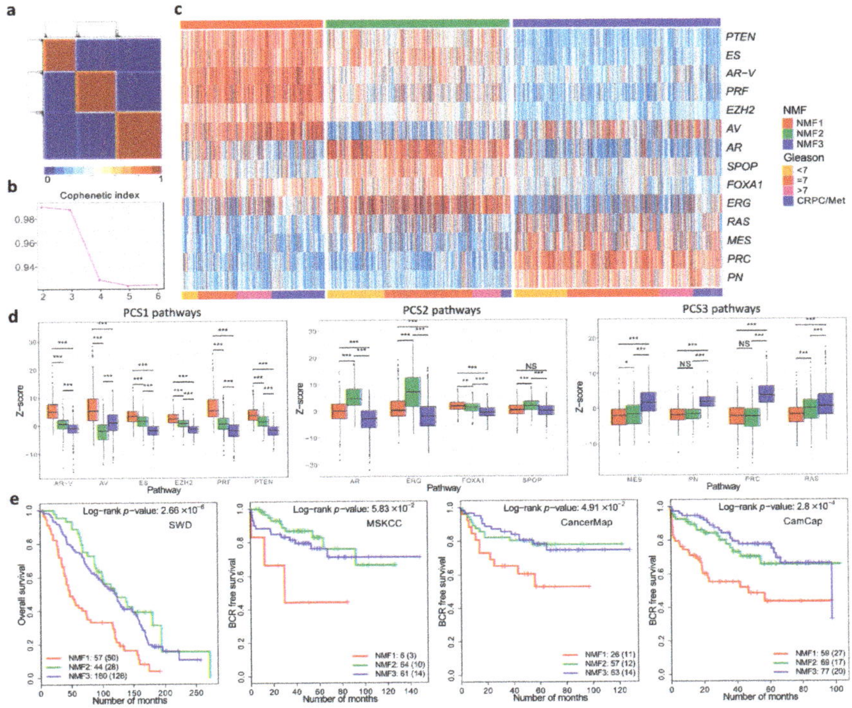

Figure 2. Non-Negative matrix factorisation of control pathways identified three prostate cancer categories. (**a**) Consensus matrix showing three cancer categories. (**b**) Cophenetic coefficient from rank 2 to 6. (**c**) Pathway activation profiles for each cancer (n = 1381) arranged according to the three cancer categories NMF1, NMF2 and NMF3. (**d**) The distribution of pathway activation scores within each cluster. The panels correspond to the three groups of pathways that are over-expressed in each cluster in the You et al. paper. (**e**) Kaplan–Meier plots for four different datasets showing clinical outcome for cancers assigned to the three different cancer categories NMF1, NMF2 and NMF3. * ≤0.05; ** ≤0.01; *** ≤0.001.

3.3. Overlaps in the Detection of Cancers at High Risk of PSA Failure

Returning to the comparisons of cancers detected by the SIG-DESNT and SIG-HES6 groups of signatures, we combined data from the CancerMap, CamCap and MSKCC datasets (n = 482 patients) with PSA failure as an end point, and then separately applied the DESNT, HES6, NMF1, and Prolaris tests. DESNT is calculated as a continuous variable designated γ representing the proportion of the analysed sample that contains the DESNT signature. Cancers were assigned as a "DESNT cancer" when this gene expression pattern had a larger γ value than any other contributing signatures. Random forest classifiers were used to detect NMF1 and Ramos-Montoya et al. high-risk cancers (Figure S2e–g). We used a published formula to calculate the Prolaris index [19] and selected the 25% of cancers exhibiting highest risk.

Based on the assumption that each of the two signature groups (SIG-DESNT and SIG-HES6) represents a separate underlying molecular mechanism, there are two predicted results. Each signature group could represent an entirely separate progression mechanism in which case two non–overlapping groups of cancers with PSA failures should be detected. Alternatively, the two underlying mechanisms may cooperate to cause cancer progression, meaning that the SIG-DESNT and SIG-HES6 predictors will detect the same or overlapping groups cancers with PSA failure.

The overlaps in memberships of each group at high risk of PSA failure are shown in Figure 3 supporting the second of these models. A total of 30 cancers were assigned as high risk by all of the prognostic makers and of these 20 had undergone PSA failure (66.7%). Of 100 cancers with 50 PSA failures assigned the SIG-DESNT signature (DESNT), 61(37 PSA failures) were also detected by at least one of the SIG-HES6 signatures (HES6, Prolaris, and/or NMF1) (Figure 3). A combination of the Ramos-Montoya et al. (SIG-HES6) and DESNT (SIG-DESNT) models predicted the majority of PSA failures present in the high-risk cancer groups (76 of 84, 90.5%), with 32 failures overlapping.

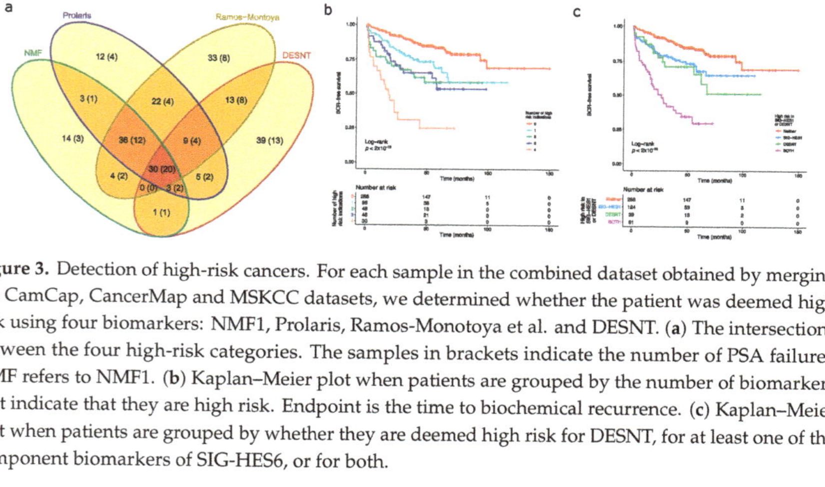

Figure 3. Detection of high-risk cancers. For each sample in the combined dataset obtained by merging the CamCap, CancerMap and MSKCC datasets, we determined whether the patient was deemed high risk using four biomarkers: NMF1, Prolaris, Ramos-Monotoya et al. and DESNT. (**a**) The intersections between the four high-risk categories. The samples in brackets indicate the number of PSA failures. NMF refers to NMF1. (**b**) Kaplan–Meier plot when patients are grouped by the number of biomarkers that indicate that they are high risk. Endpoint is the time to biochemical recurrence. (**c**) Kaplan–Meier plot when patients are grouped by whether they are deemed high risk for DESNT, for at least one of the component biomarkers of SIG-HES6, or for both.

Kaplan–Meier analyses were preformed to investigate the interactions between the high-risk categories. A particularly poor outcome was observed for patients designated as high risk by all four biomarkers (Figure 3b). Looking at the interactions between SIG-HES6 and SIG-DESNT biomarkers, we found intermediate rates of progression for patients deemed high risk either for DESNT (p = 0.0037 Benjamini–Hochberg adjusted for multiple testing (BH); pairwise comparison between DESNT only and neither; 26.9 vs. 93.0 months to 25% events) or for at least one of the SIG-HES6 biomarkers (Prolaris and/or PCS1 and/or Ramos-Monotoya et al.) (BH p = 0.0039; 42.5 vs. 93.0 months to 25% events; Figure 3c). However, when patients were designated as high risk both by DESNT and by at least one of the SIG-HES6 biomarkers, their outcome was considerably worse (Time to 25% events = 8.0 months;

BH $p < 2 \times 10^{-16}$ both vs. neither; Figure 3c). This observation is consistent with our hypothesis that two underlying mechanisms represented by SIG-HES6 and SIG-DESNT are interacting to cause cancer progression.

Upon investigation of whether poor outcome was simply determined by Gleason Score, we found that the number of signatures that indicated that a patient was at high risk was an independent prognostic indicator when Gleason was included as a covariate (IQR HR = 1.98; 95% CI 1.54–2.55; $p = 1.01 \times 10^{-7}$; Cox proportional hazards regression model). In additional, the combination of a high risk defined by at least one member of SIG-HES6 and DESNT is an independent prognostic indicator when Gleason is included as a covariate [HR = 3.86 (95% CI 2.41–6.19)]. This compares to DESNT only [HR = 1.85 (0.99–3.46)], and SIG-HES6 only [HR = 1.61 (95% CI 1.02–2.52)] (Cox proportional hazards regression models). These results show that designation as high risk provides additional prognostic information to that determined by Gleason Score.

3.4. Comparison of DESNT and Non-Negative Matrix Factorisation Subgroups

We wished to further investigate the relationship between the DESNT and NMF1 poor prognosis groups. To achieve this, the MSKCC, CancerMap, CamCap, and TCGA datasets were combined and DESNT was plotted as a continuous variable (DESNT γ), as described in Luca et al. [27]. DESNT γ was significantly higher in NMF1 cancers compared to NMF2 and NMF3 cancers (Figure 4a) and the results of Gene Set Enrichment Analysis (GSEA) analysis show a highly significant association ($p < 1 \times 10^{-6}$), giving an enrichment score of 0.61 (Figure 4b).

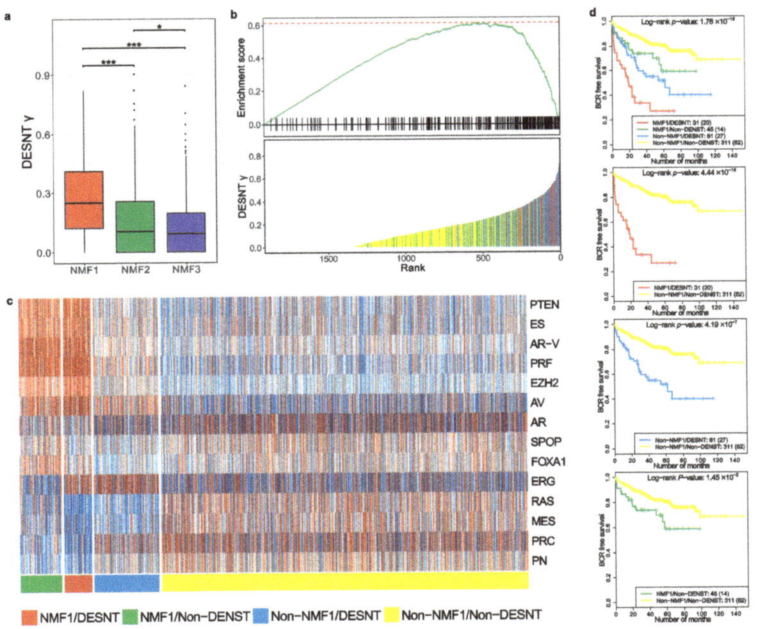

Figure 4. Comparison of DESNT and non-negative matrix classifications. (**a**) Distribution of DESNT γ for cancers assigned to NMF1, NMF2 and NMF3. (**b**) Gene Set Enrichment Analysis. Cancers were ranked according to DESNT γ (Lower Panel). The enrichment for cancers assigned to the NMF1 high-risk group (vertical lines) is shown (Upper Panel). (**c**) Pathway activation profiles for each cancer arranged according to DESNT and NMF1 subgroup status. The key is shown at the bottom of the figure. (**d**) Kaplan–Meir plots for the different cancer categories. The outcome used is time to biochemical recurrence post prostatectomy. * ≤0.05; ** ≤0.01; *** ≤0.001.

We next calculated pathway status (*z*-scores, as shown in Figure 2b) for all samples in the MSKCC, CancerMap, CancerMap and TCGA datasets and grouped the samples according to NMF1 and DESNT status. The results are shown in Figure 4c. Cancers assigned both as DESNT and NMF1 had the strongest association with time to progression (Figure 4d, $p = 4.4 \times 10^{-16}$, Log-rank test) followed by DESNT-non-NMF1 cancers ($p = 4.19 \times 10^{-7}$) and non-DESNT-NMF1 cancers ($p = 1.45 \times 10^{-2}$). Membership of DESNT accounted for 36% (31/86) of NMF1 cancers in this series but 59% (31/45) of its PSA failures. Notably, activation of the PTEN, ES, AR-V, PRF and EZH2 pathways, a feature of NMF1 cancers, was not present in DESNT-non-NMF1 cancers.

We conclude from these studies that NMF1 and DESNT are overlapping but distinct cancer categories.

4. Discussion

A number of critical observations arise from the presented studies. Signatures derived by comparisons of expression profiles to clinical features (e.g., to Gleason Score and to PSA failure) exhibited only modest overlaps in gene lists. This was exactly as predicted from previous analyses of breast cancer datasets [29]. When normal cells change into a cancer cells or when the clinical state of a cancer is altered, many thousands of genes may exhibit altered expression levels and multiple control pathways modulated. Based only on the analyses performed to identify these biomarkers, it is not possible to determine whether the genes identified are central to cancer development or represent secondary events. Nonetheless, when biomarker analyses are combined with additional studies, useful individual genes are highlighted. For example, HOXB13 was identified in expression array studies as a gene highly upregulated in prostate cancer [48] but its central importance to cancer development was not established until the analyses of cancer families were performed [49]. AMACR was first identified as a gene upregulated in three of four expression array datasets from prostate cancer, but its importance as a cancer marker was not recognised until immunohistochemical studies of tissue sections were performed [50].

When significant overlaps do occur between predictive gene lists developed using entirely different approaches, it is our belief that this indicates genes fundamental for controlling cancer progression. The observation that HES6-signature reported by Ramos-Montoya et al. [22] overlaps with the PCS1 [26] and Prolaris [19] signatures supports this view. HES6 drives castration-resistant tumour growth by enhancing the transcriptional activity of the androgen receptor, while the Prolaris signature contains many genes known to be critical for cell cycle control—both processes already known to be essential for prostate cancer growth. A second overlap occurred between downregulated DESNT genes [23,27] and a set of genes overexpressed in PCS3 [26]. We propose that genes from these two categories are also involved in processes fundamental to the development of prostate cancer. The precise mechanism is currently unknown, although possible but different models were suggested both by Luca et al. [23] and by You et al. [26].

Support for this model was obtained from analyses of the impact of DESNT and SIG-HES6 signatures on clinical outcome. When patients were designated as poor prognosis by DESNT and by one or more of the SIG-HES6 signatures (Prolaris, PCS1, Ramos-Montoya et al.), a considerably worse outcome was observed compared with use of DESNT or SIG-HES6 signatures alone, consistent with interaction. This observation also has implications for patient management, indicating that use of DESNT classification together with, for example, the Prolaris biomarker or the Ramos-Montoya et al. biomarker could greatly increase the ability to predict whether a patient with organ-confined prostate cancer will progress following treatment. This would allow targeting of treatment to the patients who need it hence avoiding the side effects, including impotence, of unnecessary treatment in men with indolent disease.

The overlapping signature DESNT, PCS3, HES6, Prolaris and PCS3 are all derived using unsupervised approaches or by investigation of biological function. It is of interest that not all signature derived using these approaches demonstrated highly significant overlaps. The derivation of the 70-gene biomarker proposed by Walker et al. [25] represent an interesting case. A 222-gene signature was originally generated using an unsupervised approach. The 70-gene signature represents a subset of these genes derived by a combination of unsupervised and supervised approaches. We failed to observe highly significant overlaps involving this signature (Figure S3). Additional signatures involving unsupervised steps that failed show gene overlaps includes those derived by Lalonde et al. [28] and by Ross Adams et al. [24].

We provide evidence that classifications based on NNMF analysis of control pathways and the DESNT classification are overlapping but distinct. The methods of clinical applications of the two tests are also different. Assignment to the PCS1 poor prognosis category is based on the use of a classifier of 37-gene classifier [26] selected from genes' differential expression between the PCS1, PCS2, and PCS3 groups. In contrast, the poor prognosis DESNT signature is only considered to be present in part of the cancer, with the exact proportion (or DESNT γ) calculated by LPD carried out on genes with the most variable levels of expression across samples [23,27]—the DESNT gene signature itself cannot be used to calculate outcome. Once calculated, the proportion of DESNT cancer can be used in a nomogram together with clinical variables to estimate likelihood of PSA failure [27]. Additionally, PCS1 and PCS3 had been assigned as having, respectively, luminal and basal phenotypes based on the expression of a set of 12 genes [26]. In contrast, we failed to find differential expression of these same genes when comparing DESNT and non-DESNT cancers (result not shown).

An important finding is that all of the highly overlapping signatures predicting poor outcome appear to be sampling from the same high-risk cancer group: the SIG-HES6-and SIG-DESNT groups of signatures are not detecting entirely separate groups of high-risk cancers. This result as well as the observed interactions between DESNT and SIG-HES6 signatures in identifying patients with poor outcome are both consistent with a model where underlying molecular processes represented by SIG-HES6 and SIG-DESNT interact, leading to aggressive disease. This observation may have relevance to approaches for therapeutic targeting. In the clinical setting, the HES6-associated signature can be pharmacologically targeted by inhibition of PLK1 with restoration of sensitivity to castration [22]. For the DESNT signature, many of the genes with downregulated expression in prostate cancer are hypermethylation [23], indicating that 5-azacytidine that could be used to enhance gene expression. Thus, a prediction of the current studies is that the combined use of inhibitors of HES6 function, androgen withdrawal and strategies for gene re-expression, would synergise in preventing the growth of castration-sensitive prostate cancer. Our results also have an implication for the use of biomarkers in general since the use of DESNT together with a SIG-HES6 biomarker may represent a much more effective method for detecting patients with aggressive disease.

5. Conclusions

To our knowledge this is the first publication to systematically analyze the relationships between multiple distinct prognostic signatures for prostate cancer. We start with the hypothesis that highly significant overlaps between signatures derived using different approaches indicates genes and processes fundamental to prostate cancer progression; leading to the identification of two sets of overlaps designated SIG-HES6 and SIG-DESNT. First, we conclude that our results support a model whereby SIG-HES6 and SIG-DESNT genes co-operated to cause cancer progression. Secondly, consistent with this model, the use of a SIG-HES6 signature in combination with DESNT provides a much better predictor of poor outcome than the use of either alone. Thirdly, for the drug treatment of patients we predict a synergy between (i) inhibitors of HES6 function, and (ii) agents, such as 5-azacytidine, that can induce re-expression of DESNT genes.

Supplementary Materials: The following are available online at http://www.mdpi.com/2073-4425/11/7/802/s1. Figure S1: MDS (Multi-Dimensional-Scaling) decomposition of the expression profiles of the DISC cohort, Figure S2: The out-of-bag(OOB) performance of NNMF random forest classifier on the DISC cohort, Figure S3: Overlaps between the gene signatures considered in this study, Figure S4: Genes identified as subtype-enriched by You et al., Figure S5: Comparison of altered genes identified by non-negative matrix factorization (NNMF) of control pathways; Table S1: Repetition of the non-negative matrix factorization (NNMF) categorisation of prostate cancer described by You et al.; Supplementary materials: Data S1: List of biomarker and signature gene used in Figure S3, Data S2: Overlaps between the DESNT, OAS-DESNT, and PCS3 gene signatures, Data S3: Overlap gene signatures for the Prolaris, PCS1, and Ramos-Montoya et al. (HES6) gene signatures; Data S4: Comparisons of differentially expressed genes, Data S5: Overlaps between the DESNT, OAS-DESNT, and NMF3 gene signatures.

Author Contributions: B.-A.L.—Conceptualization, Writing—original draft, Visualization, Formal analysis, Data curation, Methodology; V.M.—Conceptualization, Supervision, Funding acquisition; C.E.—Formal analysis. S.P.C.: Writing—review & editing, Formal analysis; D.S.B.—Conceptualization, Writing—original draft, Supervision, Methodology; C.S.C.—Conceptualization, Writing—original draft, Supervision, Funding acquisition. All authors have read and agreed to the published version of the manuscript.

Funding: This work was funded by the Bob Champion Cancer Trust, The Masonic Charitable Foundation successor to The Grand Charity, The King Family, The Hargrave Foundation and The University of East Anglia. We acknowledge support from Movember, from Prostate Cancer UK, The Big C Cancer Charity, Callum Barton and from The Andy Ripley Memorial Fund.

Acknowledgments: The research presented in this paper was carried out on the High Performance Computing Cluster supported by the Research and Specialist Computing Support service at the University of East Anglia. These authors contributed equally: Bogdan-Alexandru Luca, Daniel S. Brewer These authors jointly supervised this work: Vincent Moulton, Daniel S. Brewer, Colin S. Cooper. We thank Charlie Massie for helpful comments.

Conflicts of Interest: Colin Cooper, Daniel Brewer, Bogdan-Alexandru Luca and Vincent Moulton are co-inventors on a patent application from the University of East Anglia on the detection of DESNT prostate cancer.

References

1. D'Amico, A.V.; Moul, J.; Carroll, P.R.; Sun, L.; Lubeck, D.; Chen, M.-H. Cancer-Specific Mortality After Surgery or Radiation for Patients With Clinically Localized Prostate Cancer Managed During the Prostate-Specific Antigen Era. *J. Clin. Oncol.* **2016**, *21*, 2163–2172.
2. Buyyounouski, M.K.; Pickles, T.; Kestin, L.L.; Allison, R.; Williams, S.G. Validating the Interval to Biochemical Failure for the Identification of Potentially Lethal Prostate Cancer. *J. Clin. Oncol.* **2012**, *30*, 1857–1863. [CrossRef] [PubMed]
3. Cucchiara, V.; Cooperberg, M.R.; Dall'Era, M.; Lin, D.W.; Montorsi, F.; Schalken, J.A.; Evans, C.P. Genomic Markers in Prostate Cancer Decision Making. *Eur. Urol.* **2018**, *73*, 572–582. [CrossRef]
4. Agell, L.; Hernández, S.; Nonell, L.; Lorenzo, M.; Puigdecanet, E.; de Muga, S.; Juanpere, N.; Bermudo, R.; Fernández, P.L.; Lorente, J.A.; et al. 12-gene expression signature is associated with aggressive histological in prostate cancer: SEC14L1 and TCEB1 genes are potential markers of progression. *Am. J. Pathol.* **2012**, *181*, 1585–1594. [CrossRef] [PubMed]
5. Bibikova, M.; Chudin, E.; Arsanjani, A.; Zhou, L.; Garcia, E.W.; Modder, J.; Kostelec, M.; Barker, D.; Downs, T.; Fan, J.-B.; et al. Expression signatures that correlated with Gleason score and relapse in prostate cancer. *Genomics* **2007**, *89*, 666–672. [CrossRef] [PubMed]
6. Bismar, T.A.; Demichelis, F.; Riva, A.; Kim, R.; Varambally, S.; He, L.; Kutok, J.; Aster, J.C.; Tang, J.; Kuefer, R.; et al. Defining aggressive prostate cancer using a 12-gene model. *Neoplasia* **2006**, *8*, 59–68. [CrossRef]
7. Long, Q.; Johnson, B.A.; Osunkoya, A.O.; Lai, Y.-H.; Zhou, W.; Abramovitz, M.; Xia, M.; Bouzyk, M.B.; Nam, R.K.; Sugar, L.; et al. Protein-coding and microRNA biomarkers of recurrence of prostate cancer following radical prostatectomy. *Am. J. Pathol.* **2011**, *179*, 46–54. [CrossRef]
8. Erho, N.; Crisan, A.; Vergara, I.A.; Mitra, A.P.; Ghadessi, M.; Buerki, C.; Bergstralh, E.J.; Kollmeyer, T.; Fink, S.; Haddad, Z.; et al. Discovery and Validation of a Prostate Cancer Genomic Classifier that Predicts Early Metastasis Following Radical Prostatectomy. *PLoS ONE* **2013**, *8*, e66855. [CrossRef]
9. Glinsky, G.V.; Glinskii, A.B.; Stephenson, A.J.; Hoffman, R.M.; Gerald, W.L. Gene expression profiling predicts clinical outcome of prostate cancer. *J. Clin. Investig.* **2004**, *113*, 913–923. [CrossRef]
10. Klein, E.A.; Cooperberg, M.R.; Magi-Galluzzi, C.; Simko, J.P.; Falzarano, S.M.; Maddala, T.; Chan, J.M.; Li, J.; Cowan, J.E.; Tsiatis, A.C.; et al. A 17-gene assay to predict prostate cancer aggressiveness in the context of Gleason grade heterogeneity, tumor multifocality, and biopsy undersampling. *Eur. Urol.* **2014**, *66*, 550–560. [CrossRef]
11. Planche, A.; Bacac, M.; Provero, P.; Fusco, C.; Delorenzi, M.; Stehle, J.-C.; Stamenkovic, I. Identification of prognostic molecular features in the reactive stroma of human breast and prostate cancer. *PLoS ONE* **2011**, *6*, e18640. [CrossRef] [PubMed]

12. Rajan, P.; Stockley, J.; Sudbery, I.M.; Fleming, J.T.; Hedley, A.; Kalna, G.; Sims, D.; Ponting, C.P.; Heger, A.; Robson, C.N.; et al. Identification of a candidate prognostic gene signature by transcriptome analysis of matched pre- and post-treatment prostatic biopsies from patients with advanced prostate cancer. *BMC Cancer* **2014**, *14*, 1–10. [CrossRef]
13. Ramaswamy, S.; Ross, K.N.; Lander, E.S.; Golub, T.R. A molecular signature of metastasis in primary solid tumors. *Nat. Genet.* **2003**, *33*, 49–54. [CrossRef] [PubMed]
14. Singh, D.; Febbo, P.G.; Ross, K.; Jackson, D.G.; Manola, J.; Ladd, C.; Tamayo, P.; Renshaw, A.A.; D'Amico, A.V.; Richie, J.P.; et al. Gene expression correlates of clinical prostate cancer behavior. *Cancer Cell* **2002**, *1*, 203–209. [CrossRef]
15. Varambally, S.; Yu, J.; Laxman, B.; Rhodes, D.R.; Mehra, R.; Tomlins, S.A.; Shah, R.B.; Chandran, U.; Monzon, F.A.; Becich, M.J.; et al. Integrative genomic and proteomic analysis of prostate cancer reveals signatures of metastatic progression. *Cancer Cell* **2005**, *8*, 393–406. [CrossRef]
16. Wu, C.-L.; Schroeder, B.E.; Ma, X.-J.; Cutie, C.J.; Wu, S.; Salunga, R.; Zhang, Y.; Kattan, M.W.; Schnabel, C.A.; Erlander, M.G.; et al. Development and validation of a 32-gene prognostic index for prostate cancer progression. *Proc. Natl. Acad. Sci. USA* **2013**, *110*, 6121–6126. [CrossRef]
17. Mo, F.; Lin, D.; Takhar, M.; Ramnarine, V.R.; Dong, X.; Bell, R.H.; Volik, S.V.; Wang, K.; Xue, H.; Wang, Y.; et al. Stromal Gene Expression is Predictive for Metastatic Primary Prostate Cancer. *Eur. Urol.* **2018**, *73*, 524–532. [CrossRef]
18. Yu, J.; Yu, J.; Rhodes, D.R.; Tomlins, S.A.; Cao, X.; Chen, G.; Mehra, R.; Wang, X.; Ghosh, D.; Shah, R.B.; et al. A polycomb repression signature in metastatic prostate cancer predicts cancer outcome. *Cancer Res.* **2007**, *67*, 10657–10663. [CrossRef]
19. Cuzick, J.; Swanson, G.P.; Fisher, G.; Brothman, A.R.; Berney, D.M.; Reid, J.E.; Mesher, D.; Speights, V.O.; Stankiewicz, E.; Foster, C.S.; et al. Prognostic value of an RNA expression signature derived from cell cycle proliferation genes in patients with prostate cancer: A retrospective study. *Lancet Oncol.* **2011**, *12*, 245–255. [CrossRef]
20. Irshad, S.; Bansal, M.; Castillo-Martin, M.; Zheng, T.; Aytes, A.; Wenske, S.; Le Magnen, C.; Guarnieri, P.; Sumazin, P.; Benson, M.C.; et al. A molecular signature predictive of indolent prostate cancer. *Sci. Transl. Med.* **2013**, *5*, 202ra122. [CrossRef]
21. Sharma, N.L.; Massie, C.E.; Ramos-Montoya, A.; Zecchini, V.; Scott, H.E.; Lamb, A.D.; MacArthur, S.; Stark, R.; Warren, A.Y.; Mills, I.G.; et al. The androgen receptor induces a distinct transcriptional program in castration-resistant prostate cancer in man. *Cancer Cell* **2013**, *23*, 35–47. [CrossRef]
22. Montoya, A.R.; Lamb, A.D.; Russell, R.; Carroll, T.; Jurmeister, S.; Dalmau, N.G.; Massie, C.E.; Boren, J.; Bon, H.; Theodorou, V.; et al. HES6 drives a critical AR transcriptional programme to induce castration-resistant prostate cancer through activation of an E2F1-mediated cell cycle network. *EMBO Mol. Med.* **2014**, *6*, 651–661. [CrossRef]
23. Luca, B.-A.; Brewer, D.S.; Edwards, D.R.; Edwards, S.; Whitaker, H.C.; Merson, S.; Dennis, N.; Cooper, R.A.; Hazell, S.; Warren, A.Y.; et al. DESNT: A Poor Prognosis Category of Human Prostate Cancer. *Eur. Urol. Focus* **2018**, *4*, 842–850. [CrossRef] [PubMed]
24. Ross-Adams, H.; Lamb, A.D.; Dunning, M.J.; Halim, S.; Lindberg, J.; Massie, C.M.; Egevad, L.A.; Russell, R.; Ramos-Montoya, A.; Vowler, S.L.; et al. CamCaP Study Group, Integration of copy number and transcriptomics provides risk stratification in prostate cancer: A discovery and validation cohort study. *EBioMedicine* **2015**, *2*, 1133–1144. [CrossRef]
25. Walker, S.M.; Knight, L.A.; McCavigan, A.M.; Logan, G.E.; Berge, V.; Sherif, A.; Pandha, H.; Warren, A.Y.; Davidson, C.; Uprichard, A.; et al. Molecular Subgroup of Primary Prostate Cancer Presenting with Metastatic Biology. *Eur. Urol.* **2017**, *72*, 509–518. [CrossRef]
26. You, S.; Knudsen, B.S.; Erho, N.; Alshalalfa, M.; Takhar, M.; Al-deen Ashab, H.; Davicioni, E.; Karnes, R.J.; Klein, E.A.; Den, R.B.; et al. Integrated Classification of Prostate Cancer Reveals a Novel Luminal Subtype with Poor Outcome. *Cancer Res.* **2016**, *76*, 4948–4958. [CrossRef]
27. Luca, B.-A.; Moulton, V.; Ellis, C.; Edwards, D.R.; Campbell, C.; Cooper, R.A.; Clark, J.; Brewer, D.S.; Cooper, C.S. A novel stratification framework for predicting outcome in patients with prostate cancer. *Br. J. Cancer* **2020**, *122*, 1467–1476. [CrossRef] [PubMed]

28. Lalonde, E.; Ishkanian, A.S.; Sykes, J.; Fraser, M.; Ross-Adams, H.; Erho, N.; Dunning, M.J.; Halim, S.; Lamb, A.D.; Moon, N.C.; et al. Tumour genomic and microenvironmental heterogeneity for integrated prediction of 5-year biochemical recurrence of prostate cancer: A retrospective cohort study. *Lancet Oncol.* **2014**, *15*, 1521–1532. [CrossRef]
29. Ein-Dor, L.; Kela, I.; Getz, G.; Givol, D.; Domany, E. Outcome signature genes in breast cancer: Is there a unique set? *Bioinformatics* **2005**, *21*, 171–178. [CrossRef] [PubMed]
30. Michiels, S.; Koscielny, S.; Hill, C. Prediction of cancer outcome with microarrays: A multiple random validation strategy. *Lancet* **2005**, *365*, 488–492. [CrossRef]
31. Ein-Dor, L.; Zuk, O.; Domany, E. Thousands of samples are needed to generate a robust gene list for predicting outcome in cancer. *Proc. Natl. Acad. Sci. USA* **2006**, *103*, 5923–5928. [CrossRef] [PubMed]
32. Van't Veer, L.J.; Dai, H.; van de Vijver, M.J.; He, Y.D.; Hart, A.A.M.; Mao, M.; Peterse, H.L.; van der Kooy, K.; Marton, M.J.; Witteveen, A.T.; et al. Gene expression profiling predicts clinical outcome of breast cancer. *Nature* **2002**, *415*, 530–536. [CrossRef]
33. Yang, Y.H.; Dudoit, S.; Luu, P.; Lin, D.M.; Peng, V.; Ngai, J.; Speed, T.P. Normalization for cDNA microarray data: A robust composite method addressing single and multiple slide systematic variation. *Nucleic Acids Res.* **2002**, *30*, e15. [CrossRef] [PubMed]
34. Bolstad, B.M.; Irizarry, R.A.; Astrand, M.; Speed, T.P. A comparison of normalization methods for high density oligonucleotide array data based on variance and bias. *Bioinformatics* **2003**, *19*, 185–193. [CrossRef]
35. Ritchie, M.E.; Phipson, B.; Wu, D.; Hu, Y. Limma powers differential expression analyses for RNA-sequencing and microarray studies. *Nucleic Acids Res.* **2015**, *43*, e47. [CrossRef]
36. Irizarry, R.A.; Hobbs, B.; Collin, F.; Beazer-Barclay, Y.D.; Antonellis, K.J.; Scherf, U.; Speed, T.P. Exploration, normalization, and summaries of high density oligonucleotide array probe level data. *Biostatistics* **2003**, *4*, 249–264. [CrossRef] [PubMed]
37. Gautier, L.; Cope, L.; Bolstad, B.M.; Irizarry, R.A. Affy-analysis of Affymetrix GeneChip data at the probe level. *Bioinformatics* **2004**, *20*, 307–315. [CrossRef]
38. Carvalho, B.S.; Irizarry, R.A. A framework for oligonucleotide microarray preprocessing. *Bioinformatics* **2010**, *26*, 2363–2367. [CrossRef]
39. Love, M.I.; Huber, W.; Anders, S. Moderated estimation of fold change and dispersion for RNA-seq data with DESeq2. *Genome Biol.* **2014**, *15*, 550. [CrossRef]
40. Durinck, S.; Moreau, Y.; Kasprzyk, A.; Davis, S.; De Moor, B.; Brazma, A.; Huber, W. BioMart and Bioconductor: A powerful link between biological databases and microarray data analysis. *Bioinformatics* **2005**, *21*, 3439–3440. [CrossRef]
41. Taylor, B.S.; Schultz, N.; Hieronymus, H.; Gopalan, A.; Xiao, Y.; Carver, B.S.; Arora, V.K.; Kaushik, P.; Cerami, E.; Reva, B.; et al. Integrative genomic profiling of human prostate cancer. *Cancer Cell* **2010**, *18*, 11–22. [CrossRef]
42. Sboner, A.; Demichelis, F.; Calza, S.; Pawitan, Y.; Setlur, S.R.; Hoshida, Y.; Perner, S.; Adami, H.-O.; Fall, K.; Mucci, L.A.; et al. Molecular sampling of prostate cancer: A dilemma for predicting disease progression. *BMC Med. Genom.* **2010**, *3*, 8–12. [CrossRef] [PubMed]
43. Levine, D.M.; Haynor, D.R.; Castle, J.C.; Stepaniants, S.B.; Pellegrini, M.; Mao, M.; Johnson, J.M. Pathway and gene-set activation measurement from mRNA expression data: The tissue distribution of human pathways. *Genome Biol.* **2006**, *7*, R93. [CrossRef]
44. Gaujoux, R.; Seoighe, C. A flexible R package for nonnegative matrix factorization. *BMC Bioinform.* **2010**, *11*, 367–369. [CrossRef] [PubMed]
45. Rogers, S.; Girolami, M.; Campbell, C.; Breitling, R. The latent process decomposition of cDNA microarray data sets. *IEEE/ACM Trans. Comput. Biol. Bioinform.* **2005**, *2*, 143–156. [CrossRef]
46. Carrivick, L.; Rogers, S.; Clark, J.; Campbell, C.; Girolami, M.; Cooper, C. Identification of prognostic signatures in breast cancer microarray data using Bayesian techniques. *J. R. Soc. Interface* **2006**, *3*, 367–381. [CrossRef] [PubMed]
47. Sergushichev, A.A. An algorithm for fast preranked gene set enrichment analysis using cumulative statistic calculation. *BioRxiv* **2016**, 060012.
48. Edwards, S.; Campbell, C.; Flohr, P.; Shipley, J.; Giddings, I.; Te-Poele, R.; Dodson, A.; Foster, C.; Clark, J.; Jhavar, S.; et al. Expression analysis onto microarrays of randomly selected cDNA clones highlights HOXB13 as a marker of human prostate cancer. *Br. J. Cancer* **2005**, *92*, 376–381. [CrossRef]

49. Ewing, C.M.; Ray, A.M.; Lange, E.M.; Zuhlke, K.A.; Robbins, C.M.; Tembe, W.D.; Wiley, K.E.; Isaacs, S.D.; Johng, D.; Wang, Y.; et al. Germline mutations in HOXB13 and prostate-cancer risk. *N. Engl. J. Med.* **2012**, *366*, 141–149. [CrossRef]
50. Rubin, M.A.; Zhou, M.; Dhanasekaran, S.M.; Varambally, S.; Barrette, T.R.; Sanda, M.G.; Pienta, K.J.; Ghosh, D.; Chinnaiyan, A.M. α-Methylacyl coenzyme A racemase as a tissue biomarker for prostate cancer. *JAMA* **2002**, *287*, 1662–1670. [CrossRef]

genes

MDPI

Article

Identification and Characterization of Alternatively Spliced Transcript Isoforms of *IRX4* in Prostate Cancer

Achala Fernando [1,2], Chamikara Liyanage [1,2], Afshin Moradi [1,2], Panchadsaram Janaththani [1,2] and Jyotsna Batra [1,2,*]

1 Faculty of Health, Institute of Health and Biomedical Innovation, School of Biomedical Sciences, Queensland University of Technology, Brisbane QLD 4059, Australia; achala.vitharanage@hdr.qut.edu.au (A.F.); chamikara.liyanage@hdr.qut.edu.au (C.L.); afshin.moradi@hdr.qut.edu.au (A.M.); panchadsaram.janaththani@qut.edu.au (P.J.)

2 Translational Research Institute, Queensland University of Technology, Brisbane QLD 4102, Australia

* Correspondence: jyotsna.batra@qut.edu.au; Tel.: +61-734437336

Abstract: Alternative splicing (AS) is tightly regulated to maintain genomic stability in humans. However, tumor growth, metastasis and therapy resistance benefit from aberrant RNA splicing. Iroquois-class homeodomain protein 4 (IRX4) is a TALE homeobox transcription factor which has been implicated in prostate cancer (PCa) as a tumor suppressor through genome-wide association studies (GWAS) and functional follow-up studies. In the current study, we characterized 12 IRX4 transcripts in PCa cell lines, including seven novel transcripts by RT-PCR and sequencing. They demonstrate unique expression profiles between androgen-responsive and nonresponsive cell lines. These transcripts were significantly overexpressed in PCa cell lines and the cancer genome atlas program (TCGA) PCa clinical specimens, suggesting their probable involvement in PCa progression. Moreover, a PCa risk-associated SNP rs12653946 genotype GG was corelated with lower IRX4 transcript levels. Using mass spectrometry analysis, we identified two IRX4 protein isoforms (54.4 kDa, 57 kDa) comprising all the functional domains and two novel isoforms (40 kDa, 8.7 kDa) lacking functional domains. These IRX4 isoforms might induce distinct functional programming that could contribute to PCa hallmarks, thus providing novel insights into diagnostic, prognostic and therapeutic significance in PCa management.

Keywords: prostate cancer; IRX4; alternative splicing; transcript; isoforms

Citation: Fernando, A.; Liyanage, C.; Moradi, A.; Janaththani, P.; Batra, J. Identification and Characterization of Alternatively Spliced Transcript Isoforms of *IRX4* in Prostate Cancer. *Genes* **2021**, *12*, 615. https://doi.org/10.3390/genes12050615

Academic Editor: Anelia Horvath

Received: 6 March 2021
Accepted: 13 April 2021
Published: 21 April 2021

Publisher's Note: MDPI stays neutral with regard to jurisdictional claims in published maps and institutional affiliations.

1. Introduction

Alternative splicing (AS) in precursor mRNA plays a vital role in the regulation of gene expression by expanding the coding capacity of genomes. Diverse combinations of splice sites and alternative promoters in pre-mRNA are chosen to produce structurally distinct mRNA and, thus, protein isoforms that range from slightly different to having the opposite functions [1]. Different mechanistic modes of AS have been identified such as exon skipping, retention of introns, alternative 5′/3′ splice sites, alternative promoters, alternative polyadenylation sites and alterations in spliceosomes [1]. Apart from the contribution to the greater diversity of the proteome, AS leads to every hallmark of cancer progression [2]. Recent study findings predict that the vast heterogeneity of human cancers may be a result of the distinct roles of protease isoforms resulting from AS [3]. AS has been reported to modify the network of protein interactions in a disruptive, non-regulated manner, thus disrupting normal cell function via mediating cancer driven pathways [2]. Additionally, AS may induce degradation of the tumor suppressor transcripts by nonsense-mediated decay or induce mutations on the splice-sites, thus having intron-retention in tumor suppressors [2]. AS can considerably alter the coding region of the drug targets of proteins, which leads to drug and therapy resistance in many cancers [4]. Although the activities of transcription factors are extremely and coordinately regulated during embryonic growth and differentiation, AS can alter the transcriptional regulation and may switch cells from physiological to pathological transformation [5].

In 2020, prostate cancer (PCa) was the second most commonly diagnosed and the fifth leading cause of cancer-related death among men worldwide [6]. Alternative splicing plays a significant role in PCa, regulating malignant progression, aggressiveness, tumor cell lineage plasticity and therapy resistance [7,8]. Although androgen receptor (AR) splicing has been well studied in PCa, the knowledge on AS of other PCa oncogenes is scarce [9]. Continuous attempts to understand and translate the expertise in AS to PCa may accelerate the discovery of novel diagnostic and therapeutic targets to improve PCa patient care.

Iroquois-class homeodomain protein IRX4, also known as Iroquois homeobox protein 4, encoded by the gene *IRX4*, is a member of the homeobox gene family [10]. Homeobox family genes mainly act as transcription factors and are found in almost all multicellular organisms [11]. They play a crucial role in the regulation of many aspects of embryonic development including pattern formation [11]. The *homo sapiens IRX* genes consist of six members and are organized as two clusters containing three genes each: *IRX1, 2* and *4* cluster on chromosome 5 and *IRX3, 5* and *6* cluster on chromosome 16, which are separated by large intergenic regions [12]. *IRX4* is located at chromosome *5p15.3* locus and has been described as the most divergent member of the IRX family [10,13]. The *IRX* homeobox gene family is unique from other homeobox genes with an extra 3-amino acid loop extension (TALE) in their homeodomain (63 amino acids) and the 9-amino acid domain (the Iro box) [12]. The *IRX4* gene is expressed in different human organs including the developing central nervous system, breast, esophagus, skin, prostate and vagina, but is predominantly expressed in the cardiac ventricles [10]. IRX4 plays a crucial role in the regulation of the ventricular chamber-specific gene expression by triggering the ventricular myosin heavy chain-1 (VMHC1) gene and suppressing the atrial myosin heavy chain-1 (AMHC1) gene [14]. IRX4 supports the maintenance of cardiac contractile function and has a protective role in cardiomyopathy, cardiac hypertrophy and congenital heart disease [15,16]. In addition, *IRX4* expression was identified in the retina as a crucial regulator of *Slit1* expression [17]. Far beyond its physiological role, the differential expression and oncogenic role of IRX4 has been suggested in recent studies. High levels of IRX4 in breast cancer plasma samples have been reported, which suggest the potential of IRX4 as a biomarker for breast cancer [18]. IRX4 expression was found to be downregulated in the mesenchymal cell population compared to epithelial cells in breast cancer [19]. Overexpression of IRX4 drives cell proliferation in non-small cell lung cancer (NSCLC) and is directly associated with the overall survival of patients [20]. The *IRX4* promoter region was found to be frequently hypermethylated in pancreatic cancer, which provides an advantage for cell growth in pancreatic cancer [21]. *IRX4* has been identified recently as a potential candidate gene in PCa after genome-wide association studies (GWAS) discovered the *5p15* locus to be associated with PCa risk. Many studies carried out by Batra et al. [22], Wang et al. [23], Lindstrom et al. [24] and Qi et al. [25] have identified the association of SNP rs12653946 at *5p15*, a cis-eQTL of *IRX4,* with PCa risk in multiethnic populations. IRX4 has been described as a tumor suppressor in PCa with the interaction of vitamin D receptor [26]. The differential roles of IRXs in a tumor microenvironment have been reported [27–29], possibly suggesting their tissue-specific role and/or their splicing, which affect the protein level, thereby modifying the functional capacity of normal cells to prompt and withstand multiple mechanisms related to tumor progression.

IRX4 is a human multi-exon transcription factor, which has been reported as alternatively spliced and highly expressed in PCa [26]. To date, five transcripts of *IRX4* have been identified, including four predicted alternative promoters [26,30]. *IRX4* has mainly two types of transcripts that differ from the presence of an additional fourth exon, further divided into five transcripts differing only from the 5′UTR region, and the expression of these transcripts have been reported in PCa [26]. The total of five transcripts are predicted to translate into two IRX4 protein isoforms that only differ from the presence of an additional 27 amino acids that is encoded by an extra fourth exon (UniProt database),functional role of which is still unknown. In this article, we present the diversity of *IRX4* transcripts and IRX4 protein expression in PCa cell lines and clinical samples, as IRX4 protein isoforms may have differential roles in PCa progression. This study enables a broad understanding of the regulation and mechanisms in prostate tumorigenesis, which is very limited at present.

2. Materials and Methods

2.1. Cell Culture

A panel of PCa cell lines (LNCaP, VCaP, DuCaP, C42B, PC3, DU145, RWPE2, 22RV1) and benign prostate (BPH1, RWPE1) cell lines were purchased from the American Type Culture Collection (ATCC, Manassas, VA, USA). RWPE-1 and RWPE-2 cell lines were grown in keratinocyte serum-free medium supplemented with 5 ng/mL recombinant human epidermal growth factor (EGF) and 50 μg/mL bovine pituitary extract (Gibco™, Invitrogen, Carlsbad, CA, USA), whereas all other cell lines were grown in RPMI1640 (1X) with no phenol red (Life Technologies, Grand Island, NY, USA) supplemented with either 5% or 10% fetal bovine serum (FBS, Life Technologies, Thornton, NSW, Australia). The cell lines were authenticated by short tandem repeat (STR) profiling and tested negative for mycoplasma. The cells were maintained at 37 °C in a 5% CO_2 humidified incubator.

2.2. RNA Isolation and cDNA Synthesis

Total RNA was extracted from PCa cells using the Isolate II RNA Mini Kit (Bioline, London, UK) according to standard protocol. RNA concentration and purity were measured using NanoDropTM1000 (Thermo Scientific, BiolaB, Scoresby, VIC, Australia). A total of 1 μg of RNA was reverse transcribed to cDNA using SensiFast™ cDNA synthesis kit (Bioline, GmbH, Luckenwalde, Germany). The cDNA was diluted to 100 μL before using it as a template for PCR reaction.

2.3. Reverse Transcription-Polymerase Chain Reaction (RT-PCR)

The primers for RT-PCR and qRT-PCR were designed using NCBI tool Primer BLAST–NCBI–NIH software. Several primer sets were designed to recognize the boundary between 2 exons (exon-exon spanning region) to specifically identify the expression of the *IRX4* transcripts. All the primer sequences are given in Table S1. RT-PCR was performed with a reaction comprising 1X PCR buffer, 1.5 mM $MgCl_2$, 0.2 mM dNTPs and 0.2 μM of each of forward and reverse primers (Sigma Aldrich, Castle Hill, NSW, Australia), 1 U Platinum™ Taq DNA Polymerase (Invitrogen, Carlsbad, CA, USA) and 1 μL of th cDNA template. PCR reaction was performed on a Mastercycler® nexus machine (Eppendorf, North Ryde, NSW, Australia). The samples mixed with loading dye (NEB #B7024S) were loaded on to 0.7–2% agarose gels (Bioline, Alexandria, NSW, Australia) prepared in Tris-borate-EDTA (TBE) buffer (89 mM Tris base, 89 mM Borate, 2 mM EDTA) containing 0.5 μg/mL ethidium bromide (Invitrogen, Carlsbad, CA, USA). Approximately 0.5 μg of 1 kb ladder (New England Biolabs, Ipswich, MA, USA) was loaded to compare the size of the DNA products. Images were captured by the gel documentation system QUANTUM ST5 (Fisher Biotec, Wembley, WA, Australia).

2.4. Relative Quantification by Real-Time Quantitative RT-PCR (qRT-PCR)

Quantitative RT-PCR was performed using the ViiA7 Real-Time PCR system (Applied Biosystems, Foster City, CA, USA). Each reaction contained 1X final concentration of SYBR Green PCR Master Mix 2X (Applied Biosystems, Foster City, CA, USA), 50 nM forward and reverse primer, 2.0 μL of diluted cDNA (1:5) and nuclease-free water at a final volume of 8 μL. The cycling parameters were 95 °C for 10 min, 40 cycles of 95 °C for 15 s and 60 °C for 1 min followed by a dissociation step. All the CT values were normalized to the expression of housekeeping gene *RPL32* (ΔCT) [31]. Relative expression compared to control was performed by the comparative CT (ΔΔCT) method.

2.5. Androgen Deprivation Assay

LNCaP, VCaP and DuCaP cells were seeded in RPMI1640 media (Life Technologies, Grand Island, NY, USA) supplemented with 5% FBS and incubated at 37 °C for 3 days. The medium was then replaced with an androgen-depleted culture medium (RPMI1640) containing 5% charcoal-stripped serum (CSS, Sigma-Aldrich, Castle Hill, Australia). After 48 h, the cells in CSS were supplemented with 10 nM dihydrotestosterone (DHT), 10 nM DHT + 10 μM anti-androgens (bicalutamide or enzalutamide) and ethanol (EtOH) control and incubated at 37 °C for 48 h.

2.6. In Silico Analysis of IRX4 Transcripts and Isoforms

The UCSC genome browser (https://genome.ucsc.edu/accessed on 24 November 2019) [30], the National Center for Biotechnology Information (NCBI) (https://www.ncbi.nlm.nih.gov/accessed on 18 October 2019) [32], GTEx portal (https://gtexportal.org/home/accessed on 10 January 2020), The Human Protein Atlas database (https://www.proteinatlas.org/accessed on 25 January 2020) and UniProt database (https://www.uniprot.org/accessed on 12 July 2020) were used to obtain *in silico* data for the expression of *IRX4* transcripts and protein isoforms.

2.7. RNA-seq and Genotype Data for cis-eQTL Analysis

The RNA-seq data (bam format) of 483 PCa patients and 49 matched controls of The Cancer Genome Atlas Program (TCGA) were used in the study [33]. RASflow was used for RNA-seq analysis [34]. HISAT2, a fast and sensitive alignment program, was selected for alignment to the transcriptome [35]; feature-count was utilized in the quantification, and Deseq2 was used in the normalization of data. Cleaned genotypes data of PCa risk SNP rs10866528 were used for cis-eQTL analysis.

2.8. Reprocessing PRIDE PCa Cell Line LC-MS/MS Data

LC-MS/MS raw data files of the LNCaP cell nuclear extractions were retrieved from the Proteomics Identification Database (PRIDE) database, belonging to the project ID: PXD003262 [36]. All files were converted to MASCOT generic format (MGF) peak files using MSConvertGUI (Version 3) [37]. Next, all peak files were searched in SearchGUI (Version 3.3.17) and PeptideShaker (Version 1.16.43) against a FASTA database comprising novel IRX4 peptide sequences merged with UniProt human reference database and contaminant proteins [38,39]. X! Tandem search algorithms were implemented using the following search parameters: precursor mass error: 10 p.p.m., fragment mass error: 0.05 Da, fixed modification: carbamidomethylation of cysteine, variable modification: oxidation of methionine defined as variable modification. Minimal peptide length was set to six amino acids. The 1%-fold discover rate (FDR) was set to identify specific tryptic peptides representing each IRX4 protein isoform.

2.9. DNA Sequencing

The PCR products were purified by Wizard® SV Gel and PCR Clean-Up System (Promega, Madison, USA) and sequenced by AGRF (Gehrmann Laboratories, Research Rd, University of Queensland, Brisbane, Australia). A total of 11 μL of the purified PCR products were sent with 1 μL primers (10 μM) in standard 1.5 mL Eppendorf tubes, and the results were obtained via the AGRF online website. For the identification of amplified DNA fragments, DNA sequences were aligned against the NCBI database [32] and UCSC genome browser [40].

2.10. LC-MS/MS Analysis of PCa Cells

Cell pellets were obtained and lysed using sodium deoxycholate (SDC) buffer (1% SDC in 1M Tris pH 8.0). The samples were then sonicated in an ultrasonic bath (Thermo Scientific™, Waltham, MA, USA) for 15 min (at 4 °C, 100% Power) to denature proteins and shear DNA. The concentration of proteins was calculated using a bicinchoninic acid assay

(BCA) with Pierce™ Bovine Serum Albumin (BSA) Standards (Thermo Scientific™). A total of 10 μg of the protein extract was denatured at 95 ° C for 5 min using a thermomixer (Eppendorf ThermoMixer® F1.5, Hamburg, Germany). Denatured protein samples were reduced by 10 mM Tris (2-carboxyethyl) phosphine (TCEP) (Sigma-Aldrich, Castle Hill, NSW, Australia), alkylated by 40 mM 2-chloroacetamide (2CAA) (Sigma-Aldrich) and incubated for 30 min in the dark at room temperature. Samples were then digested overnight at 37 °C by adding trypsin (Sigma-Aldrich) at a 1:50 enzyme-protein ratio. Peptides were desalted using Pierce™ C18 Spin Tips (Thermofisher, Waltham, MA, USA), washed in 0.1% TFA, and peptides were dissolved in 80% acetonitrile (ACN) (HPLC grade, Sigma-Aldrich) elution buffer. Solvents were evaporated in a SpeedVac centrifuge (Savant Speed Vac, SPD121P-230, Thermo Electron Corporation, Milford, MA, USA) at 35 °C and re-suspended using iRT calibration mix including 2% ACN ans 0.1% TFA (Biognosys AG, Schlieren, Switzerland). Samples were prepared in 3 biological replicates and analyzed by a sequential window acquisition of all theoretical mass spectra (SWATH-MS) approach, following our previously published protocol [41].

Data-dependent acquisitions were imported into ProteinPilotTM software (Version 5.0.1, AB SCIEX) and searched using the Paragon™ algorithm against the FASTA database consisting of novel IRX4 peptide sequences merged with the UniProt human reference database and contaminant proteins. The following search parameters were used: Sample type: Identification; Cys Alkylation: Iodoacetamide; Digestion: Trypsin; Instrument: TripleTOF 5600+; Species: None; Search effort: Thorough ID; Results Quality: Detected protein threshold [Unused ProtScore (Conf)] ≥ 0.05 with FDR. Generated ion library was imported into PeakView® SWATH micro app (Version 2.1, AB SCIEX) and saved in text format and cleaned using the iSwathX tool (Version 2.0). The curated ion library was imported into Skyline software (Version 1.1) [42]. The following peptide and transition settings were followed: Enzyme: Trypsin [KR | P]; Max missed cleavages: 1; Min length: 6; Max length: 35, Precursor charges: 2+,3+,4+; Ion Charges: 1+,2+,3+; Ion types:y/b; Ion match tolerance: 0.5. Data-independent acquisitions were imported, and peptide quantification was performed using MSstats R-based statistical tool (Version 2.0) [43]. Peptides only specific to each IRX4 protein isoforms were used for the quantification, and normalized peptide intensities were used to calculate the relative fold expression.

2.11. Statistical Analysis

All statistical data were analyzed by GraphPad Prism 9.0.0 (121). The comparison was analyzed by paired t-test (two groups) and Kruskal–Wallis test with Dunn's multiple comparisons (more than two groups). The results were considered statistically significant if * $p < 0.05$ at a 95% confidence interval. All the experiments were performed in 3 biological replicates.

3. Results

3.1. In silico identification and characterisation of human IRX4 transcripts

The human *IRX4* gene at *5p15.33* locus comprises six coding exons, seven introns and 3′UTR and 5′UTR regions, which spans from 1,877,413 to 1,887,236 bp (hg38) [32,40]. Although not fully characterized, the IRX4 protein consists of a homeodomain or DNA binding domain, transactivation domain and Iro box (Figure 1a). Five *IRX4* transcripts have been reported for *IRX4* to date in NCBI [32]. The transcript 1(NM_001278632.1) and the 3(NM_001278634.2) and 5(NM_016358.3) transcripts differ from the 2(NM_001278633.1) and 4(NM_001278635.2) transcripts since the latter transcripts consist of an additional fourth exon (78 bp). *IRX4* transcripts 1, 3 and 5 are similar in their coding regions but differ from each other with their diverse 5′UTR region, and also the 2 and 4 transcripts are similar in their coding regions and contrast with each other with their different 5′UTR region (Figure 1b) [30]. According to the alternative splicing graph of *IRX4* by the Swiss Institute of Bioinformatics (Figure 1c), there is clear evidence that *IRX4* has been alternatively spliced and has a probability to encode different transcripts (Figure 1c) [30].

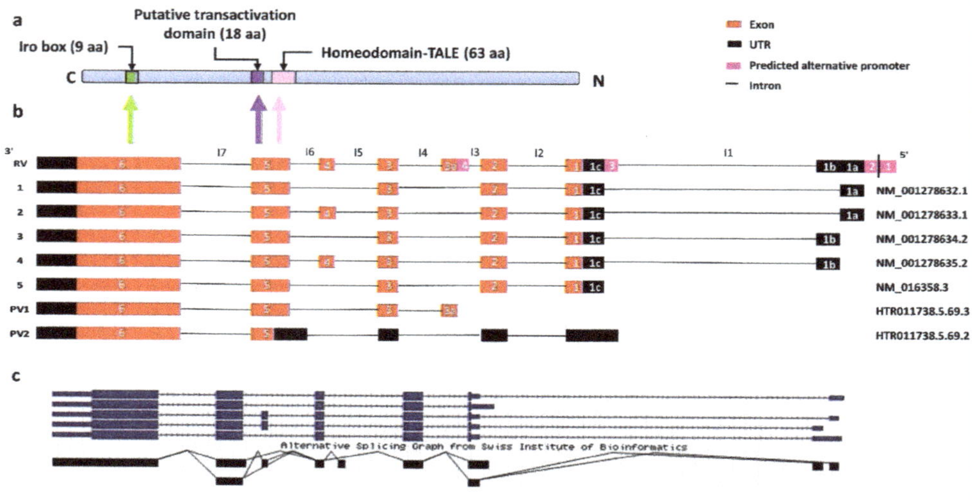

d

Cell	Cell type	Characteristics
BPH-1	Benign prostatic hyperplasia	Benign cell line
RWPE-1	Prostate epithelial	Benign cell line
RWPE-2	Tumorigenic	Derived from RWPE-1 using Ki-ras
DuCaP	Metastasis (Androgen responsive)	Originated from dura mater metastasis
VCaP	Metastasis (Androgen responsive)	Originated from lumbar vertebral metastasis
LNCaP	Metastasis (Androgen responsive)	Derived from left supraclavicular lymph node
C42B	Metastasis (Androgen nonresponsive)	Derived from LNCaP cells, Castration-resistant PCa
22RV1	Metastasis (Androgen nonresponsive)	Derived from human prostatic carcinoma xenograft, CWR22R
PC3	Metastasis (Androgen nonresponsive)	Derived from bone metastasis
DU145	Metastasis (Androgen nonresponsive)	Derived from brain metastasis
WPMY-1	Prostate fibroblast	Immortalized prostate stromal cell line

Figure 1. (**a**) Schematic representation of IRX4 domains. The predicted localization of the homeodomain, putative transactivation domain, irobox and the number of amino acids (aa) in different domains of the IRX4 proteins. The different domain encoding regions from the *IRX4* gene have been shown with the arrows, C-C-terminal, N-N-terminal regions. (**b**) The predicted *IRX4* transcripts according to gene databases. The five known *IRX4* transcripts: transcript 1(NM_001278632.1), transcript 2 (NM_001278633.1), transcript 3 (NM_001278634.2), transcript 4 (NM_001278635.2) and transcript 5 (NM_016358.3) and two predicted *IRX4* transcripts (PV1: HTR011738.5.69.3 and PV2: HTR011738.5.69.2) according to the Swiss Institute of Bioinformatics gene predictions and the novel putative exon (exon 3a) are presented with alignment with the Reference Variant (RV). Introns are labelled from I1 to I7. (**c**) Alternative splicing graph detailing alternative splicing (AS) events of the *IRX4* gene by the Swiss Institute of Bioinformatics, adapted from the UCSC genome browser. Lines on the plot show the exon–exon junctions. The overlapping lines denote the two types of exon–exon junction, suggestive of the presence of transcripts of *IRX4*. (**d**) The characteristics of PCa cell lines used in the study.

Four alternative promoters were predicted to be localized upstream of the 5′UTR of *IRX4* by the UCSC genome browser: the alternative promoter 1 (chr5:1887187–1887336), the alternative promoter 2 (chr5:1887169–1887318), the alternative promoter 3 (chr5:1882876–1883025) and the alternative promoter 4 (chr5:1881044–1881193), as shown in the Figure 1b [30]. The predicted 1 and 2 alternative promoters and the presence of an EST (AI246240) in the 5′UTR region facilitated the identification of novel 5′UTR exons (exon 1a and 1b)

(Figure 1b). Nguyen et al. suggests the presence of a diverse 5′UTR region of *IRX4*, later confirmed by Northern blot analysis and RACE [26]. Further, few additional transcripts of *IRX4* have been predicted by the Swiss Institute of Bioinformatics gene predictions based on the mRNA and EST expression [30]. A novel transcript (PV1) (HTR011738.5.69.3) with a putative novel exon (exon 3a, 89 bp) localized between the second and third exons has been predicted (Figure 1b). The presence of this putative exon is also reported for *IRX4* in an EST dataset (BY799479) and the alternative splicing prediction plot of *IRX4* by Swiss Institute of Bioinformatics (Figure 1c). The expression of this exon has been reported in the skin with no sun exposure and esophagus mucosa according to the GTEx portal data. Moreover, the presence of an alternative promoter region (alternative promoter 4) has been predicted in this region, which further confirms a new starting point for this transcript (Figure 1b). Another short transcript (PV2) (HTR011738.5.69.2) starting from a novel open reading frame (ORF) from exon 5 of the *IRX4* has also been reported (Figure 1b) [30]. The 5′UTR region of this transcript consists of four exons, slightly differing from the other transcript exons by retention of part of the intron 6 (I6) and longer first UTR region. The 3′UTR end of both predicted transcripts is similar to that of other transcripts (Figure 1b). The *IRX4* 1-4 transcripts may deploy the alternative promoter 1 and 2; the transcript 5 utilizes the alternative promoter 3, while PV1 makes use of alternative promoter 4. The existence of alternative promoters reflects the possibility of multiple transcripts for the *IRX4* gene that differ in their size and regulate separately at their transcription level.

Expression of most of the *IRX4* exons f was observed across several tissue types such as skin, prostate, salivary glands, esophagus, vagina and heart, but the last exon (exon 6) was abundantly expressed in most of the tissues studied (GTEx portal data). IRX4 proteins were mainly localized in the nucleus and vesicles. Enhanced expression of IRX4 has been detected in some human cell lines, such as brain cancer (BEWO), skin immortalized (HaCaT), lung immortalized (HBEC3-KT) and breast cancer (MCF7) cell lines (the Human Protein Atlas database).

Even though IRX4 has been identified for its transcriptional role in the human heart and tumor suppressor role in PCa, the IRX4 proteins are so far not fully structurally and functionally characterized for their interactions and the functional domains. Three domains have been predicted to date: the homeodomain (DNA binding domain) (63aa), the putative transactivation domain (18aa) and Iro box (9aa) for the IRX4 protein (Figure 1a). The homeodomain and the transactivation domain of IRX4 are encoded from part of the fifth exon of the transcript, and the iro box region is encoded from part of the sixth exon [14]. The IRX4 protein isoform 1 (Uniprot ID- P78413-1 length: 519 aa; mass: 54.4 kDa) is encoded from the transcripts 1, 3 and 5 and differs only at 26 amino acids with IRX4 protein isoform 2 (Uniprot ID- P78413-2 length: 545 aa; mass: 57.0 kDa), which is encoded from transcripts 2 and 4. The addition of 26 amino acids in isoform 2 may result in a structural change that may affect the function. A panel of PCa cell lines representing the benign and metastatic nature has been used in the study. All the features corresponding to each PCa cell line have been characterized in Figure 1d.

3.2. Identification of Human IRX4 Transcripts in PCa Cell Lines

Firstly, we tried to identify the expression of five known *IRX4* transcripts in a panel of PCa cell lines. Two main bands (369 bp and 291 bp) on the agarose gel were detected in RT-PCR with the primer set 1, designed to amplify the region from complementary to exon 2 to 5, confirming the presence of two transcripts, where one transcript contains a fourth exon and the other does not (Figure 2a,f). The sequencing analysis of the resulting bands confirmed the presence of at least two *IRX4* transcripts with and without exon 4 in PCa cell lines The upper band may correspond to either of the transcripts 2 and 4 (Figure 2a). The lower band may correspond to either of transcripts 1, 3 and 5.

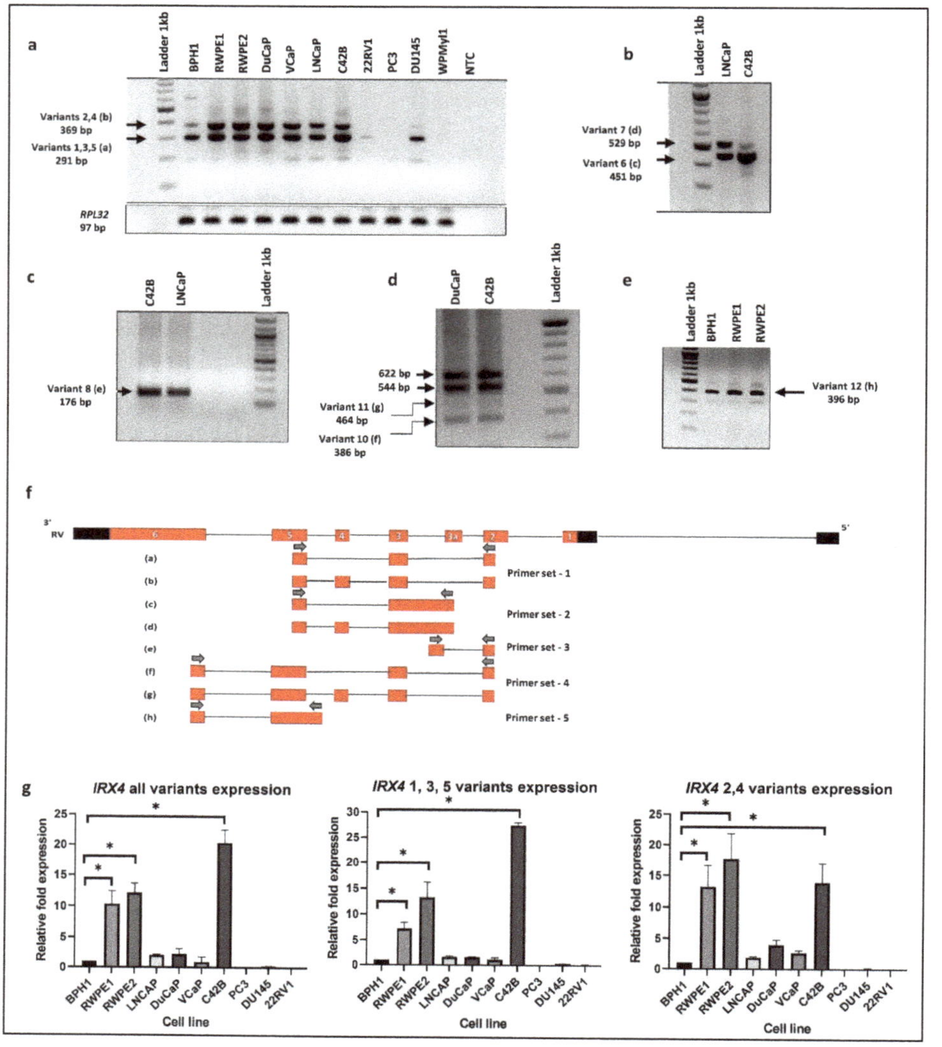

Figure 2. *IRX4* transcripts expression in a panel of PCa cell lines. (**a**) *IRX4* 1-5 transcripts expression (RT-PCR) from a panel of cell lines representing benign prostate (BPH1, RWPE1), PCa (RWPE2, DuCaP, VCaP, LNCaP, C42B, 22RV1, PC3, DU145) and immortalized prostate stromal cell line (WPMYl1). 1 Lane: ladder 1kb; 2–12 Lanes: cell lines; 13 Lanes: non-template control (NTC). *RPL32* was used as the endogenous housekeeping control. (**b**) *IRX4* predicted identification of transcript 6 and 7 in LNCaP and C42B cells (451 bp and 529 bp). (**c**) *IRX4* predicted identification of transcript 8 (176 bp) in C42B and LNCaP cells. (**d**) *IRX4* predicted identification of transcripts 10 and 11 in DuCaP and C42B cells (386 bp and 464 bp). (**e**) *IRX4* predicted identification of transcript 12 in BPH1, RWPE1 and RWPE2 cells (396 bp). (**f**) The sequencing BLAT results of gel bands (a) to (h) from Figure a to Figure e from the UCSC genome browser and the localization of primers re shown with arrows. The sequencing BLAT results for each transcript are presented with the alignment with the Reference Variant (RV). (**g**). The quantitative expression of *IRX4* 1-5 transcripts across a panel of prostate cell lines (BPH1, RWPE1, RWPE2, DuCaP, VCaP, LNCaP, C42B, 22RV1, PC3, DU145). *RPL32* was used as the endogenous housekeeping control. The relative fold expression was determined using the ΔΔCT method with respect to BPH1 (N = 3 biological and technical replicates, Mean ± SD, Kruskal–Wallis test with Dunn's multiple comparisons, * $p < 0.05$).

Next, we investigated the expression of the predicted novel transcript 1 (PV1) (HTR011738.5.69.3, transcript 6) in C42B and LNCaP cells. The presence of this transcript was identified by RT-PCR after designing a specific primer set that targets the novel exon (exon 3a) and fifth exon (primer set 2). Unexpectedly, two bands higher (about 451 bp and 529 bp) than the expected size (281 bp) were detected (Figure 2b). The Sanger sequencing analysis of the purified PCR product identified the intron retention (I4) between the novel exon (exon3a) and the third exon. The Sanger sequencing analysis of the purified product of the upper band confirmed the existence of the additional fourth exon suggestive of an AS event of transcript 6 (Figure 2b,f). For subsequent identification, the transcript that corresponds to the upper band was termed transcript 7 (Figure 3).

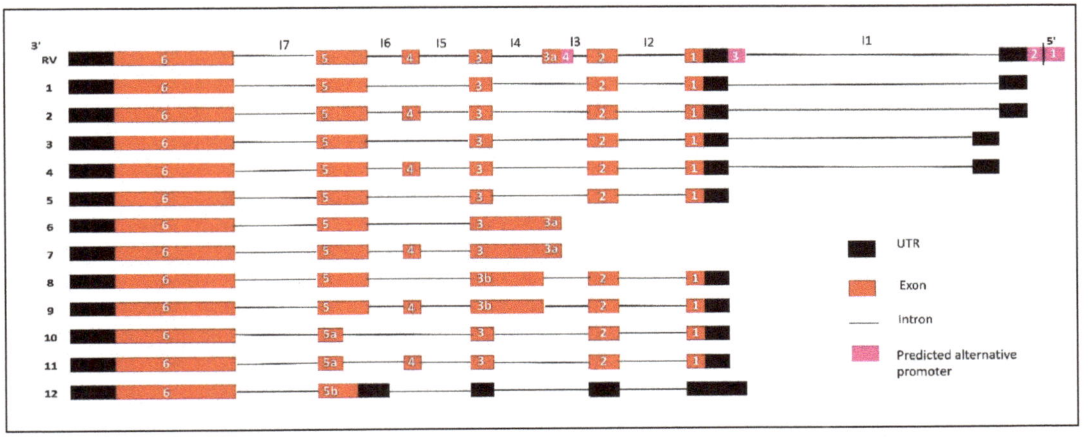

Figure 3. Summary of all identified *IRX4* predicted transcripts in PCa cells. All the exons and predicted UTRs are shown with red and black boxes, respectively, while introns are shown with black lines. The pink boxes indicate the location of predicted alternative promoters. Introns are labelled from I1 to I7. Each *IRX4* transcript is presented with the alignment with the Reference Variant (RV).

Next, we were interested to know whether the novel exon (exon 3a) is the first exon of transcripts 6 and 7. We designed a specific primer set complementary to the second exon and the novel intron region (I4) (primer set 3) and checked the expression in C42B and LNCaP cells. Interestingly, we found the presence of a band around 176 bp (Figure 2c) in PCR, but below the expected size (265 bp). Sequencing analysis confirmed a new *IRX4* transcript (which designated as variant 8) corresponding to this band (Figure 2c, f). This transcript lacks the novel 3a exon but has the intron 4 (I4) region; the intron 4 and the third exon appear as a single exon, which we called exon 3b. We saw that this transcript has an AS event with the fourth exon and appears as a new transcript with the additional fourth exon, which we termed as transcript 9 (Figure 3).

Interestingly, two new transcripts were identified with primers in the second and sixth exon in DuCaP and C42B cells. Two unexpected lower bands (386 bp and 464 bp) were detected with the primer set 4, in addition to the two expected bands (544 bp and 622 bp, Figure 2d). The sequencing results confirmed that the lower bands lack part of the exon 5 (exon 5a) of *IRX4* (Figure 2d,f). The two lower bands differ from each other with the additional fourth exon. The transcript that lacks the part of exon 5 and the fourth exon is termed transcript 10, and the transcript that lacks part of the exon 5 but retains the fourth exon is termed transcript 11 (Figure 3).

We were then able to confirm the presence of the second predicted transcript (PV2) by the Swiss Institute of Bioinformatics. The transcript has an extended fifth exon (exon 5b) in its 5′UTR region; the forward primer was designed specifically for this region to exclude the expression of other transcripts (primer set 5). The expression was detected in this transcript at 396 bp as expected in few PCa cell lines (Figure 2e). This short transcript, which we denoted as transcript 12, was predicted to have a coding region starting from the fifth exon, and all other starting exons have been predicted as 5′UTR regions (Figure 3).

The relative *IRX4* transcripts 1–5 expression levels were measured by qRT-PCR with three different primer sets, corresponding to all transcripts (primer set-12), transcripts 1, 3 and 5 (primer set-13) and transcripts 2 and 4 (primer set-14) (Figure 2g). The castration-resistant PCa cell line, C42B, had the highest expression of *IRX4* (more than 20-fold) according to the primers capturing all *IRX4* transcripts (Figure 2g) compared to the expression in BPH1. The RWPE-1 and RWPE-2 prostate epithelial cell lines expressed a significantly higher expression of all *IRX4* transcripts compared to BPH1. However, a poor correlation between the intensity of gel bands (Figure 2a) and qPCR graphs (Figure 2g) was detected due to different primer efficiencies used in PCRs. Compared to androgen-responsive cell lines LNCaP, DuCaP and VCaP, androgen-nonresponsive PCa cell lines PC3, 22RV1 and DU145 had a minimal expression of *IRX4*. Interestingly, overall *IRX4* expression was significantly higher in the castration-resistant C42B cell line (a derivative of LNCaP) compared to LNCaP, indicating a role of IRX4 in castration-resistant PCa progression.

The lower band that corresponds to transcripts 1, 3 and 5 expression was also prominently detected in C42B cells (more than 25-fold) compared to BPH1, but there was a comparatively low expression in 22RV1 and PC3 cell lines (Figure 2g) consistent with the RT-PCR results. *IRX4* transcripts 2 and 4 were prominently detected in RWPE1 and 2, but were weakly detected in 22RV1, DU145 and PC3 cells, similar to the RT-PCR upper band of the gel (Figure 2g). Overall, the results of the expression of *IRX4* transcripts showed a higher disparity between PCa cell lines in relation to their androgen responsiveness, which may be suggestive of differential regulation of *IRX4* transcripts by androgen in PCa cells. Our results indicate that human *IRX4* is extremely alternatively spliced, and many *IRX4* transcripts expressions were able to be identified in most of the PCa cell lines. The expression of *IRX4* transcripts still have not been completely annotated in gene databases. All the transcripts identified for *IRX4* in PCa cells are summarized in Figure 3. The sequence alignment for transcripts 6, 8, 10 and 12 is shown in the Supplementary Data (Figure S1) concerning their altered regions.

3.3. Characterisation of Human Novel IRX4 Transcripts in a Panel of PCa Cell Lines

Differential expression of *IRX4* transcripts 1–5 expression was detected in a panel of PCa cell lines (Figure 2a). The RT-PCR primers (grey) were designed specifically to target one transcript at one time excluding all the other transcripts (Figure 4a). However, the expression of transcripts 6 and 7 (primer set 15) and transcripts 8 and 9 (primer set 16) was not distinguishable designing specific qRT-PCR primers (green), but transcripts 10, 11 and 12 expression was individually measured with specific primer sets (primer set 17, 18 and 19) to obtain a relative quantification (Figure 4b). All the bands obtained were confirmed for their sequence with Sanger sequencing.

The highest expression of transcripts 6 and 7 was detected in DuCaP cells (~15 fold) compared to BPH1, consistent with the RT-PCR results (Figure 4b). However, C42B, RWPE1 and RWPE2 cells showed a significantly higher expression of transcripts 6 and 7 compared to BPH1 (Figure 4). The highest expression of transcripts 8, 9, 10, 11 and 12 was detected in C42B cells, and it reached statistical significance compared to BPH1, in line with the intensity of gel bands suggestive of a role of IRX4 in metastatic progression of PCa. (Figure 4b). The identified novel transcripts showed a minimal or no expression in 22RV1, PC3 and DU145 cell lines (Figure 4b), except for transcript 12 which has a comparable higher expression in DU145 cells. The PC3 cell line has a minimal expression of all identified novel transcripts. The results showed a high disparity of expression of *IRX4* novel transcripts between PCa cell lines. The results suggest that the identified novel transcripts significantly contribute to the overall overexpression of *IRX4* in PCa and may have a role in PCa progression.

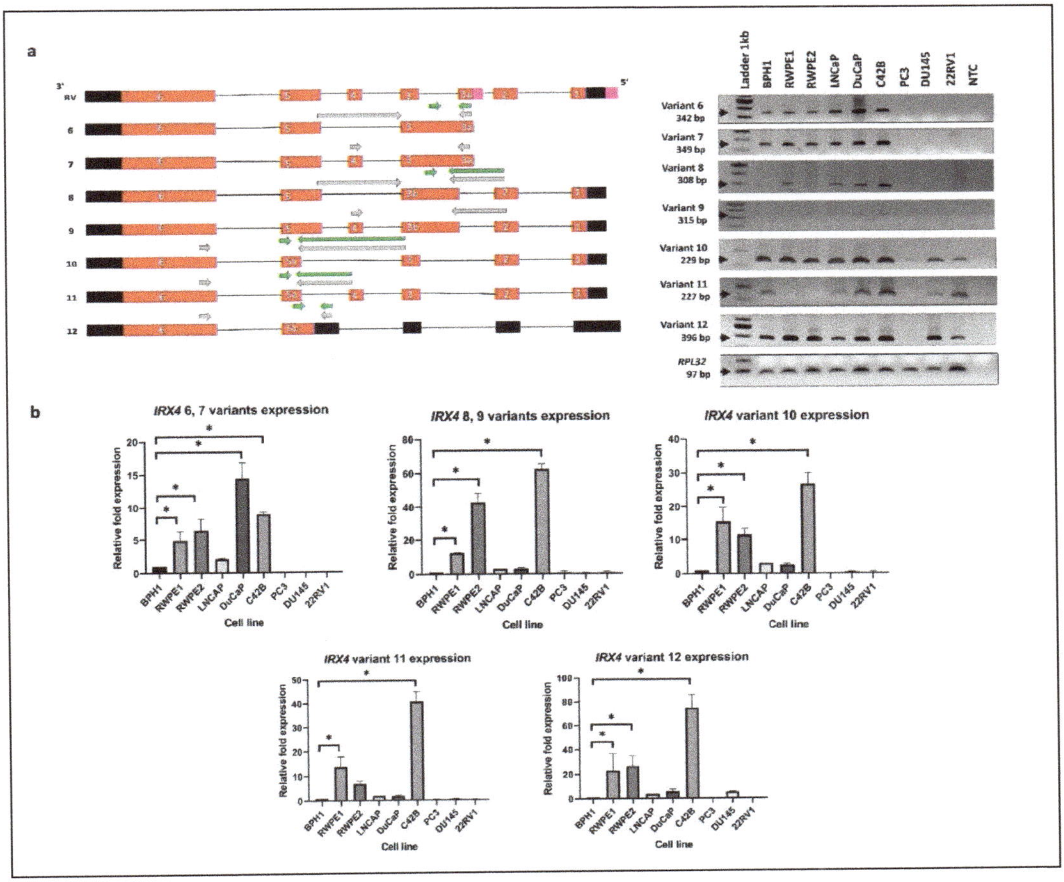

Figure 4. *IRX4* novel transcripts 6–12. (**a**) Qualitative and (**b**) quantitative expression across a panel of PCa cell lines. Expression analysis of each novel *IRX4* transcript with RT-PCR using transcript-specific primer sets (shown in grey arrows) and qRT-PCR using specific primer sets (shown in green arrows) in a panel of cell lines (BPH1, RWPE1, RWPE2, DuCaP, LNCaP, C42B, 22RV1, PC3, DU145). 1 Lane: ladder 1kb; 2–10 Lanes: cell lines; 11 Lanes: non-template control (NTC), *RPL32* was used as the endogenous control. The relative fold expression was determined using the ΔΔCT method with respect to BPH1 (N = 3 biological and technical replicates, Mean ± SD, Kruskal–Wallis test with Dunn's multiple comparisons, * $p < 0.05$).

3.4. Androgen Regulation of IRX4 Transcripts in PCa Cell Lines

Relative expression of *IRX4* transcripts in PCa cell lines with comparatively higher expression in androgen-responsive cell lines and low expression in androgen-nonresponsive cell lines may indicate that the expression of *IRX4* transcripts could be regulated by androgens and may have a role in therapy resistance. Thus, the expression of each *IRX4* transcript was determined in androgen-responsive cell lines (LNCaP, VCaP and DuCaP) with androgen (DHT) and antiandrogen (Bicalutamide and Enzalutamide) treatment. *KLK3* (PSA) expression was used as a positive control to validate both the treatments. *KLK3* was overexpressed with DHT treatment in all three cell lines, VCaP ~500-fold, DuCaP >10-fold and LNCaP ~40-fold and downregulated with antiandrogen treatment (Figure 5) compared to the ethanol (EtOH) control. Almost all *IRX4* transcripts expressions were upregulated with DHT treatment in VCaP and DuCaP cells compared to the EtOH control, and the upregulation of transcript 10 and 11 was prominent among other transcripts. Transcript 10 is overexpressed more than 300-fold and transcript 11 by more than 200-fold compared to EtOH with DHT in VCaP cells, and by 15-fold and 12-fold compared to EtOH in DuCaP cells, respectively, consistent with the treatment effect. Interestingly this expression pattern is not observed in LNCaP cells even with 40-fold *KLK3* overexpression (comparatively higher as compared to DuCaP) (Figure 5). The bicalutamide and enzalutamide treatment effectively reduced the androgen mediated expression of all identified *IRX4* transcripts in VCaP and DuCaP cells except for transcripts 6 amd 7 in DuCaP compared to DHT. This suggests that the antiandrogen treatment is not effective to reduce the expression transcripts 6 and 7, which show the highest expression in DuCaP cells among other cell lines. Although the expression of *KLK3* was reduced in LNCaP cells with antiandrogen treatment, a significant upregulation or downregulation of *IRX4* transcripts was not observed in LNCaP cells (Figure 5).

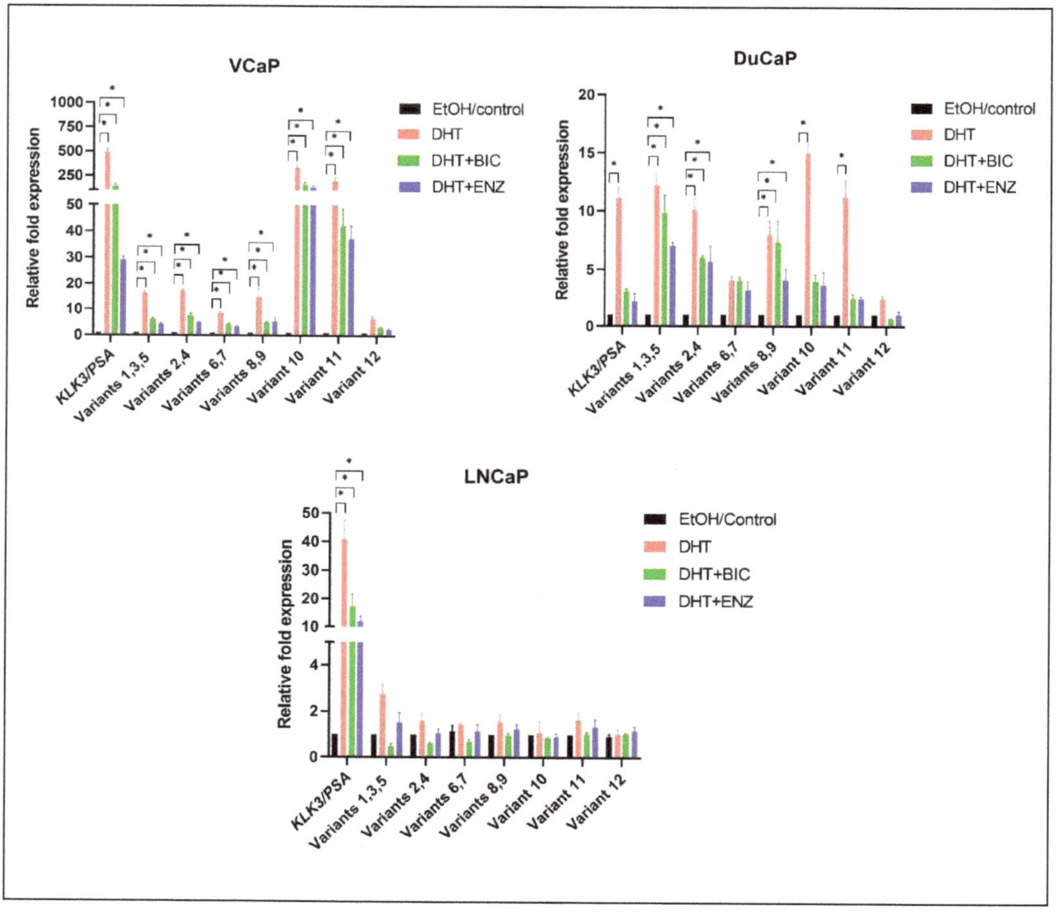

Figure 5. Regulation of *IRX4* transcripts expression by androgen and anti-androgen treatment in PCa cells. VCaP, DuCaP and LNCaP PCa cell lines were treated with 10 nM of androgens (DHT) or DHT + 10 μM bicalutamide (BIC) or DHT + 10 μM enzalutamide (ENZ) or EtOH/control. Relative fold expression of *IRX4* transcripts compared to EtOH/control expression was measured using the ΔΔCT method using *RPL32* as the endogenous control. (N = 3 biological and technical replicates, Kruskal–Wallis test with Dunn's multiple comparisons, Mean ± SD, * $p < 0.05$).

3.5. Identification and Characterisation of IRX4 Protein Isoforms by Mass Spectrometry

All the presented known and novel *IRX4* transcripts were characterized by their ORF for having the potential to encode novel protein isoforms of IRX4 using the ExPASy translate tool. The two protein isoforms, IRX4 isoform 1 and 2 encoded from transcripts 1 to 5, have been reported in the UniProt database (P78413-1 and P78413-2), and the expression has been validated in the human heart. The ORF coding site ranges from the first exon to exon 6 in *IRX4* transcripts 1 to 5 and encodes full-length proteins (54.4 kDa and 57 kDa). For most of the newly identified transcripts, the coding frame shifts from the main transcripts, and novel stop codons are incorporated (Table 1). Therefore, new transcripts produce truncated proteins. There is more than one coding frame for a few transcripts such as transcripts 6, 7, 8 and 9. The proteins generated from those novel transcripts differ from the main two protein isoforms and may lack the essential domains for their functional activity. Splice deletion of exon 1 and 2 and insertion of intron 4 (I4) produces transcripts 6 and 7 and thus has the capability to produce truncated proteins with two probable coding frames (ORFs), exon 3a–exon 5 and exon 5–exon 6 (Table 1). Moreover, transcripts 8 and 9 have intron retention between exons 3 and 3a, excluding exon 3a, and change the whole coding frame to encode a short protein isoform (11 kDa) due to the insertion of a new stop codon in their intron region (I4). The predicted second frame for these two transcripts is similar to the frame of exon 5–exon 6. Among the 12 *IRX4* SVs, two transcripts had a deletion of exon 5 (transcript 10 and 11), thus inserting a new stop codon at exon 6 which is predicted to encode two protein isoforms (21 kDa and 23.6 kDa), respectively. Transcript 12 presents the coding frame from exon 5 to exon 6 that predicted the encoding of a novel protein isoform (40 kDa). The transcripts 6, 7, 8, 9 and 12 also have the potential to encode this protein isoform. Then, we tried to identify the exact coding frame for each *IRX4* transcript by looking at their protein expression level. The details of the ExPASy tool analysis for probable protein coding frames of the identified *IRX4* transcripts and the probability to obtain their predicted domains are mentioned in Table 1.

Table 1. ExPASy tool analysis for predicted proteins from identified transcripts.

Transcript	Coding Frames	No. of Amino Acids	Predicted Protein Size	The Presence of Predicted Domains		
				Homeodomain	Transactivation Domain	Irobox
1,3,5	Exon 1–Exon 6	519	54.4 kDa (Isoform 1)	✓	✓	✓
2,4	Exon 1–Exon 6	545	57 kDa (Isoform 2)	✓	✓	✓
6,7	Exon 3a–Exon 5	114	8.7 kDa (Isoform 4)	-	-	-
	Exon 5–Exon 6	380	40 kDa (Isoform 3)	✓	✓	✓
8,9	Exon 1–Exon 3	105	11 kDa	-	-	-
	Exon 5–Exon 6	380	40 kDa (Isoform 3)	✓	✓	✓
10	Exon 1–Exon 6	204	21 kDa	✓	✓	-
11	Exon 1–Exon 6	230	23.6 kDa	✓	✓	-
12	Exon 5–Exon 6	380	40 kDa (Isoform 3)	✓	✓	✓

✓ = Presence of predicted domain.

Due to the lack of availability of IRX4 protein isoform-specific antibodies, we analyzed the predicted protein coding frames of all identified *IRX4* transcripts with the available proteomic identificationdatabase of PCa cell lines (PRIDE) with peptide shaker and seachGUI software. We could detect the MS/MS fragmentation ion spectrums of peptides that correspond to the coding frames of four IRX4 protein isoforms in LNCaP cells (Figure S2). They are the following: the first isoform (54.4 kDa) encoded by transcripts 1, 3 and 5, the second isoform (57 kDa) encoded by transcripts 2 and 4, the third isoform (40 kDa) encoded by transcripts 12, 8 or 9 and the fourth isoform (8.7 kDa) encoded transcripts 6 and 7. A peptide sequence that is specific to isoform 2 and a specific peptide sequence that is specific to isoform 4 were detected in the mass spectrometry data individually (Figure S2).

We also detected the common peptides related to isoforms 1 and 3 but cannot exactly differentiate these isoforms since they share a similar sequence identity with isoform 1. Unfortunately, we did not see the expression of peptides generated from the coding frames of transcript 10 (21 kDa) and 11 (23.6 kDa) and the short coding frame of transcripts 8 and 9 (11 kDa). Thus, the fate of these short proteins is still unknown, which may be suggestive of nonsense-mediated decay of these transcripts or might be temporal. The identified peptides in the IRX4 protein isoforms and the MS/MS fragmentation ion spectrum files of the peptides are presented in Figure S2.

Following the identification of the IRX4 protein isoforms in the published PRIDE data, we intended to quantify the isoform-specific peptide expression in PCa cell lines and BPH1 using the SWATH-MS/MS approach. Specific peptides used to determine isoform-specific expression in PCa cell lines were mentioned in Figure 6. Expression of protein isoform 3 was excluded from the analysis since it does not contain a specific peptide to differentiate from other isoforms. According to the analysis, the expression of IRX4 protein isoform 1 showed significantly higher expression in the VCaP cell line (~10 fold) and LNCaP cell line (~6 fold) compared to the expression in BPH1 (Figure 6a). Although not statistically significant, we observed a moderately higher expression of protein isoform 2 in LNCaP and DuCaP cell lines (Figure 6b). The expression of IRX4 protein isoform 4 showed significantly higher expression in the LNCaP cell line (~2 fold) compared to the expression in BPH1 (Figure 6c). In summary, IRX4 protein isoforms showed a marked expression in androgen-responsive cell lines (LNCaP, DuCaP and VCaP) compared to androgen-nonresponsive cell lines (22RV1 and DU145). However, all three IRX4 protein isoforms were found highly expressed in the castration resistant C42B cell line compared to other androgen-nonresponsive cell lines used for the analysis.

3.6. Validation of Expression of IRX4 Transcripts in PCa Patients

Since the expression of individual *IRX4* transcripts has not been determined in patients with PCa earlier, the expression of the identified *IRX4* transcripts was analysed in 49 PCa patient sample tumors ($N = 49$) and their matched non-malignant tissues ($N = 49$) from the TCGA database. We observed significant overexpression of *IRX4* transcripts 3, 5 (protein isoform 1) and 6 (protein isoform 4) in PCa tissues compared with their non-malignant tissues (Figure 7). However, the expression of *IRX4* transcripts 2 and 4 (protein isoform 2) in PCa tumor samples did not reach statistical significance compared to their normal tissues. The number of samples that express *IRX4* transcript 1 was quite low in the TCGA database and was thus excluded from the study. Since the identified novel transcripts are still not annotated with precise IDs, we could not extract the data related to the expression levels from the RNAseq data of the TCGA patient samples.

3.7. Association between PCa Risk SNP rs12653946 and Expression Levels of IRX4 Transcripts

The numerous prostate cancer-associated SNPs have been identified by GWAS, and the regulation effect of these SNPs on the expression of the nearby genes is investigated by expression quantitative trait loci (eQTLs) methods [44]. Those SNPs that can influence the expression of nearby genes are called cis-eQTL. Xu *et al.* identified fifty-nine SNPs from 39 distinct PCa risk loci, among them rs12653946 SNP was found to be a cis-eQTL which has the strongest association with *IRX4* [44]. The risk-associated genotype (GG) was associated with lower IRX4 levels in PCa [44]. In view of this, we were interested to see whether the expression pattern of the *IRX4* transcript varies with PCa patients' genotypes. rs10866528 is a tag SNP for PCa-risk SNP rs12653946, which is in linkage disequilibrium (LD), and it was later found that this SNP itself can act as a PCa-risk SNP [45]. We extracted the *IRX4* transcripts expression data of483 patients from the TCGA database and correlated them with the patients' SNP rs10866528 genotype. According to the analyzed expression of *IRX4* transcripts 2, 4, 5, and 6. we found that the GG PCa risk-associated genotype is associated with lower levels of all the analyzed *IRX4* transcripts (Figure 8). Although we did not see a significant lower expression of *IRX4* transcript 3 with the genotype GG, the sample number that showed the expression of transcript 3 was quite low.

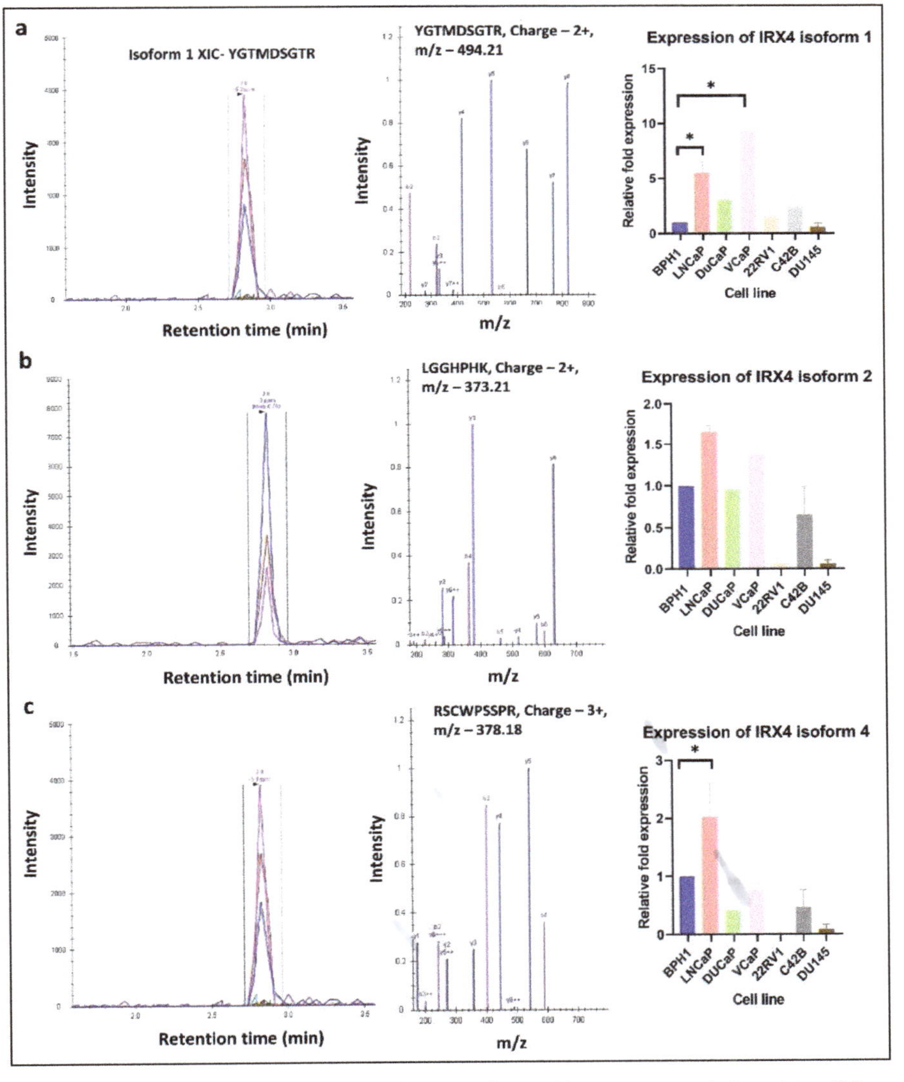

Figure 6. Identification and quantification of IRX4 isoform-specific peptides across a panel of prostate cell lines. Figures demonstrate the extracted ion chromatogram (XIC), MS/MS spectrum, charge and mass/charge ratio (m/z) of (**a**) IRX4 protein isoform 1, (**b**) IRX4 protein isoform 2 and (**c**) IRX4 protein isoform 4-specific peptides. The expression of each isoform-specific peptide was measured in BPH1, DuCaP, VCaP, LNCaP, C42B, 22RV1 and DU145 cell lines. The relative fold expression was measured compared to the BPH1 cell line. (N = 3 biological replicates, Mean ± SD, Kruskal–Wallis test with Dunn's multiple comparisons, * $p < 0.05$).

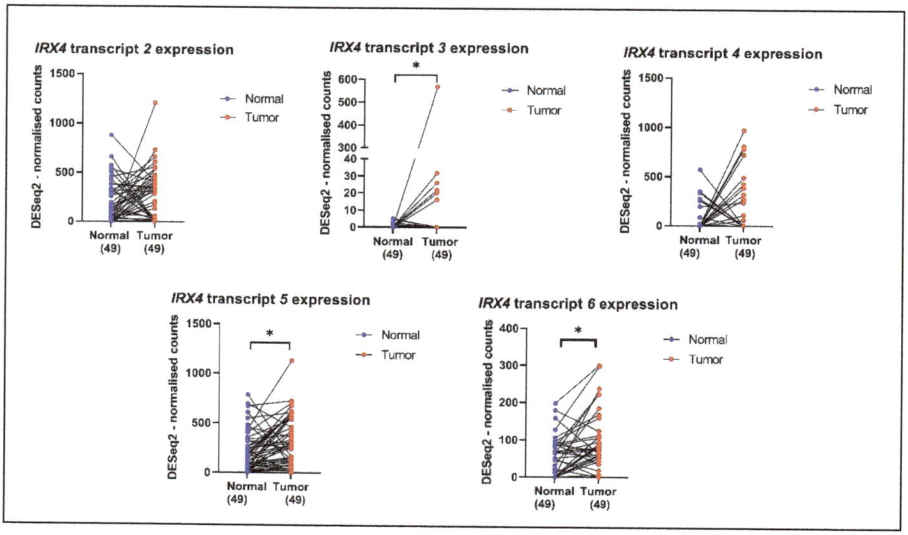

Figure 7. The expression of *IRX4* transcripts 2, 3, 4, 5 and 6 in PCa patient tumors and their matched normal tumor samples according to the TCGA database. A single patient is shown by a blue dot (the normal tissue) and a red dot (prostate tumor tissue), and the expression is matched with the line joining the dots. The number of patients in the normal and tumor categories is mentioned under each graph. (paired *t*-test, * $p < 0.05$).

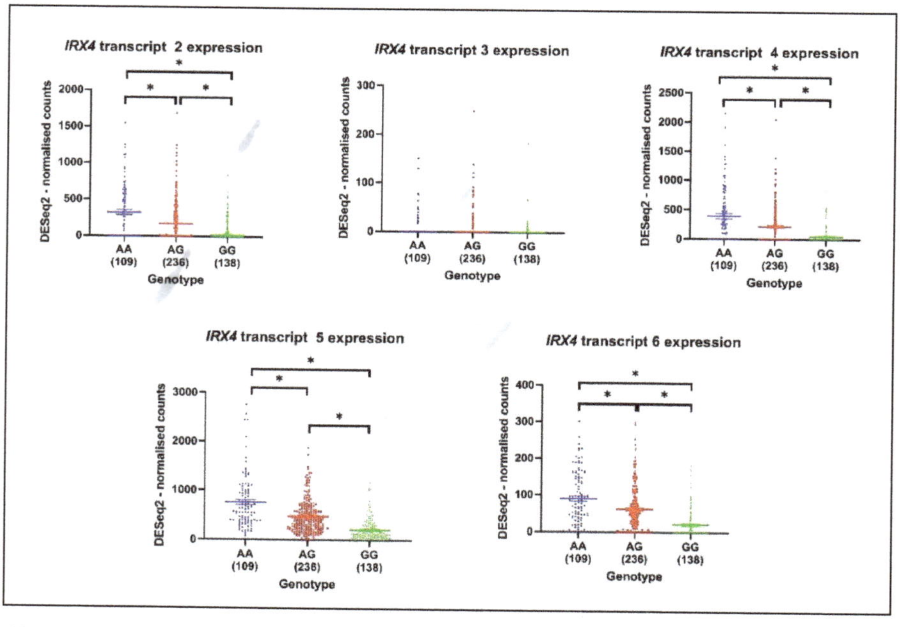

Figure 8. The expression of *IRX4* transcripts (mapped on DESeq2-normalized counts) by genotypes (AA, AG, GG) related to risk SNP rs10866528 in PCa. The number of patients in each genotype is mentioned under each genotype of the graph. (Mean ± SEM, Kruskal–Wallis test with Dunn's multiple comparisons, * $p < 0.05$).

4. Discussion

Novel research insights have proved that protein isoforms are sophisticated expression-based biomarkers in cancer prediction and progression [46–51]. Systematic sequencing of the human genome and transcriptome has revealed that more than 90% of genes express multiple mRNAs via AS events, suggesting a major impact on the functional diversity of proteins [52]. However, cancer cells frequently exhibit abnormalities in RNA splicing to survive, grow and progress to therapeutic resistance [53]. Many studies have highlighted that alternative RNA splicing is a common intrinsic mechanism leading to therapy and drug resistance in PCa [8,9,54]. For example, constitutively active ARtranscript 7 (*AR-V7*) in prostate tumor cells confers a primary or an acquired resistance to androgen deprivation therapy [55]. In addition to *AR*, several other genes undergoing AS, such as *FGFR, VEGF, Bcl-x, SH3GLB1* and *CCDN1*, were found to be associated with PCa development and progression [8]. Homeodomain transcription factors are shown to frequently be alternatively spliced [56]. For example, the homeodomain gene *HNF1B*, which encodes for three protein isoforms A, B and C, has been shown to have different functions with respect to the transcripts, as HNF1B A and B protein isoforms act as transcriptional activators, while HNF1B C protein isoform that lacks the transactivation domain functions as a transcription repressor [56–58].

Nguyen et al. have identified four novel transcripts of *IRX4* in PCa cells [26]. Although for these four *IRX4* transcripts, the sequences of exons 1 to 6 were highly conserved, the sequences of their upstream exons, which encode the 5′UTRs, are diverse in sequence and length [26]. We now reveal considerably greater diversity in human *IRX4* transcripts than previously realized. In the present study, we identified by RT-PCR analysis 12 *IRX4* transcripts, including seven novel transcripts in PCa cells. This diversity of the *IRX4* gene mainly arises from alternative promoter usage, intron retention, exon skipping and alternative 3′ and 5′ splice site usage. The 5′UTRs of mRNAs play important roles in the posttranscriptional regulation of gene expression [59]. Nguyen et al. have identified additional exons of the *IRX4* 5′UTR region by Northern blot analysis of *IRX4* transcripts [26]. The identified novel *IRX4* transcripts are yet to be characterized for their UTR regions. This complex isoform diversity of *IRX4* gene may not be constrained to PCa but can deviate to other cancers with significant expression of IRX4; thus, it is worth exploring the pathological and/or physiological impact of them.

Androgens and AR play a critical role in PCa pathogenesis. Androgen deprivation therapy (ADT) has been the mainstay of management for advanced PCa. Despite the initial strong responses to androgen deprivation therapy, the majority of patients with advanced PCa relapse with fatal castration-resistant PCa (CRPC) [60]. According to our data, *IRX4* and its isoforms are differentially expressed depending on the cell lines' androgen responsiveness. Castration-resistant cell line C42B showed the highest expression of most of the *IRX4* transcripts at the mRNA level. However, RNA expression does not correlate with proteomic data and complicates the precise understanding of the castration resistance and its relationship with IRX4. *IRX4* transcriptomic datasets in PCa showed a poor correlation with the proteomic data in line with recent research insights in PCa [41,61,62] due to varied discordant regulation between the transcriptional and translational regulation, post-transcriptional modifications associated with translation regulation, lack of temporal synchronization between transcription and translation and kinetic changes between protein generation and turnover in complex biological samples [63]. However, in consonance with *IRX4* RNA expression, the expression level of IRX4 protein isoforms was more prominent in the androgen-responsive cell lines than androgen-nonresponsive cell lines. Overall, the results suggest that IRX4 protein isoforms are androgen regulated. The simultaneous overexpression of *IRX4* transcripts together with *KLK3* with androgen may indicate a role of IRX4 in the early stages of PCa progression, and the antiandrogen treatment with bicalutamide and enzalutamide is not completely effective in eliminating the overexpression of *IRX4* transcripts in PCa cells. The effect of androgens and antiandrogens in the regulation of *IRX4* transcripts compared to VCaP and DuCaP cells is lower in LNCaP cells, suggestive of differential androgen regulation of *IRX4* in LNCaP cells, which needs to be further elucidated. Thus, IRX4 will be a better therapeutic target in combination with anti-androgen therapy for the treatment of PCa.

The lack of response to antiandrogen treatment of *IRX4* transcripts 6 and 7 was seen in DuCaP cells. IRX4 protein isoform 4 is encoded by transcripts 6 and 7, and these two transcripts use an alternative promoter distinct from other transcripts that are located immediately upstream of exon 3a. This may critically affect and alter the androgen regulation of IRX4 protein isoform 4 in PCa cells and can lead to therapeutic resistance to antiandrogen therapy in PCa patients. Although the function of IRX4 is tightly regulated at both transcriptional and post-transcriptional levels in human heart ventricles [64], the transcriptional regulation of IRX4 is still not completely clear in PCa. Nguyen et al. have explored the tumor suppressor role of IRX4 through the interaction with vitamin D receptors in PCa, but the isoform-specific roles are unknown [26]. Since IRX4 protein isoform 4 lacks both an N-terminus and C-terminus and predicted functional domains compared to full-length IRX4 proteins, we suspect the possibility of this isoform to act as a transcription factor. Thus, this isoform 4 may act distinctly compared to full-length isoforms by various mechanisms. Recent studies have elucidated that cancer-associated AS of transcription factors generates isoforms with altered activity, opposite transcription or antagonistic functions that severely impact tumor initiation and progression [3]. As Belluti et al. summarized, the lack of binding domains in transcription factors can show opposite functions in cancer progression [65]. For instance, the AP-2B isoform produced by AS of *AP-2* lacks a DNA binding domain and shows an inhibitor effect of the transactivation of *AP2* and leads to increased tumorigenicity, anchorage-independent growth, invasiveness and angiogenesis in melanoma [66,67]. In addition, functional domains lacking transcription factor isoforms can be mislocalized within the cancer cell and exert a dominant-negative activity as a result of the excessive expression of the non-functional isoform over functional or cytoplasmic titration of the functional isoform or regulation of full-length isoforms by non-functional isoforms directly or indirectly [65]. For example, the HELIOS-V1 isoform lacks exon 6 for the nuclear localization signal and therefore has a cytoplasmic localization in human leukemic T-cell lines and contributes to T-cell growth and survival. Further, the expression of HELIOS isoforms triggers the deregulation of various downstream target genes in T-cells compared to full-length isoforms [68]. The isoform 3 encoded from transcripts 12, 8 or 9 is expected to have the essential domains such as the transactivation domain and homeodomain, which are essential for DNA binding and act as transcription factors. However, compared to full-length IRX4 isoforms, this isoform 3 lacks an N-terminal region that may affect the structure and function in PCa, which needs to be further elucidated. Unfortunately, we could not see any protein expression in mass spectrometry data in several transcripts (10 and 11) identified in PCa cell lines that may suggest a nonsense-mediated decay of these transcripts. In view of this, although the potential importance of IRX4 isoforms towards PCa progression is still unknown, the diagnostic and therapeutic value of these isoforms cannot be ignored. Additional functional studies with isoform-specific overexpression and knockdown models are essential to prove the differential roles of IRX4 isoforms in PCa.

GWAS have identified over 160 PCa risk loci, some of which act as cis-eQTL regulatory elements which modulate the expression of nearby genes. The PCa risk SNP rs12653946 identified in the *5p15* locus was found to have a strong relationship with the *IRX4* gene ($P = 4.91 \times 10^{-5}$, FDR = 0.00468) [22,44]. rs10866528 has been used to tag the SNP rs12653946 and has been found to have a lower *IRX4* expression with the PCa high-risk homozygous genotype GG than with the common heterozygous AG genotype in a 50-patient sample cohort [44]. Similar to the reported results, with rs10866528 SNP we observed low levels of expression of *IRX4* transcripts in common AA genotypes and the PCa risk-associated GG genotype in a large sample cohort (483 samples). All the analyzed *IRX4* transcripts showed a similar pattern, and we were not able to observe a disparity between the analyzed transcripts. This suggests that *IRX4* transcripts equally contribute to PCa risk and one cannot ignore the individual value in the progression of PCa.

Isoform discovery is a challenging task in cancer research. Although many transcripts are identified at a cellular level, only a small proportion of transcripts encode isoforms in context-dependent manner [63]. Most of the isoforms share overlapping regions; therefore, the identification of unique isoforms is extremely difficult at a lab-based assays and computational level. The commonly used techniques, such as, PCR, RNA-seq and mass spectroscopy, have their own limitations in identifying and quantifying low abundant isoforms. We have selected a small number of PCa cell lines to quantify the expression of IRX4 isoforms by mass spectrometry, but these needed to be measured in a large sample cohort including clinical specimens. Low reproducibility, low sensitivity, high variability and high data noise of the instruments are real challenges in isoform annotation. Moreover, the limited correlation between the transcriptomic and proteomic data makes it difficult to interpret differences in isoform expression between cancer cell lines.

Analyzing isoform-specific expression in prostate tumor progression can provide clear insights to develop drugs against specific isoforms that promote tumors. Besides PCa, the diversity of the *IRX4* gene can be a hallmark for other cancers; thus, the characterization of these isoforms would augment our knowledge for the development of specific therapeutic strategies. Considering the IRX4 isoform-specific expression, our study suggests that IRX4 can have distinct roles in PCa, which suggests the clinical importance of isoform-specific therapeutic targets in improving PCa patient care.

5. Conclusions

In summary, identifying isoform-specific expression is a challenging but critically important task, especially in cancer progression. We identified a subset of *IRX4* transcripts in PCa whose mRNA and encoding isoforms show distinct expression profiles across a panel of androgen-responsive and nonresponsive PCa cell lines. Apart from the prominent *IRX4* 1–5 transcripts, a large contribution has been imparted by other *IRX4* transcripts to the overall *IRX4* gene expression in PCa. Some transcripts are not only lacking in regulatory and essential coding regions but also possess additional sequence features via intron retention, which suggests that the functional roles of their encoding isoforms may be distinct from the primary full-length isoforms. Given the experimental evidence associated with *IRX4* AS, our results prioritize those splicing events that can show a clear expression signal of functional importance. Thus, this highlights the importance of exploring the differential roles of potential isoforms in cancer and how cells obtain the benefit of AS in tumor progression, therapy and drug resistance. Therefore, understanding the AS of *IRX4* in PCa could provide insights into tumor development and lead to the development of new therapeutic targets.

Supplementary Materials: The following are available online at https://www.mdpi.com/article/10.3390/genes12050615/s1, Figure S1: Alignment of predicted *IRX4* transcripts 6, 8, 10 and 12 sequences, Figure S2: Identification of IRX4 isoform-specific peptides by reprocessing LNCaP LC-MS/MS data. Table S1: Primer sequences for *IRX4* transcripts used in the study.

Author Contributions: Conceptualization, A.F., P.J., and J.B.; methodology, A.F., P.J., C.L., A.M; investigation, A.F., C.L., A.M.; formal analysis A.F., C.L., A.M.; validation, A.F. and J.B.; writing—original draft preparation, A.F.; writing—review and editing, C.L., A.M., P.J. and J.B.; supervision, P.J. and J.B.; project administration, J.B.; funding acquisition, J.B. All authors have read and agreed to the published version of the manuscript.

Funding: This project was supported by grant 1124950 awarded through the 2016 Priority-driven Collaborative Cancer Research Scheme and funded by Cancer Australia awarded to J. Batra. J. Batra was supported by a National Health and Medical Research Council (NHMRC) Career Development Fellowship and Advance Qld Industry Research Fellowship. Achala Fernando acknowledges the Research Training Stipend (RTP) and QUT HDR Tuition Fee Sponsorship. Chamikara Liyanage and

Afshin Moradi acknowledge the QUT Postgraduate Research Award (QUTPRA) and QUT HDR Tuition Fee Sponsorship.

Institutional Review Board Statement: The cell line-based studies are approved by the QUT ethics committee (QUT Ethics approval number: 1500001082).

Informed Consent Statement: Not applicable.

Data Availability Statement: The data presented in this study are available on request from the corresponding author.

Acknowledgments: The authors acknowledge Srilakshmi Srinivasn for her guidance in the experiments and Mr Adil Malik for his contribution to PRIDE data reprocessing. Pawel Sadowski and Raj Gupta are acknowledged for carrying out the LC-MS/MS analysis at the Central Analytical Research Facility (CARF), operated by the Institute for Future Environments at QUT.

Conflicts of Interest: The authors declare no conflict of interest.

References

1. Wang, Y.; Liu, J.; Huang, B.O.; Xu, Y.M.; Li, J.; Huang, L.F.; Lin, J.; Zhang, J.; Min, Q.H.; Yang, W.M.; et al. Mechanism of alternative splicing and its regulation. *Biomed. Rep.* **2015**, *3*, 152–158. [CrossRef] [PubMed]
2. Singh, B.; Eyras, E. The role of alternative splicing in cancer. *Transcription* **2017**, *8*, 91–98. [CrossRef]
3. Liyanage, C.; Fernando, A.; Batra, J. Differential roles of protease isoforms in the tumor microenvironment. *Cancer Metastasis Rev.* **2019**, *38*, 389–415. [CrossRef]
4. Siegfried, Z.; Karni, R. The role of alternative splicing in cancer drug resistance. *Curr. Opin. Genet. Dev.* **2018**, *48*, 16–21. [CrossRef]
5. Biamonti, G.; Infantino, L.; Gaglio, D.; Amato, A. An Intricate Connection between Alternative Splicing and Phenotypic Plasticity in Development and Cancer. *Cells* **2019**, *9*, 34. [CrossRef]
6. Bray, F.; Ferlay, J.; Soerjomataram, I.; Siegel, R.L.; Torre, L.A.; Jemal, A. Global cancer statistics 2018: GLOBOCAN estimates of incidence and mortality worldwide for 36 cancers in 185 countries. *CA Cancer J. Clin.* **2018**, *68*, 394–424. [CrossRef]
7. Blee, A.M.; Huang, H. Lineage plasticity-mediated therapy resistance in prostate cancer. *Asian J. Androl.* **2019**, *21*, 241–248. [CrossRef]
8. Olender, J.; Lee, N.H. Role of Alternative Splicing in Prostate Cancer Aggressiveness and Drug Resistance in African Americans. *Adv. Exp. Med. Biol.* **2019**, *1164*, 119–139. [CrossRef] [PubMed]
9. Paschalis, A.; Sharp, A.; Welti, J.C.; Neeb, A.; Raj, G.V.; Luo, J.; Plymate, S.R.; de Bono, J.S. Alternative splicing in prostate cancer. *Nat. Rev. Clin. Oncol.* **2018**, *15*, 663–675. [CrossRef]
10. Bruneau, B.G.; Bao, Z.Z.; Tanaka, M.; Schott, J.J.; Izumo, S.; Cepko, C.L.; Seidman, J.G.; Seidman, C.E. Cardiac expression of the ventricle-specific homeobox gene Irx4 is modulated by Nkx2-5 and dHand. *Dev. Biol.* **2000**, *217*, 266–277. [CrossRef]
11. Pearson, J.C.; Lemons, D.; McGinnis, W. Modulating Hox gene functions during animal body patterning. *Nat. Rev. Genet.* **2005**, *6*, 893–904. [CrossRef]
12. Cavodeassi, F.; Modolell, J.; Gomez-Skarmeta, J.L. The Iroquois family of genes: From body building to neural patterning. *Development* **2001**, *128*, 2847–2855.
13. Kim, K.H.; Rosen, A.; Bruneau, B.G.; Hui, C.C.; Backx, P.H. Iroquois homeodomain transcription factors in heart development and function. *Circ. Res.* **2012**, *110*, 1513–1524. [CrossRef]
14. Bao, Z.Z.; Bruneau, B.G.; Seidman, J.G.; Seidman, C.E.; Cepko, C.L. Regulation of chamber-specific gene expression in the developing heart by Irx4. *Science* **1999**, *283*, 1161–1164. [CrossRef]
15. Bruneau, B.G.; Bao, Z.Z.; Fatkin, D.; Xavier-Neto, J.; Georgakopoulos, D.; Maguire, C.T.; Berul, C.I.; Kass, D.A.; Kuroski-de Bold, M.L.; de Bold, A.J.; et al. Cardiomyopathy in Irx4-deficient mice is preceded by abnormal ventricular gene expression. *Mol. Cell Biol.* **2001**, *21*, 1730–1736. [CrossRef] [PubMed]
16. Cheng, Z.; Wang, J.; Su, D.; Pan, H.; Huang, G.; Li, X.; Li, Z.; Shen, A.; Xie, X.; Wang, B.; et al. Two novel mutations of the IRX4 gene in patients with congenital heart disease. *Hum. Genet.* **2011**, *130*, 657–662. [CrossRef]
17. Jin, Z.; Zhang, J.; Klar, A.; Chedotal, A.; Rao, Y.; Cepko, C.L.; Bao, Z.Z. Irx4-mediated regulation of Slit1 expression contributes to the definition of early axonal paths inside the retina. *Development* **2003**, *130*, 1037–1048. [CrossRef]
18. Correa, S.; Panis, C.; Binato, R.; Herrera, A.C.; Pizzatti, L.; Abdelhay, E. Identifying Potential Markers in Breast Cancer Subtypes Using Plasma Label-Free Proteomics. *J. Proteom.* **2017**, *151*, 33–42. [CrossRef]
19. Ding, L.; Su, Y.; Fassl, A.; Hinohara, K.; Qiu, X.; Harper, N.W.; Huh, S.J.; Bloushtain-Qimron, N.; Jovanovic, B.; Ekram, M.; et al. Perturbed myoepithelial cell differentiation in BRCA mutation carriers and in ductal carcinoma in situ. *Nat. Commun.* **2019**, *10*, 4182. [CrossRef]
20. Zhang, D.L.; Qu, L.W.; Ma, L.; Zhou, Y.C.; Wang, G.Z.; Zhao, X.C.; Zhang, C.; Zhang, Y.F.; Wang, M.; Zhang, M.Y.; et al. Genome-wide identification of transcription factors that are critical to non-small cell lung cancer. *Cancer Lett.* **2018**, *434*, 132–143. [CrossRef]

21. Chakma, K.; Gu, Z.D.; Motoi, F.; Unno, M.; Horii, A.; Fukushige, S. DNA hypermethylation of IRX4 is a frequent event that may confer growth advantage to pancreatic cancer cells. *Cancer Res.* **2019**, 821–821. [CrossRef]
22. Batra, J.; Lose, F.; Chambers, S.; Gardiner, R.A.; Aitken, J.; Yaxley, J.; Clements, J.A.; Spurdle, A.B.; Australian Prostate Cancer BioResource. A replication study examining novel common single nucleotide polymorphisms identified through a prostate cancer genome-wide association study in a Japanese population. *Am. J. Epidemiol.* **2011**, *174*, 1391–1395. [CrossRef]
23. Wang, M.; Liu, F.; Hsing, A.W.; Wang, X.; Shao, Q.; Qi, J.; Ye, Y.; Wang, Z.; Chen, H.; Gao, X.; et al. Replication and cumulative effects of GWAS-identified genetic variations for prostate cancer in Asians: A case-control study in the ChinaPCa consortium. *Carcinogenesis* **2012**, *33*, 356–360. [CrossRef]
24. Lindstrom, S.; Schumacher, F.R.; Campa, D.; Albanes, D.; Andriole, G.; Berndt, S.I.; Bueno-de-Mesquita, H.B.; Chanock, S.J.; Diver, W.R.; Ganziano, J.M.; et al. Replication of five prostate cancer loci identified in an Asian population—Results from the NCI Breast and Prostate Cancer Cohort Consortium (BPC3). *Cancer Epidemiol. Biomark. Prev.* **2012**, *21*, 212–216. [CrossRef]
25. Qi, J.; Tian, L.; Chen, Z.; Wang, L.; Tao, S.; Gu, X.; Na, R.; Jiao, Y.; Kang, J.; Zheng, S.; et al. Genetic variants in 2q31 and 5p15 are associated with aggressive benign prostatic hyperplasia in a Chinese population. *Prostate* **2013**, *73*, 1182–1190. [CrossRef] [PubMed]
26. Nguyen, H.H.; Takata, R.; Akamatsu, S.; Shigemizu, D.; Tsunoda, T.; Furihata, M.; Takahashi, A.; Kubo, M.; Kamatani, N.; Ogawa, O.; et al. IRX4 at 5p15 suppresses prostate cancer growth through the interaction with vitamin D receptor, conferring prostate cancer susceptibility. *Hum. Mol. Genet.* **2012**, *21*, 2076–2085. [CrossRef]
27. Holmquist Mengelbier, L.; Lindell-Munther, S.; Yasui, H.; Jansson, C.; Esfandyari, J.; Karlsson, J.; Lau, K.; Hui, C.C.; Bexell, D.; Hopyan, S.; et al. The Iroquois homeobox proteins IRX3 and IRX5 have distinct roles in Wilms tumour development and human nephrogenesis. *J. Pathol.* **2019**, *247*, 86–98. [CrossRef]
28. Wang, P.; Zhuang, C.; Huang, D.; Xu, K. Downregulation of miR-377 contributes to IRX3 deregulation in hepatocellular carcinoma. *Oncol. Rep.* **2016**, *36*, 247–252. [CrossRef]
29. Morey, S.R.; Smiraglia, D.J.; James, S.R.; Yu, J.; Moser, M.T.; Foster, B.A.; Karpf, A.R. DNA methylation pathway alterations in an autochthonous murine model of prostate cancer. *Cancer Res.* **2006**, *66*, 11659–11667. [CrossRef] [PubMed]
30. Kent, W.J.; Sugnet, C.W.; Furey, T.S.; Roskin, K.M.; Pringle, T.H.; Zahler, A.M.; Haussler, D. The human genome browser at UCSC. *Genome Res.* **2002**, *12*, 996–1006. [CrossRef]
31. Schmittgen, T.D.; Livak, K.J. Analyzing real-time PCR data by the comparative C(T) method. *Nat. Protoc.* **2008**, *3*, 1101–1108. [CrossRef]
32. Sayers, E.W.; Beck, J.; Brister, J.R.; Bolton, E.E.; Canese, K.; Comeau, D.C.; Funk, K.; Ketter, A.; Kim, S.; Kimchi, A.; et al. Database resources of the National Center for Biotechnology Information. *Nucleic Acids Res.* **2019**, *48*, D9–D16. [CrossRef]
33. The Cancer Genome Atlas Program. Available online: https://www.cancer.gov/tcga (accessed on 24 February 2021).
34. Zhang, X.; Jonassen, I. RASflow: An RNA-Seq analysis workflow with Snakemake. *BMC Bioinform.* **2020**, *21*, 110. [CrossRef] [PubMed]
35. Kim, D.; Paggi, J.M.; Park, C.; Bennett, C.; Salzberg, S.L. Graph-based genome alignment and genotyping with HISAT2 and HISAT-genotype. *Nat. Biotechnol.* **2019**, *37*, 907–915. [CrossRef] [PubMed]
36. Metzger, E.; Willmann, D.; McMillan, J.; Forne, I.; Metzger, P.; Gerhardt, S.; Petroll, K.; von Maessenhausen, A.; Urban, S.; Schott, A.K.; et al. Assembly of methylated KDM1A and CHD1 drives androgen receptor-dependent transcription and translocation. *Nat. Struct. Mol. Biol.* **2016**, *23*, 132–139. [CrossRef]
37. Chambers, M.C.; Maclean, B.; Burke, R.; Amodei, D.; Ruderman, D.L.; Neumann, S.; Gatto, L.; Fischer, B.; Pratt, B.; Egertson, J.; et al. A cross-platform toolkit for mass spectrometry and proteomics. *Nat. Biotechnol.* **2012**, *30*, 918–920. [CrossRef]
38. Barsnes, H.; Vaudel, M. SearchGUI: A Highly Adaptable Common Interface for Proteomics Search and de Novo Engines. *J. Proteome Res.* **2018**, *17*, 2552–2555. [CrossRef]
39. Vaudel, M.; Burkhart, J.M.; Zahedi, R.P.; Oveland, E.; Berven, F.S.; Sickmann, A.; Martens, L.; Barsnes, H. PeptideShaker enables reanalysis of MS-derived proteomics data sets. *Nat. Biotechnol.* **2015**, *33*, 22–24. [CrossRef]
40. Kent, W.J. BLAT—The BLAST-like alignment tool. *Genome Res.* **2002**, *12*, 656–664. [CrossRef]
41. Liyanage, C.; Adil, M.; Abeysinghe, P.; Clements, J.; Batra, J. SWATH-MS Based Proteomic Profiling of Prostate Cancer Cells Reveals Adaptive Molecular Mechanisms in Response to Anti-Androgen Therapy. *Cancers* **2021**, *13*, 715. [CrossRef]
42. MacLean, B.; Tomazela, D.M.; Shulman, N.; Chambers, M.; Finney, G.L.; Frewen, B.; Kern, R.; Tabb, D.L.; Liebler, D.C.; MacCoss, M.J. Skyline: An open source document editor for creating and analyzing targeted proteomics experiments. *Bioinformatics* **2010**, *26*, 966–968. [CrossRef] [PubMed]
43. Choi, M.; Chang, C.Y.; Clough, T.; Broudy, D.; Killeen, T.; MacLean, B.; Vitek, O. MSstats: An R package for statistical analysis of quantitative mass spectrometry-based proteomic experiments. *Bioinformatics* **2014**, *30*, 2524–2526. [CrossRef] [PubMed]
44. Xu, X.; Hussain, W.M.; Vijai, J.; Offit, K.; Rubin, M.A.; Demichelis, F.; Klein, R.J. Variants at IRX4 as prostate cancer expression quantitative trait loci. *Eur. J. Hum. Genet.* **2014**, *22*, 558–563. [CrossRef]
45. Conti, D.V.; Darst, B.F.; Moss, L.C.; Saunders, E.J.; Sheng, X.; Chou, A.; Schumacher, F.R.; Olama, A.A.A.; Benlloch, S.; Dadaev, T.; et al. Publisher Correction: Trans-ancestry genome-wide association meta-analysis of prostate cancer identifies new susceptibility loci and informs genetic risk prediction. *Nat. Genet.* **2021**, *53*, 413. [CrossRef]

46. Feng, H.; Jin, Z.; Liu, K.; Peng, Y.; Jiang, S.; Wang, C.; Hu, J.; Shen, X.; Qiu, W.; Cheng, X.; et al. Identification and validation of critical alternative splicing events and splicing factors in gastric cancer progression. *J. Cell Mol. Med.* **2020**, *24*, 12667–12680. [CrossRef] [PubMed]
47. Kitamura, K.; Matsushita, K.; Kobayashi, S.; Ishige, T.; Semba, T.; Kimura, A.; Kazami, T.; Ohyama, M.; Itoga, S.; Beppu, M.; et al. Alternative Splicing Detection as a Biomarker for Cancer Diagnosis: A Novel Progressive Mechanism of Acute Lymphoblastic Leukemia with Alternative Splicing as a Biomarker Candidate. *Rinsho Byori* **2015**, *63*, 1091–1102.
48. Le, K.Q.; Prabhakar, B.S.; Hong, W.J.; Li, L.C. Alternative splicing as a biomarker and potential target for drug discovery. *Acta Pharmacol. Sin.* **2015**, *36*, 1212–1218. [CrossRef]
49. Matsushita, K.; Itoga, S.; Nishimura, M.; Furuta, K.; Nomura, F. Alternative Splicing Detection as Biomarker Candidates for Cancer Diagnosis and Treatment with Establishment of Clinical Biobank in Chiba University. *Rinsho Byori* **2015**, *63*, 347–360.
50. Wu, P.; Zhou, D.; Wang, Y.; Lin, W.; Sun, A.; Wei, H.; Fang, Y.; Cong, X.; Jiang, Y. Identification and validation of alternative splicing isoforms as novel biomarker candidates in hepatocellular carcinoma. *Oncol. Rep.* **2019**, *41*, 1929–1937. [CrossRef]
51. Zhao, X.; Si, S.; Li, X.; Sun, W.; Cui, L. Identification and validation of an alternative splicing-based prognostic signature for head and neck squamous cell carcinoma. *J. Cancer* **2020**, *11*, 4571–4580. [CrossRef]
52. Narasimhan, A.; Greiner, R.; Bathe, O.F.; Baracos, V.; Damaraju, S. Differentially expressed alternatively spliced genes in skeletal muscle from cancer patients with cachexia. *J. Cachexia Sarcopenia Muscle* **2018**, *9*, 60–70. [CrossRef]
53. Wang, B.D.; Lee, N.H. Aberrant RNA Splicing in Cancer and Drug Resistance. *Cancers* **2018**, *10*, 458. [CrossRef]
54. Wang, B.D.; Ceniccola, K.; Hwang, S.; Andrawis, R.; Horvath, A.; Freedman, J.A.; Olender, J.; Knapp, S.; Ching, T.; Garmire, L.; et al. Alternative splicing promotes tumour aggressiveness and drug resistance in African American prostate cancer. *Nat. Commun.* **2017**, *8*, 15921. [CrossRef] [PubMed]
55. Anand, A.U.; Bjartell, A. Re: AR-V7 and resistance to enzalutamide and abiraterone in prostate cancer. *Eur. Urol.* **2015**, *67*, 349–350. [CrossRef]
56. Harries, L.W.; Perry, J.R.; McCullagh, P.; Crundwell, M. Alterations in LMTK2, MSMB and HNF1B gene expression are associated with the development of prostate cancer. *BMC Cancer* **2010**, *10*, 315. [CrossRef]
57. Harries, L.W. Alternate mRNA processing of the hepatocyte nuclear factor genes and its role in monogenic diabetes. *Expert Rev. Endocrinol. Metab.* **2006**, *1*, 715–726. [CrossRef] [PubMed]
58. Chandra, S.; Srinivasan, S.; Batra, J. Hepatocyte nuclear factor 1 beta: A perspective in cancer. *Cancer Med.* **2021**, *10*, 1791–1804. [CrossRef] [PubMed]
59. Ghiasvand, S.; Bakhshinejad, B.; Mowla, S.J.; Sadeghizadeh, M. Potential roles of 5 UTR and 3 UTR regions in post-trans-criptional regulation of mouse Oct4 gene in BMSC and P19 cells. *Iran. J. Basic Med. Sci.* **2014**, *17*, 490–496.
60. Lee, I.I.; Kuznik, N.C.; Rottenberg, J.T.; Brown, M.; Cato, A.C.B. BAG1L: A promising therapeutic target for androgen receptor-dependent prostate cancer. *J. Mol. Endocrinol.* **2019**, *62*, R289–R299. [CrossRef]
61. Ferrari, N.; Granata, I.; Capaia, M.; Piccirillo, M.; Guarracino, M.R.; Vene, R.; Brizzolara, A.; Petretto, A.; Inglese, E.; Morini, M.; et al. Adaptive phenotype drives resistance to androgen deprivation therapy in prostate cancer. *Cell Commun. Signal.* **2017**, *15*, 51. [CrossRef]
62. Iglesias-Gato, D.; Wikstrom, P.; Tyanova, S.; Lavallee, C.; Thysell, E.; Carlsson, J.; Hagglof, C.; Cox, J.; Andren, O.; Stattin, P.; et al. The Proteome of Primary Prostate Cancer. *Eur. Urol.* **2016**, *69*, 942–952. [CrossRef]
63. Liu, Y.; Beyer, A.; Aebersold, R. On the Dependency of Cellular Protein Levels on mRNA Abundance. *Cell* **2016**, *165*, 535–550. [CrossRef]
64. Nelson, D.O.; Jin, D.X.; Downs, K.M.; Kamp, T.J.; Lyons, G.E. Irx4 identifies a chamber-specific cell population that contributes to ventricular myocardium development. *Dev. Dyn.* **2014**, *243*, 381–392. [CrossRef]
65. Belluti, S.; Rigillo, G.; Imbriano, C. Transcription Factors in Cancer: When Alternative Splicing Determines Opposite Cell Fates. *Cells* **2020**, *9*, 760. [CrossRef] [PubMed]
66. Gershenwald, J.E.; Sumner, W.; Calderone, T.; Wang, Z.; Huang, S.; Bar-Eli, M. Dominant-negative transcription factor AP-2 augments SB-2 melanoma tumor growth in vivo. *Oncogene* **2001**, *20*, 3363–3375. [CrossRef] [PubMed]
67. Buettner, R.; Kannan, P.; Imhof, A.; Bauer, R.; Yim, S.O.; Glockshuber, R.; Van Dyke, M.W.; Tainsky, M.A. An alternatively spliced mRNA from the AP-2 gene encodes a negative regulator of transcriptional activation by AP-2. *Mol. Cell Biol.* **1993**, *13*, 4174–4185. [CrossRef] [PubMed]
68. Zhao, S.; Liu, W.; Li, Y.; Liu, P.; Li, S.; Dou, D.; Wang, Y.; Yang, R.; Xiang, R.; Liu, F. Alternative Splice Variants Modulates Dominant-Negative Function of Helios in T-Cell Leukemia. *PLoS ONE* **2016**, *11*, e0163328. [CrossRef] [PubMed]

genes

MDPI

Article

Characterization of Hormone-Dependent Pathways in Six Human Prostate-Cancer Cell Lines: A Gene-Expression Study

Andras Franko [1,2,3], Lucia Berti [2,3,*], Alke Guirguis [4], Jörg Hennenlotter [5], Robert Wagner [1,2,3], Marcus O. Scharpf [6], Martin Hrabě de Angelis [3,7], Katharina Wißmiller [8,9,10], Heiko Lickert [3,8,9,10], Arnulf Stenzl [5], Andreas L. Birkenfeld [1,2,3], Andreas Peter [2,3,4], Hans-Ulrich Häring [1,2,3], Stefan Z. Lutz [1,11,†] and Martin Heni [1,2,3,4,†]

1 Department of Internal Medicine IV, Division of Diabetology, Endocrinology, and Nephrology, University Hospital Tübingen, 72076 Tübingen, Germany; andras.franko@med.uni-tuebingen.de (A.F.); robert.wagner@med.uni-tuebingen.de (R.W.); andreas.birkenfeld@med.uni-tuebingen.de (A.L.B.); hans-ulrich.haering@med.uni-tuebingen.de (H.-U.H.); s.lutz@bad-sebastiansweiler.de (S.Z.L.); martin.heni@med.uni-tuebingen.de (M.H.)
2 Institute for Diabetes Research and Metabolic Diseases of the Helmholtz Centre Munich at the University of Tübingen, 72076 Tübingen, Germany; andreas.peter@med.uni-tuebingen.de
3 German Center for Diabetes Research (DZD), 85764 Neuherberg, Germany; hrabe@helmholtz-muenchen.de (M.H.d.A.); heiko.lickert@helmholtz-muenchen.de (H.L.)
4 Department for Diagnostic Laboratory Medicine, Institute for Clinical Chemistry and Pathobiochemistry, University Hospital Tübingen, 72076 Tübingen, Germany; alke.guirguis@med.uni-tuebingen.de
5 Department of Urology, University Hospital Tübingen, 72076 Tübingen, Germany; joerg.hennenlotter@med.uni-tuebingen.de (J.H.); arnulf.stenzl@med.uni-tuebingen.de (A.S.)
6 Institute of Pathology, University Hospital Tübingen, 72076 Tübingen, Germany; Marcus.Scharpf@med.uni-tuebingen.de
7 Institute of Experimental Genetics, Helmholtz Zentrum München, 85764 Neuherberg, Germany
8 Institute of Diabetes and Regeneration Research, Helmholtz Zentrum München, 85764 Neuherberg, Germany; katharina.wissmiller@helmholtz-muenchen.de
9 Institute of Stem Cell Research, Helmholtz Zentrum München, 85764 Neuherberg, Germany
10 Department of Medicine, Technical University of Munich, 81675 München, Germany
11 Clinic for Geriatric and Orthopedic Rehabilitation Bad Sebastiansweiler, 72116 Mössingen, Germany
* Correspondence: lucia.berti@helmholtz-muenchen.de; Tel.: +49-89-3187-4432
† These authors contributed equally to this work.

Received: 25 August 2020; Accepted: 30 September 2020; Published: 7 October 2020

Abstract: Prostate cancer (PCa), the most incident cancer in men, is tightly regulated by endocrine signals. A number of different PCa cell lines are commonly used for in vitro experiments, but these are of diverse origin, and have very different cell-proliferation rates and hormone-response capacities. By analyzing the gene-expression pattern of main hormone pathways, we systematically compared six PCa cell lines and parental primary cells. We compared these cell lines (i) with each other and (ii) with PCa tissue samples from 11 patients. We found major differences in the gene-expression levels of androgen, insulin, estrogen, and oxysterol signaling between PCa tissue and cell lines, and between different cell lines. Our systematic characterization gives researchers a solid basis to choose the appropriate PCa cell model for the hormone pathway of interest.

Keywords: androgen receptor; estrogen receptor; gene expression; insulin receptor; prostate cancer

1. Introduction

Genetic and metabolic alterations can both drive the development of prostate cancer (PCa) [1,2]. Metabolic alterations can influence autocrine and paracrine signaling pathways, which are crucial in the carcinogenic processes of PCa [3]. A tumor is a complex milieu and, in addition to prostate-cancer cells, it contains fibroblasts, endothelial cells, and mesenchymal stem cells [4]. These various cells communicate with PCa cells via paracrine factors that drive carcinogenesis [4,5]. For instance, paracrine factors secreted by PCa cells were shown to stimulate osteoblasts, which led to excess bone formation [6]. Metabolic alterations, including obesity, metabolic syndrome, insulin resistance, and diabetes, are known to exacerbate PCa [7,8]. These conditions mediate the aggressivity of PCa via endocrine signaling [1]. The major hormone pathways in PCa are androgen, insulin, estrogen, and oxysterol signaling [1,9,10]. In vitro studies use different cell models that are of diverse origin, and have very different cell-proliferation rates and hormone-response capacities [11–13]. Although all these cell lines represent valuable tools to study the development of PCa *in vitro*, whether these cell lines are suitable for studying endocrine pathways, which regulate PCa development in humans, is poorly understood. Therefore, it is vital to use a suitable in vitro model that resembles the physiology of endocrine pathways in human PCa. To characterize the major hormone pathways in human PCa cell lines, we systematically compared six commonly applied PCa cell lines and parental primary cells with (i) each other and (ii) PCa tissue samples of patients with PCa by analyzing their gene-expression patterns.

2. Materials and Methods

2.1. Cell Culture

Human prostate adenocarcinoma cell lines PC3 and LNCaP, isolated from bone and lymph-node metastasis respectively, were purchased from CLS-Cell line services (Eppelheim, Baden-Württemberg, Germany). DU145 cells (isolated from brain metastasis) were obtained from the DSMZ-German collection of microorganisms and cell cultures (Braunschweig, Niedersachsen, Germany). NCI-H660 cells (isolated from lymph-node metastasis) and MDA-PCa-2b (isolated from bone metastasis) were purchased from the ATCC-American type culture collection (Manassas, VA, USA). CWR-R1ca cells (isolated from human xenograft tumors) and human prostate epithelial cells (HPEC) were purchased from Merck (Darmstadt, Hessen, Germany). HPEC cells were used at passages 3 and 4. All cell lines were propagated according to the supplier's instructions and maintained in a medium supplemented with 100 IU/mL penicillin, 0.1 mg/mL streptomycin, and 2 mM glutamine (Gibco/Thermo Fisher Scientific, Karlsruhe, Baden-Württemberg, Germany) in a 5% CO_2 humidified atmosphere at 37 °C. LNCaP cells for 14 days of androgen deprivation were cultured in a basal medium (ATCC #CRL-1740, ThermoFisher Scientific #21875034) containing 2 mM L-glutamine and 10% FBS Good (Pan-Biotech P40-37500, Aidenbach, Freistaat Bayern, Germany). The medium used for androgen deprivation (ThermoFisher Scientific #11835030) was supplemented with 10% charcoal-stripped FBS (Sigma-Aldrich, F6765, Munich, Freistaat Bayern, Germany) and 10 nM dihydrotestosterone (Sigma-Aldrich, D-073) for 14 days. Detailed information regarding culture medium and supplements is given in Table 1. Cells were routinely cultivated until 90% confluency before harvesting.

Table 1. Characteristics of the cell lines used in the study. If not otherwise specified, fetal bovine serum (FBS) was purchased from Bio&SELL (FBS Superior, #S 0615, Feucht, Germany).

Cell Line	Origin	Catalog Nr	Medium	Supplements (Final Concentration)
PC3	Adult, male Caucasian. Prostate, from metastatic site: bone. Adenocarcinoma, Grade IV	CLS, cell line service #300312	ThermoFisher Scientific #11039047	2 mM L-glutamine, 5% FBS
LNCaP	Adult, male Caucasian. Prostate, from metastatic site: left supraclavicular lymph node. Carcinoma	CLS, cell line service #300265	ThermoFisher Scientific #51200046	2 mM L-glutamine, 10% FBS, 0.1 mM nonessential amino acids, 1 mM sodium pyruvate
MDA-PCa-2b	Adult, male African American. Prostate, from metastatic site: bone	ATCC #CRL-2422	ATCC #30-2004	20% FBS, 25 ng/mL cholera toxin, 10 ng/mL mouse EGF, 0.005 mM phosphoethanolamine, 100 pg/mL hydrocortisone, 45 nM sodium selenite, 0.005 mg/mL human recombinant insulin
CWR-R1ca	Fibroblast-free cell line derived from the castration-resistant or recurrent CWR-R1 human prostate-cancer cell line	Merck-Millipore #SCC118	ThermoFisher Scientific #32404014	2 mM L-glutamine, 10% FBS
NCI-H660	Adult, male Caucasian. Prostate, from metastatic site: lymph node	ATCC #CRL-5813	ThermoFisher Scientific #32404014	4 mM L-glutamine, 5% FBS, 0.005 mg/mL insulin, 0.01 mg/mL transferrin, 30 nM sodium selenite, 10 nM hydrocortisone, 10 nM β-estradiol
DU145	Adult, male Caucasian. Prostate, from metastatic site: brain	DSMZ #ACC 261	ThermoFisher Scientific #32404014	2 mM L-glutamine, 10% FBS
HPEC	Adult, normal	Merck-Millipore #SCCE019	Merck-Millipore #SCMP001 (Kit)	6 mM L-glutamine, 0.4% epiFactor P, 1 µM epinephrine, 0.5 ng/mL rh TGF-α, 100 ng/mL hydrocortisone hemisuccinate, 5 µg/mL human recombinant insulin, 5 µg/mL apo-transferrin

2.2. Human Samples

Newly diagnosed PCa patients who had not received treatment before surgery were recruited prior to radical prostatectomy. Tissue sampling was performed by an experienced uropathologist. PCa tissues were immediately snap-frozen in liquid nitrogen and stored at −80 °C. Hematoxylin and eosin staining was performed on paraffinized samples for histological confirmation. Histopathological features were assessed, and pT- and postoperative Gleason scores were determined [14,15]. Informed written consent was obtained from all participants, and the Ethics Committee of the University of Tübingen (575/2018BO2) approved the protocol according to the Declaration of Helsinki.

2.3. Real-Time qPCR

Total RNA was isolated using an AllPrep DNA/RNA/miRNA kit (Qiagen, Hilden, Germany) according to the manufacturer's description, and cDNA was synthesized (Roche, Basel, Switzerland). Real-time PCR was performed with LightCycler 480 Probes Master (Roche) with universal probe library using LightCycler 480 (Roche) [16]. Delta–delta crossing-point (Cp) values were calculated, and data were normalized to ubiquitin c (*UBC*) [8,15]. For real-time PCR analysis, the following primers and probes were applied: *ESR1* 5′-TCCTAACTTGCTCTTGGACAGG and 3′-GTAGCCAGCAGCATGTCG (probe nr 22), *ESRRA* 5′-GTGGGCGGCAGAAGTACA and 3′-TCAACCACCAGCAGATGAGA (probe nr 3), *CYP46A1* 5′-GCTGGACAACTTCGTCACCT and 3′-CATCACTGTGAACGCCAAGT (probe nr 53), *CCND1* 5′-GACCTTCGTTGCCCTCTGT and 3′-GGTTCAGGCCTTGCACTG (probe nr 87), *TP53* 5′-AAGTCTAGAGCCACCGTCCA and 3′-AGTCTGGCTGCCAATCCA (probe nr 3), *GLUT1* 5′-GGTTGTGCCATACTCATGACC and

3′-CAGATAGGACATCCAGGGTAGC (probe nr 67), *GLUT12* 5′-TGCTGCTTTTCAATTGGTCT and 3′-AGGAAAGATCTCGCTGAGCA (probe nr 37), *IRS1* 5′-GCCTATGCCAGCATCAGTTT and 3′-TTGCTGAGGTCATTTAGGTCTTC (probe nr 71), *IRS2* 5′-TGACTTCTTGTCCCACCACTT and 3′-CATCCTGGTGATAAAGCCAGA (probe nr 49), *FGFR1* 5′-GGCAGCATCAACCACACATA and 3′-TACCCAGGGCCACTGTTTT (probe nr 42), *CH25H* 5′-CCTTCTTCCCGGTCATCTTC and 3′-GATATCCAGGACCACGAAGG (probe nr 9), *CDKN1A* 5′-CCGAAGTCAGTTCCTTGTGG and 3′-CATGGGTTCTGACGGACAT (probe nr 82), *CDKN1B* 5′-GAGAGCCAGGATGTCAGCG and 3′-TTGTTTTGAGTAGAAGAATCGTCGGT (probe CCTTTAATTGGGGCTCCGGCTAACT), *NOS2* 5′-GCTCAAATCTCGGCAGAATC and 3′-GCCATCCTCACAGGAGAGTT (probe nr 42), *SEPP1* 5′-GGAGCTGCCAGAGTAAAGCA and 3′-ACATTGCTGGGGTTGTCAC (probe nr 38). Primer sequences for *AR, CYP27A1, CYP7B1, ESR2, FOLH1, IGF1R, INSRA, INSRB, KLK3, UBC* and *MKI67* are given in [9]. Primer sequences for *HIF1A, RELA* and *BIRC5* are given in [14]. For the human PCa samples, the gene-expression levels of *GLUT1, HIF1A, BIRC5, NOS2, RELA* and *MKI67* were previously studied [15].

2.4. Western Blot

Total cellular protein was extracted using RIPA buffer (50 mmol/L Tris-HCl pH 7.4, 150 mmol/L NaCl, 1% NP-40, 0.25% Na-deoxycholate, 1 mmol/L phenyl-methyl-sulfonyl-fluoride, 1 mmol/L dithiothreitol) containing a protease and phosphatase inhibitor cocktail (Roche Molecular Biochemicals, Mannheim, Baden-Württemberg, Germany), and cleared by centrifugation. Protein concentration was determined using a BCA protein assay from Bio-Rad (Hercules, CA, USA). The 20 µg protein lysates were separated on a 4–12% Bis-Tris gel (Invitrogen, Carlsbad, CA, USA). After electrophoresis, proteins were transferred using nitrocellulose ministacks and the iBlot dry-blotting system (Invitrogen, Carlsbad, CA, USA). Membranes were blocked for two hours in Odyssey Blocking Buffer (LiCor, Lincoln, NE, USA) and further incubated with antibodies against androgen receptor (AR, ab133273), prostate-specific antigen (PSA, KLK3, ab53774), β-tubulin (ab21057), prostate-specific membrane antigen (PSMA, FOLH1, ab19071) (Abcam, Cambridge, Cambridgeshire, UK), IGF1R β subunit (D23H3, #9750), and insulin receptor β subunit (L55B10, #3020) (Cell Signaling Technologies, Danvers, MA, USA). IRDye® or AlexaFluor® secondary antibodies (LiCor or Abcam, Cambridge, UK) were used, and signals were detected and quantified using the iBright device (Invitrogen, Carlsbad, CA, USA).

3. Results

For our comprehensive analysis, we chose six commonly investigated human PCa cell lines (CWR-R1ca, DU145, LNCaP, NCI-H660, MDA-PCa-2b, and PC3). Human prostate epithelial cells (HPEC) were included as parental, nontumorous primary prostate cells. To compare gene expression for hormone pathways in the PCa cell lines to the human situation, we analyzed 11 PCa samples isolated from patients who underwent radical prostatectomy due to their tumor. Histopathological screening confirmed the presence of prostate cancer in the collected tissues. As prostate-cancer metabolism could be different at various tumor stages, we specifically selected patients at a similar tumor stage with comparable Gleason scores (7a and 7b) and without lymph-node metastasis (Table 2). Data for the 11 human samples are shown as pooled values (mean ± standard deviation) in Figures 1–3.

As a first step, we quantified the transcripts of the main hormone receptors. The expression levels of androgen receptor (*AR*) and its target genes, prostate-specific antigen (*PSA, KLK3*), and prostate-specific membrane antigen (*PSMA, FOLH1*) were most pronounced in LNCaP and MDA-PCa-2b cells (Figure 1A–C). The insulin receptor isoform A (*INSRA*)/insulin receptor isoform B (*INSRB*) ratio was highest in MDA-PCa-2b cells (Figure 1D–F), suggesting the activation of a mitogenic insulin pathway [17]. PC3 cells demonstrated the largest insulin receptor substrate (IRS) IRS1/IRS2 ratio (Figure 1G–I).

Table 2. Patient characteristics. Abbreviations: BMI: body-mass index, PSA: prostate-specific antigen, N: number of patients, Stdev: standard deviation. pT and Gleason scores represent prostate-cancer (PCa) pathological stages; pN denotes lymph-node status.

Patient Characteristics	Mean	Stdev	N
Age (y)	65	8	11
BMI (kg/m^2)	26.5	3.2	11
PSA (ng/mL)	12.1	8.1	10
pT–2c			6
pT–3a			4
pT–3b			1
pN			0
GLEASON 7a			7
GLEASON 7b			4

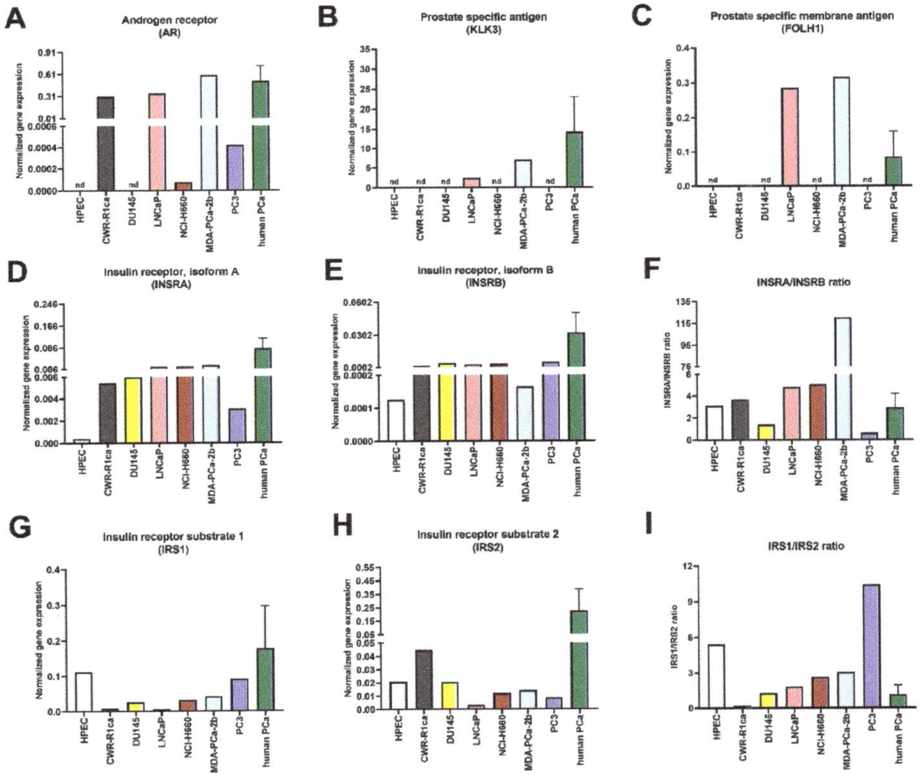

Figure 1. Transcript levels of hormone receptors and downstream substrates in prostate-cancer cell lines and in human prostate-cancer tissue. Transcript levels of indicated genes measured using real-time PCR. (**A**) *AR*, (**B**) *KLK3*, (**C**) *FOLH1*, (**D**) *INSRA*, (**E**) *INSRB*, (**F**) *INSRA/INSRB* ratio, (**G**) *IRS1*, (**H**) *IRS2*, (**I**) *IRS1/IRS2* ratio. PCa: prostate cancer, HPEC: parental primary prostate cells, CWR-R1ca: xenograft PCa cells, DU145: brain metastasis PCa cells, LNCaP: lymph-node metastasis PCa cells, NCI-H660: lymph-node metastasis PCa cells, MDA-PCa-2b: bone metastasis PCa cells, PC3: bone metastasis PCa cells, nd: not detected. For human PCa samples, data shown as pooled samples: mean ± standard deviation (n = 11).

The highest gene levels of estrogen receptors α (*ESR1*) and β (*ESR2*) were observed in NCI-H660 and PC3 cells, whereas the expression of estrogen-related receptor α (*ESRRA*) was the highest in LNCaP cells (Figure 2A–C). The gene-expression levels of insulin-like growth factor 1 receptor (*IGF1R*) were the highest in HPEC and DU145, whereas fibroblast growth factor receptor 1 (*FGFR1*) showed the strongest expression in CWR-R1ca and DU145 cells (Figure 2D–E).

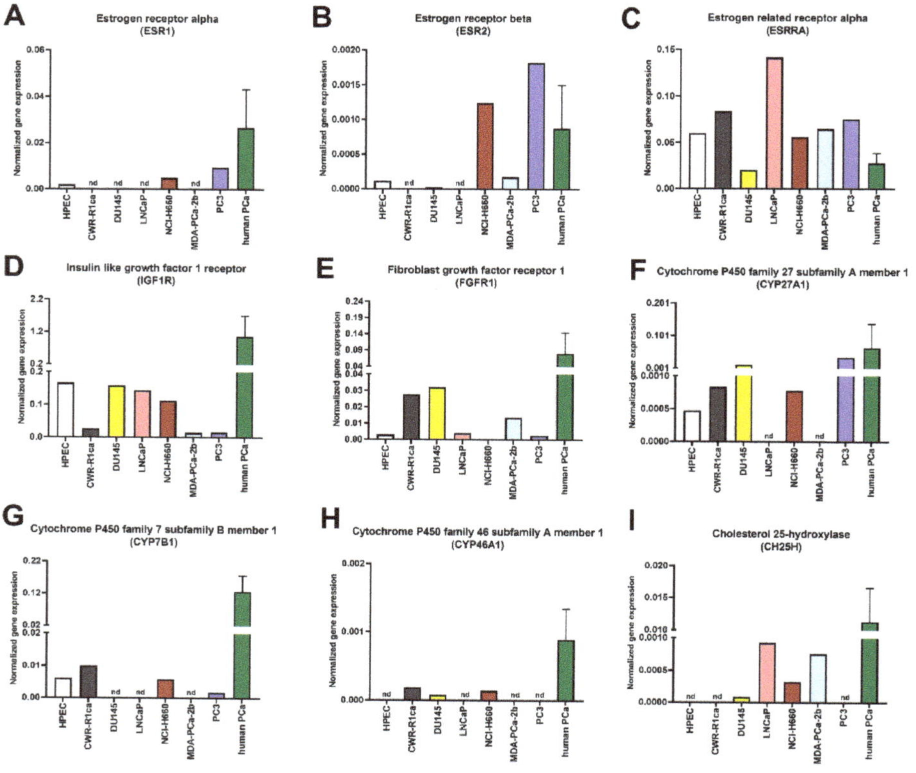

Figure 2. Transcript levels of hormone receptors and potential oncogenic mediators in prostate-cancer cell lines and in human prostate-cancer tissue. Transcript levels of indicated genes were measured using real-time PCR. (**A**) *ESR1*, (**B**) *ESR2*, (**C**) *ESRRA*, (**D**) *IGF1R*, (**E**) *FGFR1*, (**F**) *CYP27A1*, (**G**) *CYP7B1*, (**H**) *CYP46A1*, (**I**) *CH25H*. PCa: prostate cancer, HPEC: parental primary prostate cells, CWR-R1ca: xenograft PCa cells, DU145: brain metastasis PCa cells, LNCaP: lymph-node metastasis PCa cells, NCI-H660: lymph-node metastasis PCa cells, MDA-PCa-2b: bone metastasis PCa cells, PC3: bone metastasis PCa cells, nd: not detected. For human PCa samples, data shown as pooled samples: mean ± standard deviation (n = 11).

As a second step, we quantified the mRNA levels of further major intracellular regulators of PCa. Among the three investigated enzymes of oxysterol metabolism, the strongest expression for cytochrome P450 family 27 subfamily A member 1 (*CYP27A1*) was observed in DU145 and PC3 cells (Figure 2F), whereas cytochrome P450 family 7 subfamily B member 1 (*CYP7B1*) was highly represented in CWR-R1ca cells (Figure 2G). The transcript for cytochrome P450 family 46 subfamily A member 1 (*CYP46A1*) was detected only in CWR-R1ca, DU145, and NCI-H660 cells (Figure 2H). In order to use a statistical test among the analyzed cells lines, we examined cell-type-dependent gene expressions in a mixed model by bundling different genes to pathways: androgen pathway: *AR*, *PSA*, and *PSMA*; insulin pathway: *INSRA, INSRB, IRS1*, and *IRS2*; estrogen pathway: *ESR1, ESR2*, and *ESRRA*; oxysterols: *CYP7B1, CYP46A1*, and *CYP27A1*. The expressed genes were considered as random effects, and the pathways, cell types, and their interactions as fixed effects. We observed a significant interaction between cell type MDA-PCa-2b and the androgen pathway, which suggested that the MDA-PCa-2b cell line has a different gene expression for this pathway. Cholesterol 25-hydroxylase (*CH25H*) was only identified in four cell lines (Figure 2I).

Next, we analyzed three genes that are involved in proliferative pathways [18,19]. The expression of Ki-67 (*MKI67*) was the highest in NCI-H660 and MDA-PCa-2b cells (Figure 3A). CWR-R1ca cells showed the strongest expression of cyclin D1 (*CCND1*) (Figure 3B). The mRNA level of tumor protein p53 (*TP53*) was high in LNCaP and MDA-PCa-2b cells (Figure 3C). Hypoxia inducible factor 1 subunit α (*HIF1A*) and RELA proto-oncogene NF-kB subunit (*RELA*), as well as the NF-kB target genes, including baculoviral IAP repeat containing 5 (*BIRC5*), nitric oxide synthase 2 (*NOS2*), and selenoprotein P (*SEPP1*), are important regulators of carcinogenic processes in the development of PCa [20–23]. *HIF1A* was strongly expressed in PC3 cells (Figure 3D). mRNA levels of *RELA* and *BIRC5* were comparable among the cell lines (Figure 3E,F). *NOS2* was detected in three cell lines (Figure 3G). The transcript levels of *SEPP1* were the largest in LNCaP and MDA-PCa-2b cells (Figure 3H). Cyclin-dependent kinases and their inhibitors are implicated in mitogenic signaling and were demonstrated to be regulated by androgen signaling [18]. Cyclin-dependent kinase inhibitor 1A (*CDKN1A*) showed strong gene expression in HPEC cells (Figure 3I). The mRNA levels of *CDKN1B* were the highest in LNCaP and MDA-PCa-2b cells (Figure 3J). Glucose transporters regulate crucial tumorigenic pathways, which have particular interest in the term of diabetes [24]. Among the analyzed glucose transporters, solute carrier family 2 member 1 (*GLUT1*) showed the strongest expression in HPEC cells (Figure 3K). The strongest gene expression for solute carrier family 2 member 12 (*GLUT12*) was found in CWR-R1ca cells (Figure 3L).

Ghandi and colleagues recently assessed the transcript pattern of five PCa cell lines using RNA sequencing [25]. The observed differences of gene-expression pattern among the analyzed cells lines in our study resemble these previous data for DU145, LNCaP, NCI-H660, MDA-PCa-2b, and PC3 cells (Table 3, https://www.cbioportal.org/, and [25]).

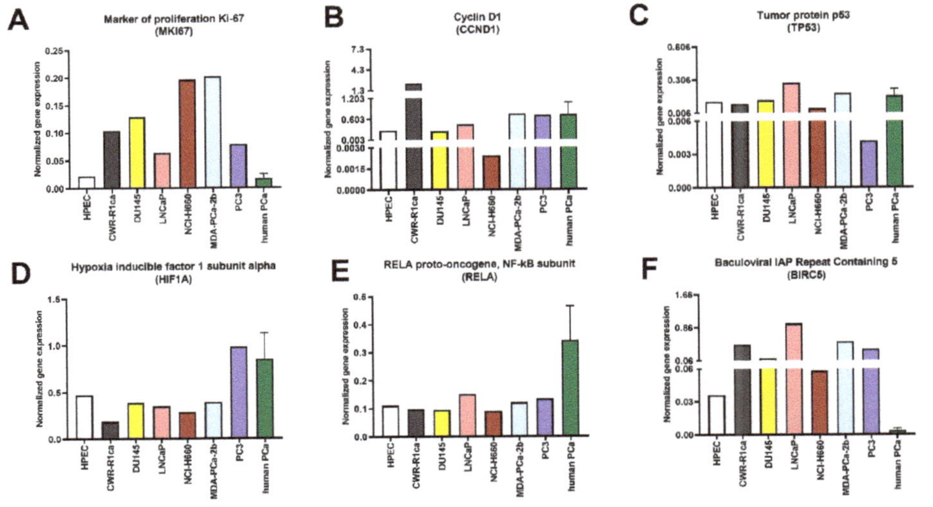

Figure 3. *Cont.*

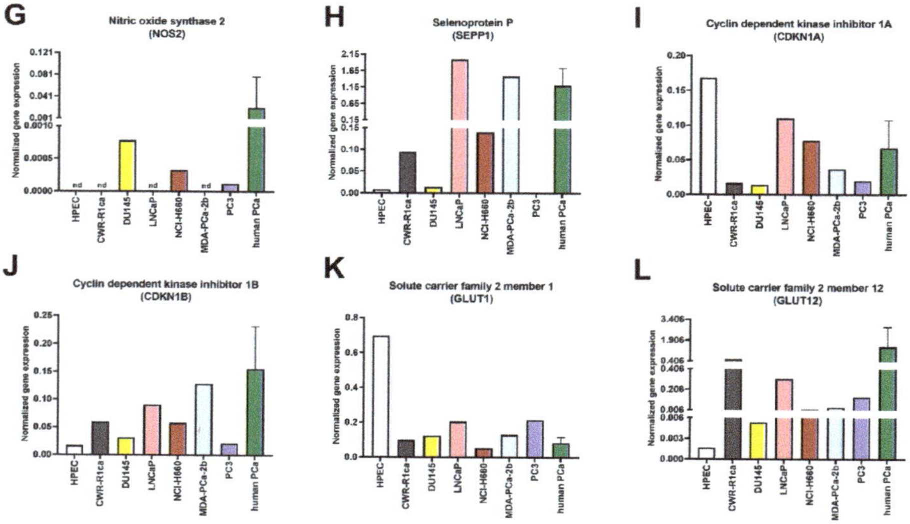

Figure 3. Transcript levels of potential oncogenic mediators in prostate-cancer cell lines and in human prostate-cancer tissue. Transcript levels of indicated genes measured using real-time PCR. (**A**) *MKI67*, (**B**) *CCND1*, (**C**) *TP53*, (**D**) *HIF1A*, (**E**) *RELA*, (**F**) *BIRC5*, (**G**) *NOS2*, (**H**) *SEPP1*, (**I**) *CDKN1A*, (**J**) *CDKN1B*, (**K**) *GLUT1*, (**L**) *GLUT12*. PCa: prostate cancer, HPEC: parental primary prostate cells, CWR-R1ca: xenograft PCa cells, DU145: brain metastasis PCa cells, LNCaP: lymph-node metastasis PCa cells, NCI-H660: lymph-node metastasis PCa cells, MDA-PCa-2b: bone metastasis PCa cells, PC3: bone metastasis PCa cells, nd: not detected. For human PCa samples, data shown as pooled samples: mean ± standard deviation ($n = 11$).

Table 3. RNA sequencing data from https://www.cbioportal.org/ [25].

Cell Line	DU145	LNCaP	NCI-H660	MDA-PCa-2b	PC3
Gene ID					
AR	0.0934	52.35	0.0104	24.77	0.0098
KLK3	0.0152	3066.78	0	3292.29	0
FOLH1	0.0447	473.17	0.0119	308.79	0.1089
INSR*	1.27	5.68	6.41	5.39	1.87
IRS1	3.59	2.50	2.15	5.40	7.05
IRS2	3.44	0.8080	2.80	6.19	0.7902
ESR1	0.0184	0.0015	0.9132	0.0016	0.2013
ESR2	0.1613	0.0867	0.5153	0.0632	0.2337
ESRRA	11.19	27.04	12.65	29.41	19.22
IGF1R	5.41	9.46	6.21	22.71	1.58
FGFR1	13.30	2.68	0.3394	3.32	5.23
CYP27A1	3.42	0.0650	0.1743	0.0264	5.67
CYP7B1	0.0550	0	0.3231	0.0092	0.7905
CYP46A1	0.0446	0.2518	0.1962	0.1349	0.1359
CH25H	0.0455	0.0151	0.0282	0	0.0157
MKI67	58.09	16.15	38.52	26.22	24.39
CCND1	129.71	110.32	0.6017	98.12	232.60
TP53	32.21	39.17	5.85	17.03	1.01
HIF1A	111.95	40.72	54.63	123.32	83.55
RELA	27.04	19.18	20.18	19.53	16.76
BIRC5	6.84	2.61	4.12	2.57	6.09

Table 3. *Cont.*

Cell Line	DU145	LNCaP	NCI-H660	MDA-PCa-2b	PC3
Gene ID					
NOS2	0.5080	0	0.0462	0	0.0081
SEPP1	0.6737	27.18	34.10	44.55	1.32
CDKN1A	18.73	92.39	52.25	40.94	4.49
CDKN1B	17.72	28.15	40.70	32.49	8.18
GLUT1	68.77	19.17	10.53	25.15	63.88
GLUT12	0.5633	26.02	0.9042	6.98	0.6577

Numbers denote median values.* Applied method did not distinguish between INSRA and INSRB isoforms, which were, however, differentiated by our qPCR analysis.

Furthermore, we measured the protein levels for AR, PSMA, IR, IGF1R, and PSA using Western blotting (Figure 4A). Most of the observed changes at the gene-expression level among HPEC, LNCaP, and PC3 cells were mirrored well on the protein level. The gene-expression profiles of CWR-R1ca, DU145, NCI-H660, and MDA-PCa-2b cells are in line with protein-expression data of previous studies [11,12,26–29]. As LNCaP cells showed profound PSA and PSMA expressions at the protein level, we treated these cells with dihydrotestosterone (DHT) after serum deprivation. DHT treatment increased the protein level of PSA (Figure 4B), suggesting that LNCaP cells are partly sensitive for hormone treatment after serum deprivation.

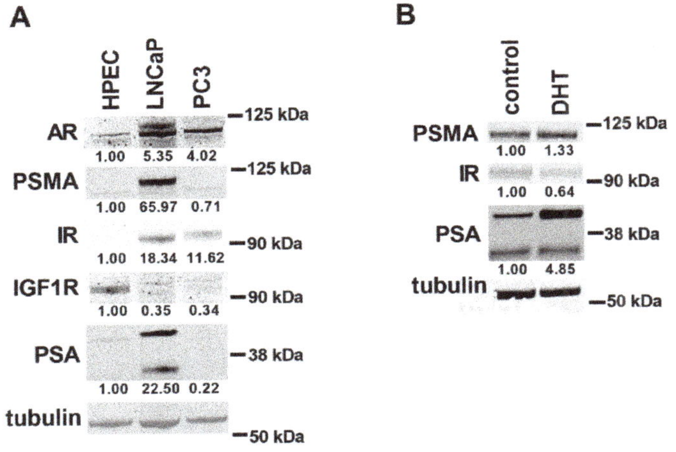

Figure 4. Protein levels of androgen receptor (AR), PSA, prostate-specific membrane antigen (PSMA), insulin receptor (IR), and insulin-like growth factor 1 receptor (IGF1R) in prostate-cancer cell lines. Proteins detected using SDS-PAGE and Western blot. (**A**) For HPEC, LNCaP, and PC3 cell lysates, AR, PSA, PSMA, IR, and IGF1R antibodies were applied. (**B**) For LNCaP cell lysates, PSA, PSMA, and IR antibodies were applied. As loading control, β-tubulin was used. Protein intensities were normalized to (**A**) β-tubulin and HPEC cells or (**B**) control condition. These relative protein intensities are under the pictures. For quantifying PSA intensity, upper protein bands were evaluated. Control: LNCaP cells grown in growth medium; DHT: dihydrotestosterone treatment of LNCaP cells after serum deprivation. Numbers on the right side represent molecular-weight markers.

4. Discussion

All analyzed PCa cell lines had lower insulin-receptor expression levels than those of PCa samples obtained from PCa patients. On the other hand, transcript levels of potential oncogenic mediators and proliferation markers in different prostate-cancer cell lines and in human prostate-cancer tissue have similar expression. These data indicate that certain PCa cell lines can serve as an appropriate model for investigating the main hormone pathways on the transcript level.

Our comprehensive data regarding the mRNA levels of endocrine-signaling components identified major differences among the six human PCa cell lines and the parental nontumorous primary cells. Androgen-receptor signaling is the best-studied pathway in PCa [1,22]. LNCaP and MDA-PCa-2b cells have very high transcript levels for androgen-signaling components; the present data suggest that, out of the six analyzed PCa cell lines, LNCaP and MDA-PCa-2b cells are probably the best choice for studies on androgen signaling (Table 4).

Table 4. Summary table.

Pathway of Interest	Proposed Useful PCa Cell Lines
Androgen signaling	LNCaP, MDA-PCa-2b
Mitogenic insulin signaling	MDA-PCa-2b
Estrogen signaling	NCI-H660, PC3
HIF1A pathway	PC3

In contrast to the androgen-signaling pathway, there is less knowledge about insulin, estrogen, and oxysterol signaling in PCa cell models. A high insulin-receptor isoform A to insulin-receptor isoform B ratio in PCa suggests the prevailing activation of the mitogenic insulin pathway [17,30]. Our analysis indicates that MDA-PCa-2b cells could be highly responsive to mitogenic insulin signaling (Table 4). On the other hand, NCI-H660 and PC3 cells appear to be suitable for analyzing estrogen signaling due to the high expression levels of *ESR1* and *ESR2* (Table 4). Cholesterol derivates, including oxysterols [31], were recently implicated in the regulation of PCa growth by modulating androgen-receptor signaling [9]. Our current data indicate that, on the mRNA level, the major enzymes of oxysterol metabolism show significant differences in the analyzed PCa cell lines.

Furthermore, the activation of HIF1A and NF-kB pathways plays a pivotal role in oncogenic processes [24,32]. In human prostate samples, we observed a positive correlation of the transcript levels of HIF1A and NF-kB pathways to the oncometabolite fumarate [14]. These real-time PCR data indicate that, for investigating HIF1A pathways *in vitro*, PC3 cells could serve as an appropriate model (Table 4). Patients with Type 2 diabetes develop more aggressive PCa [33,34] and show increased risk of PCa mortality [35]; therefore, altered glucose metabolism and the upregulation of glucose transporters could be involved in the progression of PCa in patients with diabetes [7,24]. HPEC cells had high *GLUT1* expression levels, whereas CWR-R1ca cells showed the highest expression for *GLUT12* transcripts, indicating that these cell lines could be used to analyze the involvement of glucose metabolism in cancer progression.

Androgen-deprivation therapy (ADT) is one of the common treatment options for localized PCa [36] and was shown to modify intracellular hormonal metabolism [37]. PCa cell lines represent in vitro models studying the consequences of ADT and hormone stimulus [13]. In LNCaP cells, PSA showed high androgen sensitivity, which is in line with previous studies [38,39]. Nevertheless, ADT has additional effects on PCa cells, as it also stimulates neuroendocrine differentiation in LNCaP cells [40]; thus, the in vitro application of hormone treatment after serum deprivation should be carefully interpreted.

The cell lines in our study were cultivated in various culture conditions, such as different media, hormonal supplements, and FBS concentrations, following suggestions for each cell line from the provider. Nevertheless, these different conditions could have created a bias on the gene expressions that we could not control.

In summary, our systematic characterization gives researchers a solid basis to choose the most appropriate PCa cell-culture model to characterize the hormone pathway of interest (Table 4).

Author Contributions: Conceptualization, S.Z.L. and H.-U.H.; methodology, A.F., L.B., J.H., A.G., K.W., and R.W.; formal analysis, A.F., L.B., and R.W.; resources, H.L., A.S., A.L.B., M.H.d.A., and H.-U.H.; writing—original-draft preparation, A.F.; writing—review and editing, L.B., A.G., J.H., R.W., M.O.S., M.H.d.A., K.W., H.L., A.S., A.L.B., A.P., H.-U.H., S.Z.L., and M.H.; All authors have read and agreed to the published version of the manuscript.

Funding: The study was supported in part by a grant from the German Federal Ministry of Education and Research (BMBF) to the German Center for Diabetes Research (DZD e.V.).

Conflicts of Interest: The authors declare no conflict of interest.

References

1. Grossmann, M.; Wittert, G. Androgens, diabetes and prostate cancer. *Endocr. Relat. Cancer* **2012**, *19*, F47–F62. [CrossRef]
2. Elia, I.; Schmieder, R.; Christen, S.; Fendt, S.M. Organ-Specific Cancer Metabolism and Its Potential for Therapy. *Handb. Exp. Pharmacol.* **2016**, *233*, 321–353. [CrossRef] [PubMed]
3. Bushman, W. Hedgehog Signaling in Prostate Development, Regeneration and Cancer. *J. Dev. Biol.* **2016**, *4*, 30. [CrossRef] [PubMed]
4. Ridge, S.M.; Sullivan, F.J.; Glynn, S.A. Mesenchymal stem cells: Key players in cancer progression. *Mol. Cancer* **2017**, *16*, 31. [CrossRef] [PubMed]
5. Ishii, K.; Takahashi, S.; Sugimura, Y.; Watanabe, M. Role of Stromal Paracrine Signals in Proliferative Diseases of the Aging Human Prostate. *J. Clin. Med.* **2018**, *7*, 68. [CrossRef] [PubMed]
6. Hensel, J.; Thalmann, G.N. Biology of Bone Metastases in Prostate Cancer. *Urology* **2016**, *92*, 6–13. [CrossRef]
7. Di Sebastiano, K.M.; Pinthus, J.H.; Duivenvoorden, W.C.M.; Mourtzakis, M. Glucose impairments and insulin resistance in prostate cancer: The role of obesity, nutrition and exercise. *Obes. Rev.* **2018**, *19*, 1008–1016. [CrossRef]
8. Franko, A.; Berti, L.; Hennenlotter, J.; Rausch, S.; Scharpf, M.; De Angelis, M.H.; Stenzl, A.; Birkenfeld, A.L.; Peter, A.; Lutz, S.Z.; et al. Prostate tissue of patients with type 2 diabetes is characterized by increased gene expressions of Matrix metalloproteinases (MMPs) and CC chemokine ligands (CCLs). **2020**. In Preparation.
9. Lutz, S.Z.; Hennenlotter, J.; Scharpf, M.O.; Sailer, C.; Fritsche, L.; Schmid, V.; Kantartzis, K.; Wagner, R.; Lehmann, R.; Berti, L.; et al. Androgen receptor overexpression in prostate cancer in type 2 diabetes. *Mol. Metab.* **2018**, *8*, 158–166. [CrossRef]
10. Nelles, J.L.; Hu, W.-Y.; Prins, G.S. Estrogen action and prostate cancer. *Expert Rev. Endocrinol. Metab.* **2011**, *6*, 437–451. [CrossRef]
11. Wu, X.; Gong, S.; Roy-Burman, P.; Lee, P.; Culig, Z. Current mouse and cell models in prostate cancer research. *Endocr. Relat. Cancer* **2013**, *20*, R155–R170. [CrossRef] [PubMed]
12. Sobel, R.E.; Sadar, M.D. Cell Lines Used in Prostate Cancer Research: A Compendium of Old and New Lines—Part 1. *J. Urol.* **2005**, *173*, 342–359. [CrossRef] [PubMed]
13. Namekawa, T.; Ikeda, K.; Horie-Inoue, K.; Inoue, S. Application of Prostate Cancer Models for Preclinical Study: Advantages and Limitations of Cell Lines, Patient-Derived Xenografts, and Three-Dimensional Culture of Patient-Derived Cells. *Cells* **2019**, *8*, 74. [CrossRef]
14. Franko, A.; Shao, Y.; Heni, M.; Hennenlotter, J.; Hoene, M.; Hu, C.; Liu, X.; Zhao, X.; Wang, Q.; Birkenfeld, A.L.; et al. Human Prostate Cancer is Characterized by an Increase in Urea Cycle Metabolites. *Cancers* **2020**, *12*, 1814. [CrossRef] [PubMed]
15. Franko, A.; Berti, L.; Hennenlotter, J.; Rausch, S.; Scharpf, M.O.; De Angelis, M.H.; Stenzl, A.; Birkenfeld, A.L.; Peter, A.; Lutz, S.Z.; et al. Transcript Levels of Aldo-Keto Reductase Family 1 Subfamily C (AKR1C) Are Increased in Prostate Tissue of Patients with Type 2 Diabetes. *J. Pers. Med.* **2020**, *10*, 124. [CrossRef] [PubMed]
16. Franko, A.; Kovarova, M.; Feil, S.; Feil, R.; Wagner, R.; Heni, M.; Königsrainer, A.; Ruoss, M.; Nussler, A.K.; Weigert, C.; et al. cGMP-dependent protein kinase I (cGKI) modulates human hepatic stellate cell activation. *Metabolism* **2018**, *88*, 22–30. [CrossRef]
17. Heni, M.; Hennenlotter, J.; Scharpf, M.; Lutz, S.Z.; Schwentner, C.; Todenhöfer, T.; Schilling, D.; Kühs, U.; Gerber, V.; Machicao, F.; et al. Insulin Receptor Isoforms A and B as well as Insulin Receptor Substrates-1 and -2 Are Differentially Expressed in Prostate Cancer. *PLoS ONE* **2012**, *7*, e50953. [CrossRef]
18. Balk, S.P.; Knudsen, K.E. AR, the cell cycle, and prostate cancer. *Nucl. Recept. Signal.* **2008**, *6*. [CrossRef]

19. Bjartell, A.; Montironi, R.; Berney, D.M.; Egevad, L. Tumour markers in prostate cancer II: Diagnostic and prognostic cellular biomarkers. *Acta Oncol.* **2011**, *50* (Suppl. 1), 76–84. [CrossRef]
20. Debelec-Butuner, B.; Alapinar, C.; Varisli, L.; Erbaykent-Tepedelen, B.; Hamid, S.M.; Gonen-Korkmaz, C.; Korkmaz, K.S. Inflammation-mediated abrogation of androgen signaling: An in vitro model of prostate cell inflammation. *Mol. Carcinog.* **2014**, *53*, 85–97. [CrossRef]
21. McIlwain, D.W.; Fishel, M.L.; Boos, A.; Kelley, M.R.; Jerde, T.J. APE1/Ref-1 redox-specific inhibition decreases survivin protein levels and induces cell cycle arrest in prostate cancer cells. *Oncotarget* **2018**, *9*, 10962–10977. [CrossRef] [PubMed]
22. Russo, M.A.; Ravenna, L.; Pellegrini, L.; Petrangeli, E.; Salvatori, L.; Magrone, T.; Fini, M.; Tafani, M. Hypoxia and Inflammation in Prostate Cancer Progression. Cross-talk with Androgen and Estrogen Receptors and Cancer Stem Cells. *Endocr. Metab. Immune Disord. Drug Targets* **2016**, *16*, 235–248. [CrossRef] [PubMed]
23. Sciacovelli, M.; Frezza, C. Oncometabolites: Unconventional triggers of oncogenic signalling cascades. *Free. Radic. Biol. Med.* **2016**, *100*, 175–181. [CrossRef] [PubMed]
24. Ramteke, P.; Deb, A.; Shepal, V.; Bhat, M.K. Hyperglycemia Associated Metabolic and Molecular Alterations in Cancer Risk, Progression, Treatment, and Mortality. *Cancers* **2019**, *11*, 1402. [CrossRef] [PubMed]
25. Ghandi, M.; Huang, F.W.; Jané-Valbuena, J.; Kryukov, G.V.; Lo, C.C.; McDonald, E.R., 3rd; Barretina, J.; Gelfand, E.T.; Bielski, C.M.; Li, H.; et al. Next-generation characterization of the Cancer Cell Line Encyclopedia. *Nature* **2019**, *569*, 503–508. [CrossRef]
26. Shourideh, M.; DePriest, A.; Mohler, J.L.; Wilson, E.M.; Koochekpour, S. Characterization of fibroblast-free CWR-R1ca castration-recurrent prostate cancer cell line. *Prostate* **2016**, *76*, 1067–1077. [CrossRef]
27. Hsieh, T.C.; Wu, J.M. Cell growth and gene modulatory activities of Yunzhi (Windsor Wunxi) from mushroom Trametes versicolor in androgen-dependent and androgen-insensitive human prostate cancer cells. *Int. J. Oncol.* **2001**, *18*, 81–88. [CrossRef]
28. Li, Y.; Chen, R.; Bowden, M.; Mo, F.; Lin, Y.Y.; Gleave, M.; Collins, C.; Dong, X. Establishment of a neuroendocrine prostate cancer model driven by the RNA splicing factor SRRM4. *Oncotarget* **2017**, *8*, 66878–66888. [CrossRef]
29. Thomas-Jardin, S.E.; Kanchwala, M.S.; Jacob, J.; Merchant, S.; Meade, R.K.; Gahnim, N.M.; Nawas, A.F.; Xing, C.; Delk, N.A. Identification of an IL-1-induced gene expression pattern in AR(+)PCa cells that mimics the molecular phenotype of AR(−)PCa cells. *Prostate* **2018**, *78*, 595–606. [CrossRef]
30. Belfiore, A.; Malaguarnera, R.; Vella, V.; Lawrence, M.C.; Sciacca, L.; Frasca, F.; Morrione, A.; Vigneri, R. Insulin Receptor Isoforms in Physiology and Disease: An Updated View. *Endocr. Rev.* **2017**, *38*, 379–431. [CrossRef]
31. Alfaqih, M.A.; Nelson, E.R.; Liu, W.; Safi, R.; Jasper, J.S.; Macias, E.; Geradts, J.; Thompson, J.W.; Dubois, L.G.; Freeman, M.R.; et al. CYP27A1 Loss Dysregulates Cholesterol Homeostasis in Prostate Cancer. *Cancer Res.* **2017**, *77*, 1662–1673. [CrossRef] [PubMed]
32. Kimbro, K.S.; Simons, J.W. Hypoxia-inducible factor-1 in human breast and prostate cancer. *Endocr. Relat. Cancer* **2006**, *13*, 739–749. [CrossRef] [PubMed]
33. Lutz, S.Z.; Todenhöfer, T.; Wagner, R.; Hennenlotter, J.; Ferchl, J.M.; Scharpf, M.O.; Martus, P.; Staiger, H.; Fritsche, A.; Stenzl, A.; et al. Higher prevalence of lymph node metastasis in prostate cancer in patients with diabetes. *Endocr. Relat. Cancer* **2018**, *25*, L19–L22. [CrossRef] [PubMed]
34. Kim, H.S.; Presti, J.C., Jr.; Aronson, W.J.; Terris, M.K.; Kane, C.J.; Amling, C.L.; Freedland, S.J. Glycemic control and prostate cancer progression: Results from the SEARCH database. *Prostate* **2010**, *70*, 1540–1546. [CrossRef] [PubMed]
35. Bensimon, L.; Yin, H.; Suissa, S.; Pollak, M.N.; Azoulay, L. Type 2 diabetes and the risk of mortality among patients with prostate cancer. *Cancer Causes Control.* **2014**, *25*, 329–338. [CrossRef]
36. Jhan, J.H.; Yeh, H.C.; Chang, Y.H.; Guu, S.J.; Wu, W.J.; Chou, Y.H.; Li, C.C. New-onset diabetes after androgen-deprivation therapy for prostate cancer: A nationwide propensity score-matched four-year longitudinal cohort study. *J. Diabetes Complicat.* **2018**, *32*, 688–692. [CrossRef]
37. Shiota, M.; Fujimoto, N.; Kashiwagi, E.; Eto, M. The Role of Nuclear Receptors in Prostate Cancer. *Cells* **2019**, *8*, 602. [CrossRef]
38. Zhao, H.; Ma, T.F.; Lin, J.; Liu, L.L.; Sun, W.J.; Guo, L.X.; Wang, S.Q.; Otecko, N.O.; Zhang, Y.P. Identification of valid reference genes for mRNA and microRNA normalisation in prostate cancer cell lines. *Sci. Rep.* **2018**, *8*, 1949. [CrossRef]

39. Olsen, J.R.; Azeem, W.; Hellem, M.R.; Marvyin, K.; Hua, Y.; Qu, Y.; Li, L.; Lin, B.; Ke, X.; Øyan, A.M.; et al. Context dependent regulatory patterns of the androgen receptor and androgen receptor target genes. *BMC Cancer* **2016**, *16*, 377. [CrossRef]
40. Yuan, T.C.; Veeramani, S.; Lin, F.F.; Kondrikou, D.; Zelivianski, S.; Igawa, T.; Karan, D.; Batra, S.K.; Lin, M.F. Androgen deprivation induces human prostate epithelial neuroendocrine differentiation of androgen-sensitive LNCaP cells. *Endocr. Relat. Cancer* **2006**, *13*, 151–167. [CrossRef]

genes

MDPI

Article

Transcriptomic Profiling Identifies Differentially Expressed Genes in Palbociclib-Resistant ER+ MCF7 Breast Cancer Cells

Lilibeth Lanceta [1], Conor O'Neill [2], Nadiia Lypova [1], Xiahong Li [3], Eric Rouchka [4], Sabine Waigel [1], Jorge G. Gomez-Gutierrez [5,6], Jason Chesney [1,6] and Yoannis Imbert-Fernandez [1,6,*]

1 Department of Medicine, School of Medicine, University of Louisville, Louisville, KY 40202, USA; Lilibeth.lanceta@louisville.edu (L.L.); nadiia.lypova@louisville.edu (N.L.); sabine.waigel@louisville.edu (S.W.); jason.chesney@louisville.edu (J.C.)
2 College of Medicine, University of Kentucky, Lexington, KY 40506, USA; cson222@uky.edu
3 Department of Anatomical Sciences and Neurobiology, Bioinformatics Core, University of Louisville, Louisville, KY 40202, USA; xiaohong.li@louisville.edu
4 Department of Computer Engineering and Computer Science, University of Louisville, Louisville, KY 40292, USA; eric.rouchka@louisville.edu
5 Department of Surgery, School of Medicine, University of Louisville, Louisville, KY 40202, USA; jgguti01@louisville.edu
6 James Graham Brown Cancer Center School of Medicine, University of Louisville, Louisville, KY 40202, USA
* Correspondence: yoannis.imbertfernandez@louisville.edu; Tel.: +1-502-852-6570

Received: 20 March 2020; Accepted: 18 April 2020; Published: 24 April 2020

Abstract: Acquired resistance to cyclin-dependent kinases 4 and 6 (CDK4/6) inhibition in estrogen receptor-positive (ER+) breast cancer remains a significant clinical challenge. Efforts to uncover the mechanisms underlying resistance are needed to establish clinically actionable targets effective against resistant tumors. In this study, we sought to identify differentially expressed genes (DEGs) associated with acquired resistance to palbociclib in ER+ breast cancer. We performed next-generation transcriptomic RNA sequencing (RNA-seq) and pathway analysis in ER+ MCF7 palbociclib-sensitive (MCF7/pS) and MCF7 palbociclib-resistant (MCF7/pR) cells. We identified 2183 up-regulated and 1548 down-regulated transcripts in MCF7/pR compared to MCF7/pS cells. Functional analysis of the DEGs using Gene Ontology (GO) and the Kyoto Encyclopedia of Genes and Genomes (KEGG) database identified several pathways associated with breast cancer, including 'cell cycle', 'DNA replication', 'DNA repair' and 'autophagy'. Additionally, Ingenuity Pathway Analysis (IPA) revealed that resistance to palbociclib is closely associated with deregulation of several key canonical and metabolic pathways. Further studies are needed to determine the utility of these DEGs and pathways as therapeutics targets against ER+ palbociclib-resistant breast cancer.

Keywords: palbociclib; estrogen receptor; breast cancer; CDK4/6; CDK4/6 inhibitors; therapy resistance; DNA repair; metabolic rewiring

1. Introduction

Breast cancer is the most frequent malignancy among women, and approximately 60–70% of cases are estrogen receptor-positive (ER+). Selective inhibition of cyclin-dependent kinases 4 and 6 (CDK4/6) and ER signaling is now standard-of-care therapy for ER+ metastatic breast cancer [1]. Three CDK4/6 inhibitors, palbociclib, ribociclib and abemaciclib, are currently used in combination with endocrine therapy given their shown improvement in progression-free survival compared to endocrine therapy alone in the metastatic setting [2]. Despite the clear benefit of this combination, approximately 10% of patients remain insensitive, whereas nearly all patients become resistant within 12 to 36 months of therapy initiation [3]. Therefore, determining the underlying mechanisms of resistance is required to design novel treatment strategies that delay or overcome clinical resistance.

Previous studies have shown that resistance to palbociclib is commonly associated with cyclin E or CDK6 amplification, CDK2 activation and loss of the retinoblastoma (Rb) protein in ER+ breast cancer cells [4–6]. Analysis of circulating tumor DNA from patients enrolled in the PALOMA-3 trial (fulvestrant or fulvestrant + palbociclib) identified an enrichment of Rb mutations, although this only occurred in 4.5% of the palbociclib-treated cohort [7,8]. Importantly, acquired alterations in *ESR* and *PIK3CA* were also observed; however, these alterations occurred in both treatment arms indicating distinct events driving resistance to palbociclib versus fulvestrant [9]. Additional studies have implicated fibroblast growth factor receptor (FGFR) or aurora kinase A amplifications, enhanced MAPK or AKT signaling and decreased DNA repair as mechanisms of resistance against CDK4/6 inhibition [10–13]. Taken together, these studies have provided rationale for the testing of CDK4/6 inhibitors in combination with MEK or PI3K inhibitors [11,14].

The major goal of this study was to identify additional mechanisms of resistance to palbociclib in ER+ breast cancer cells through transcriptomic analyses. We previously demonstrated that ER+ palbociclib-resistant cells exhibit a marked decrease in the cellular antiviral interferon (IFN) response [6], and thus we expected that other drivers of resistance remained to be identified. Here, we determined the transcriptional landscape of ER+ MCF7 palbociclib-sensitive (MCF7/pS) and palbociclib-resistant (MCF7/pR) breast cancer cells via next-generation transcriptomic RNA sequencing (RNA-seq). Gene expression profile and pathway analysis identified significant canonical pathways associated with resistance to palbociclib including cell cycle regulation, immune responses and DNA damage repair (DDR) among others. Importantly, we identified several metabolic pathways uniquely enriched in palbociclib-resistant cells compared to palbociclib-sensitive cells. These studies provide a mechanistic base for the further validation of these pathways in mediating resistance to palbociclib.

2. Materials and Methods

2.1. Cell Culture, Generation of Palbociclib-Resistant Cells and Palbociclib Treatment

MCF7 (HTB-22) cells were purchased from the American Type Culture Collection (ATCC) and maintained at 37 °C with 5% CO_2. MCF7 cells were cultured in IMEM (Corning) supplemented with 10% fetal bovine serum (FBS, Invitrogen). Drug-resistant MCF7 cells were established by culturing in media containing palbociclib (0.1–4 μM). Drug was replenished every 3 days. Cells were subcultured every 1–2 weeks with 25% increments in drug concentration. The resistant cells were established after 6 months and maintained in the presence of 1 μM palbociclib. Cells were authenticated by the short tandem repeat (STR) assay (Genetica).

2.2. RNA Extraction and Next-Generation Sequencing

MCF7/pS and MCF7/pR cells were seeded in 10 cm^2 dishes at a density of 2×10^6 cells and allowed to incubate overnight prior to RNA extraction using the RNeasy kit (Qiagen) for a total of three independent replicates per cell line. Libraries were prepared simultaneously for all replicates and cell lines using the TruSeq Stranded mRNA LT Sample Prep Kit - Set A (Cat# RS-122-2101) with poly-A enrichment. Sequencing was performed on the University of Louisville Center for Genetics and Molecular Medicine's (CGeMM) Illumina NextSeq 500 using the NextSeq 500/550 1 × 75 cycle High Output Kit v2 (Cat# FC-404-2005). A second run was performed on all samples to achieve an average of 45 million reads per sample.

2.3. DEG Analysis

The resulting samples were downloaded from Illumina's BaseSpace [15] (https://basespace.illumina.com/). Sequences were directly aligned to the Homo sapiens hg38 reference genome assembly (hg38.fa) using tophat2 (version 2.0.13), generating alignment files in bam format. DEGs were identified for the pairwise comparison MCF7/pS versus MCF7/pR using the tuxedo suite programs including cufflinks-cuffdiff2 (VERSION2.2.1). A total of 60,603 ENSEMBL genes were considered. Of these, 26,837 showed no gene expression and were excluded. A q-value cutoff ≤ 0.05 with $|\log_2 FC| \geq 1$ and gene expression greater than 1 in at least one replicate was used to determine differential expression. RNA-seq data are available (GEO accession number GSE130437). Gene Ontology Biological Processes (GO:BP) and KEGG pathway analysis was performed by using CategoryCompare [16].

2.4. In Silico Ingenuity Network Analysis

Pathway and biological processes analysis of all differentially expressed genes was performed using Ingenuity Pathway Analysis (Qiagen).

2.5. GFP-LC3 Visualization

Plasmid vector containing green fluorescent protein linked to microtubule-associated protein 1 LC3 was used to detect autophagosome formation in MCF7/pS and MCF7/pR cell lines [17]. Cells were treated with either vehicle control or palbociclib after 24 h of transfection. The expression of GFP was monitored by fluorescence microscopy 48 h after treatment. Cells were classified as having a predominantly diffuse GFP stain or having numerous punctate structures representing autophagosomes. Images were taken at 40× magnification with the EVOS FL Imaging System (Thermo Fisher Scientific, Waltham, MA, USA) under 357/44 and 447/60 nanometers (nm) excitation and emission visualization, respectively.

3. Results

3.1. RNA-Seq Profiling Reveals a Distinct Transcriptomic Profiling in Palbociclib-Resistant Cells

To characterize transcriptional alterations driven by acquired resistance to palbociclib, we performed gene expression profiling in MCF7/pS and MCF7/pR cells. These cells were developed by our group and were previously shown to be resistant to palbociclib [6]. Hierarchical clustering based on differentially expressed RNA transcripts revealed a distinct transcriptomic profile in MCF7/pR cells compared to MCF7/pS (Figure 1). Using a q-value cutoff ≤ 0.05 with $|\log_2 FC| \geq 1$, we identified 2183 up-regulated genes and 1548 down-regulated transcripts in MCF7/pR cells. Table 1 shows the top 20 up-regulated and down-regulated genes in MCF7/pR cells compared to MCF7/pS cells.

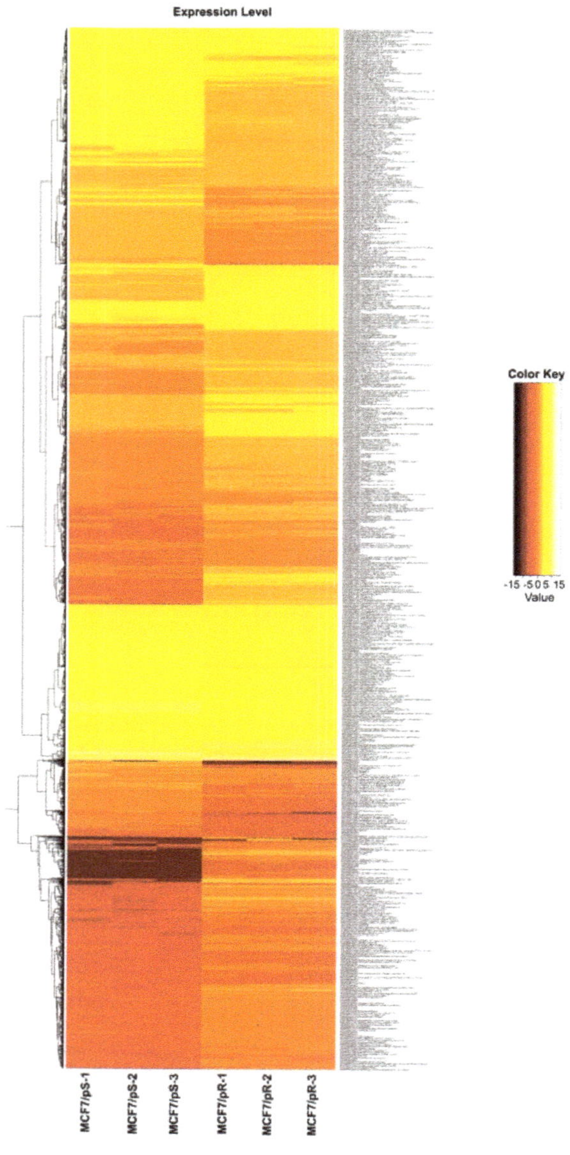

Figure 1. Differential expression heatmap of estrogen receptor-positive (ER+) MCF7 palbociclib-sensitive (MCF7/pS) compared to MCF7 palbociclib-resistant (MCF7/pR) cells. Next-generation transcriptomic RNA sequencing (RNA-seq) was performed and the raw expression of genes is shown as a heatmap. Replicate samples are clustered. Red and yellow indicate lower and higher gene expression, respectively.

Table 1. Top 20 up-regulated and down-regulated genes between MCF7/pS and MCF7/pR ranked by p-value (pval ≤ 0.05; qval ≤ 0.05; $|\log_2 FC| \geq 1$).

Ensembl ID	Up-Regulated Gene Symbol|Description	Log_2FC	p-Value	Q-value	Ensembl ID	Down-Regulated Gene Symbol|Description	Log_2FC	p-Value	Q-Value
ENSG00000011465	DCN|decorin	18.6778	5×10^{-5}	0.00026	ENSG00000211816	TRAV38-1|	−16.6898	5×10^{-5}	0.00026
ENSG00000011677	GABRA3|gamma-aminobutyric acid (GABA) A receptor, alpha 3	18.6778	5×10^{-5}	0.00026	ENSG00000228340	MIR646HG|MIR646 host gene	−16.6898	5×10^{-5}	0.00026
ENSG00000012223	LTF|lactotransferrin	18.6778	5×10^{-5}	0.00026	ENSG00000259761		−16.6898	5×10^{-5}	0.00026
ENSG00000070729	CNGB1|cyclic nucleotide gated channel beta 1	18.6778	5×10^{-5}	0.00026	ENSG00000082482	KCNK2|potassium channel, two pore domain subfamily K, member 2	−7.09413	5×10^{-5}	0.00026
ENSG00000079689	SCGN|secretagogin, EF-hand calcium binding protein	18.6778	5×10^{-5}	0.00026	ENSG00000182836	PLCXD3|phosphatidylinositol-specific phospholipase C, X domain containing 3	−6.53477	5×10^{-5}	0.00026
ENSG00000083782	EPYC|epiphycan	18.6778	5×10^{-5}	0.00026	ENSG00000253293	HOXA10|homeobox A10	−6.24454	5×10^{-5}	0.00026
ENSG00000086205	FOLH1|folate hydrolase (prostate-specific membrane antigen) 1///FOLH1|folate hydrolase 1B	18.6778	5×10^{-5}	0.00026	ENSG00000259527	LINC00052|long intergenic non-protein coding RNA 52	−6.17543	5×10^{-5}	0.00026
ENSG00000101441	CST4|cystatin S	18.6778	5×10^{-5}	0.00026	ENSG00000006747	SCIN|scinderin	−5.91229	5×10^{-5}	0.00026
ENSG00000101460	MAP1LC3A|microtubule-associated protein 1 light chain 3 alpha	18.6778	5×10^{-5}	0.00026	ENSG00000110492	MDK|midkine (neurite growth-promoting factor 2)	−5.60036	5×10^{-5}	0.00026
ENSG00000102854	MSLN|mesothelin	18.6778	5×10^{-5}	0.00026	ENSG00000140538	NTRK3|neurotrophic tyrosine kinase, receptor, type 3	−5.39223	5×10^{-5}	0.00026
ENSG00000104313	EYA1|EYA transcriptional coactivator and phosphatase 1	18.6778	5×10^{-5}	0.00026	ENSG00000102271	KLHL4|kelch-like family member 4	−5.2236	5×10^{-5}	0.00026
ENSG00000109846	CRYAB|crystallin, alpha B	18.6778	5×10^{-5}	0.00026	ENSG00000105974	CAV1|caveolin 1, caveolae protein, 22kDa	−5.18256	5×10^{-5}	0.00026
ENSG00000111863	ADTRP|androgen-dependent TFPI-regulating protein	18.6778	5×10^{-5}	0.00026	ENSG00000112175	BMP5|bone morphogenetic protein 5	−5.08812	5×10^{-5}	0.00026
ENSG00000113805	CNTN3|contactin 3 (plasmacytoma associated)	18.6778	5×10^{-5}	0.00026	ENSG00000159618	ADGRG5|adhesion G protein-coupled receptor G5	−4.97744	5×10^{-5}	0.00026
ENSG00000115009	CCL20|chemokine (C-C motif) ligand 20	18.6778	5×10^{-5}	0.00026	ENSG00000174498	IGDCC3|immunoglobulin superfamily, DCC subclass, member 3	−4.96488	5×10^{-5}	0.00026
ENSG00000125788	DEFB126|defensin, beta 126	18.6778	5×10^{-5}	0.00026	ENSG00000168269	FOXI1|forkhead box I1	−4.95038	5×10^{-5}	0.00026
ENSG00000131095	GFAP|glial fibrillary acidic protein	18.6778	5×10^{-5}	0.00026	ENSG00000087077	TRIP6|thyroid hormone receptor interactor 6	−4.80366	5×10^{-5}	0.00026
ENSG00000139329	LUM|lumican	18.6778	5×10^{-5}	0.00026	ENSG00000134193	REG4|regenerating islet-derived family, member 4	−4.705	5×10^{-5}	0.00026
ENSG00000140287	HDC|histidine decarboxylase	18.6778	5×10^{-5}	0.00026	ENSG00000003436	TFPI|tissue factor pathway inhibitor (lipoprotein-associated coagulation inhibitor)	−4.62227	5×10^{-5}	0.00026
ENSG00000146233	CYP39A1|cytochrome P450, family 39, subfamily A, polypeptide 1	18.6778	5×10^{-5}	0.00026	ENSG00000064205	WISP2|WNT1 inducible signaling pathway protein 2	−4.57396	5×10^{-5}	0.00026

3.2. KEGG Annotation of DEG and Enriched Biological Processes Analysis

To gain insight into the molecular mechanisms underlying palbociclib resistance, we performed KEGG pathway analysis of all DEGs identified using CategoryCompare [16]. Table 2 lists the enriched KEGG pathways identified in MCF7/pS vs. MCF7/pR cells (false discovery rate (FDR) ≤ 0.05 and p-value ≤ 0.001). The KEGG terms associated with resistance to palbociclib included 'cell cycle', 'DNA replication', 'mismatch repair' and 'phagosome'. Subsequent analysis of GO:BP identified many enriched biological processes that correlated with palbociclib resistance (Figure 2). Importantly, we observed distinct groups of nodes such as DNA replication, cell cycle transition, mitosis, protein–DNA assembly and organization and response to virus revealing multiple functional 'themes' associated with resistance to palbociclib.

Table 2. Top enriched KEGG terms between MCF7/pS and MCF7/pR ranked by p-value. (pval ≤ 0.05; qval ≤ 0.05; $|\log_2 FC| \geq 1$).

KEGG ID	Description	p-value	FDR
4110	Cell cycle	8.73×10^{-8}	0
3030	DNA replication	1.20×10^{-7}	0
4114	Oocyte meiosis	6.34×10^{-5}	0
4914	Progesterone-mediated oocyte maturation	8.36×10^{-5}	0
4360	Axon guidance	8.76×10^{-5}	0
3430	Mismatch repair	5.20×10^{-4}	0.01
4115	p53 signaling pathway	1.11×10^{-3}	0.01143
4010	MAPK signaling pathway	2.91×10^{-3}	0.03333
600	Sphingolipid metabolism	3.88×10^{-3}	0.044
4145	Phagosome	4.78×10^{-3}	0.05

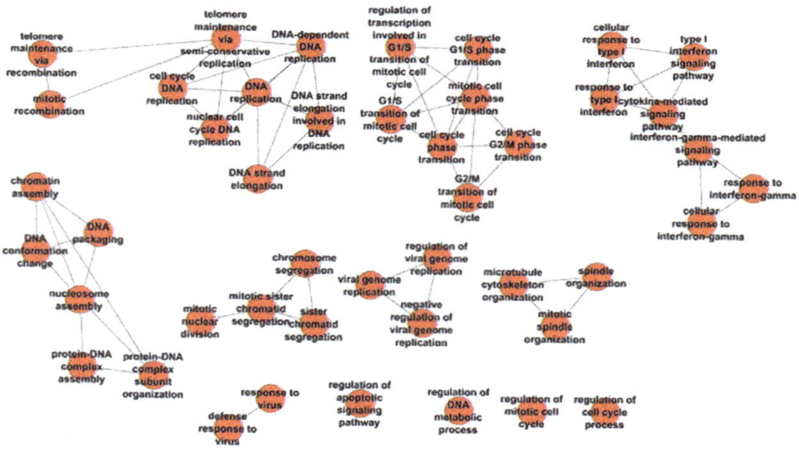

Figure 2. Enriched biological processes (BP) analysis of ER+ palbociclib-resistant breast cancer cells.

3.3. Resistance to Palbociclib Is Associated with Increased Autophagosome Formation

Characterization of MCF7/pR cells by KEGG pathway analysis revealed an enrichment in genes associated with phagosomes (Table 2). Given a previous observation suggesting a crosstalk between phagocytosis and autophagy, we sought to investigate autophagy levels in the context of resistance to palbociclib [18]. We performed hierarchical clustering of autophagy-related genes in MCF7/pS and MCF7/pR cells (Figure 3A). Using a p-value cutoff ≤ 0.05, we identified a significant number of autophagy-related genes differentially expressed in MCF7/pR compared to MCF7/pS cells. Next, we measured autophagosome formation by monitoring the conversion of cytoplasm-diffuse GFP-LC3-I to punctate forms of membrane-associated GFP-LC3-II, which indicates LC3-II incorporation into the autophagosomes. We observed that MCF7/pR cells displayed a significant increase in autophagosome formation compared to MCF7/pS and that the addition of palbociclib led to a marked increase in autophagosome formation in both MCF7/pS and MCF7/pR cells (Figure 3B). These results confirmed an increase in autophagy in MCF7/pR cells and are in line with previous studies [19]. Numerous studies have demonstrated that autophagy contributes to the resistance of breast cancer cells to targeted therapies by promoting tumor cell survival and blocking apoptosis [20–22]. Recently, it has been shown that autophagy inhibitors synergize with palbociclib in ER+ MCF7 and T47D breast cancer cells resulting in a significant increase in growth inhibition [19]. Our results provide rationale for the use of autophagy inhibitors to treat palbociclib-resistant cells in addition to palbociclib-sensitive cells. Future studies will test the efficacy of this combination against CDK4/6 inhibition in the resistance setting and determine the molecular mechanisms driving the increase in autophagy in resistant tumors.

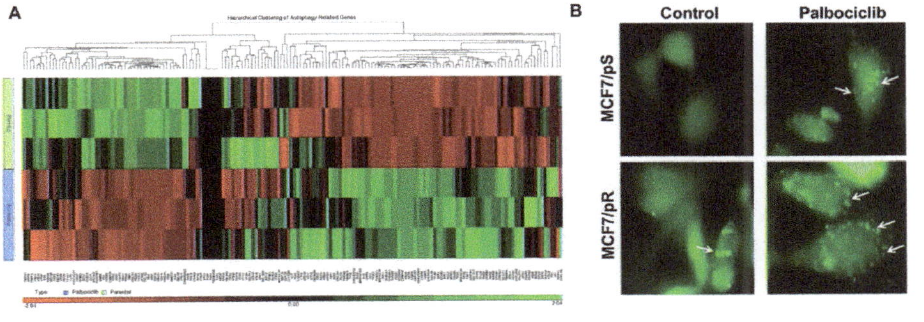

Figure 3. Increased autophagy is associated with palbociclib resistance in ER+ MCF7 cells. (**A**) Hierarchical clustering of autophagy-related genes performed by MetaCore analysis. (**B**) Cells were transfected with a pEGFP-LC3 plasmid and treated with either vehicle control (0.5% water) or 500 nM palbociclib for 24 h. Formation of autophagosomes is depicted by punctate structures (arrows). Images were taken at 40× magnification with an EVOS microscope.

3.4. Pathway Enrichment Analysis of DEG

To identify potential targetable pathways, all altered transcripts were mapped to known pathways using Ingenuity Pathway Analysis (IPA). We observed significant enrichment of several canonical pathways including four pathways involved in cell cycle regulation ('Estrogen-mediated S-phase entry', 'Cell cycle control of chromosomal replication', 'Mitotic roles of Polo-Like Kinase' and 'Role of CHK proteins in cell cycle checkpoint control'), four involved in DDR ('ATM signaling', 'Role of BRCA in DNA damage response', 'Mismatch repair in eukaryotes' and 'G2/M DNA damage checkpoint regulation'), eight involved in immune responses ('IL-17A signaling', 'Interferon signaling', 'STAT3 pathway', 'April mediated signaling', 'Tec Kinase signaling', 'Antigen presentation pathway', 'Production of nitric oxide and reactive oxygen species in macrophages' and 'IL-15 production') among other pathways (Figure 4).

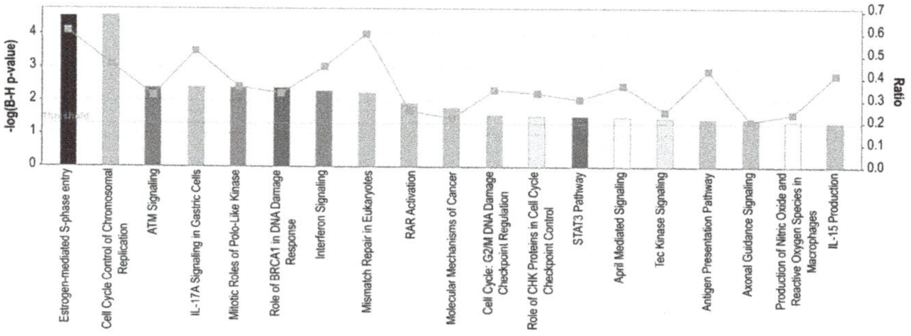

Figure 4. Canonical pathway analysis of ER+ palbociclib-resistant breast cancer cells. A higher–log(B-H p-value) shown on the left Y axis represents more significant pathways. The ratio (right Y axis) refers to the number of genes from the data set that map to the pathway divided by the total number of genes that map the canonical pathway from the Ingenuity Pathway Analysis (IPA) database. pval ≤ 0.05; qval ≤ 0.05; $|\log_2 FC| \geq 1$.

3.5. Metabolic Pathways Associated with Resistance to Palbociclib

Previous reports have indicated that cellular metabolism is a downstream target of CDK4/6 inhibition. Specifically, it has been shown that palbociclib administration increases glucose utilization in cancer, whereas cyclin D3-CDK6 can directly phosphorylate and inhibit the activity of two key enzymes in the glycolytic pathway [23,24]. To identify metabolic pathways associated with resistance to palbociclib, we performed metabolic pathway analysis of all DEGs using IPA (Figure 5). We observed an enrichment of several metabolic pathways including three pathways involved in ribonucleotides synthesis ('Pyrimidine deoxyribonucleotides de novo biosynthesis I', 'dTMP de novo biosynthesis' and 'Salvage pathway of pyrimidine ribonucleotides'), six pathways involved in inositol metabolism ('3-Phosphoinositide biosynthesis', '3-Phosphoinositide degradation', 'D-myo-inositol(1,4,5,6)-tetrakisphosphate biosynthesis', 'D-myo-inositol-5-phosphate metabolism' and 'Superpathway of inositol phosphate compounds'). Among other pathways, we found 'Glycerol-3-phosphate shuttle', 'Asparagine degradation' and 'NAD biosynthesis II (from tryptophan)' to be enriched in our dataset. These results indicate that deregulated metabolism may play an essential role in mediating resistance to palbociclib.

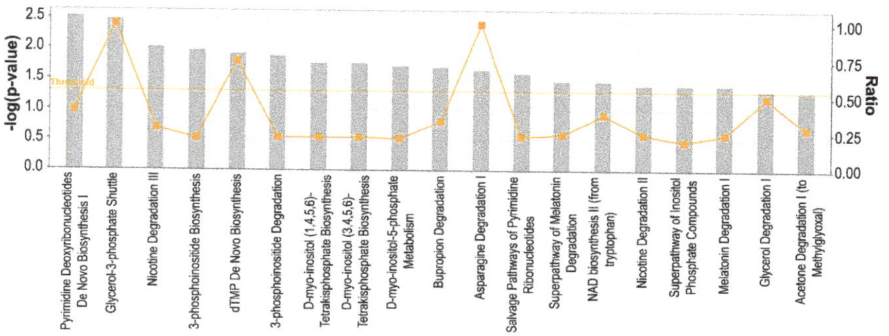

Figure 5. Metabolic pathway analysis of ER+ palbociclib-resistant breast cancer cells. A higher–log(p-value) shown on the left Y axis represents more significant pathways. The ratio (right Y axis) refers to the number of genes from the dataset that map to the pathway divided by the total number of genes that map the canonical pathway from the IPA database. pval ≤ 0.05; qval ≤ 0.05; $|\log_2 FC| \geq 1$.

4. Discussion

Three orally available inhibitors of CDK4/6 are currently used in combination with endocrine therapy (ET) as first-line therapy ER+ metastatic breast cancer patients [25]. Although initially beneficial, resistance to CDK4/6 inhibition arises in almost all patients within two years thus limiting durable responses. Currently, there are no biomarkers that can predict treatment response or early resistance [26]. Here, we identified a number of clinically relevant pathways that are associated with resistance to palbociclib, largely focusing on metabolic alterations and oncogenic signaling such as nucleotide metabolism, inositol metabolism, cell cycle, immune regulation and DDR.

Previous efforts to identify mechanisms of resistance to CDK4/6 inhibition have found that lack of Rb protein, increased cyclin E expression, IL6/STAT3 pathway activation and decreased DNA repair are some of the underlying mechanisms of resistance in ER+ breast cancer cells [6,13,19,27,28]. Analysis of ctDNA or tumor mRNA from patients enrolled in the PALOMA-3, NeoPalAna and MONALEESA-3 trials have identified Rb mutations, activating mutations in PIK3CA and ESR1, increased cyclin E1 and activation of the PDK1-AKT axis as some of the drivers of resistance [7,9,11]. Consistent with previous findings, we observed a significant enrichment in pathways involved in DDR [13]. Furthermore, we observed an increased in autophagy in MCF7/pR cells which is consistent with the previously described increase in autophagy driven by CDK4/6 inhibition in palbociclib-sensitive ER+ breast cancer cells [19]. Previous studies have shown that resistance to CDK4/6 inhibition is associated with a loss of ER/progesterone receptor (PR) expression in tumor biopsies of patients treated with the CDK4/6 inhibitor abemaciclib [5]. Notably, we observed a significant decrease in PR expression in palbociclib-resistant cells (Supplementary File 1). This finding is relevant given that unliganded PR sustains ER expression levels by maintaining a low methylation status of the ER gene [29]. Taken together, these observations suggest that PR loss may drive breast cancer cells to escape CDK4/6 inhibition by altering ER methylation thereby resulting in the down-regulation of ER expression. Additionally, our results highlight that ER methylation status can potentially be used to predict acquired resistance to CDK4/6 inhibition.

While our findings are in line with previously identified mechanisms of resistance, our analysis uncovered additional potential mechanisms of resistance such as deregulation of 'Polo-Like Kinase (PLK)', 'April mediated signaling' and 'Tec Kinase signaling'. Of these, targeting PLK1 is of high relevance due to its role as a master regulator of the G2-M phase and DNA replication [30,31]. Importantly, PLK1 has been shown to play a role in mediating tamoxifen resistance in ER+ breast cancer cells, and thus we will conduct additional studies evaluating the role of PLK1 as a novel target for the ER+ breast cancer resistant to CDK4/6 inhibition [32]. Importantly, a potent PLK1 inhibitor, volasertib (BI6727), has been recently approved for the treatment of acute myeloid leukemia and would be a promising therapeutic agent against palbociclib-resistant breast cancer [33,34].

Close examination of the DEGs revealed significant expression changes in many genes involved in tumorigenesis and chemoresistance. For example, up-regulation of three of the small leucine-rich family of proteoglycans (SLRP), decorin, epiphycan and lumican, was observed in our dataset (Table 1). These proteoglycans are known for their ability to regulate cell signaling, adhesion, migration, proliferation and apoptosis in many types of cancer [35,36]. Notably, accumulated evidence supports a role for both decorin and lumican in mediating drug resistance, and thus our data suggest a potential role for these proteoglycans in mediating resistance to palbociclib [37–41]. Other promising genes that were shown to be up-regulated in our dataset were cystatin S and alpha B-crystallin. Elevated blood levels of cystatin-C have been detected in women with breast cancer and are shown to correlate with cancer progression [42,43]. Alpha B-crystallin expression has been associated with high metastatic potential, poor clinical outcome and drug resistance in breast cancer [44–46]. Our findings raise the possibility of the potential use of alpha B-crystallin and cystatin-C as biomarkers of sensitivity to CDK4/6 inhibition. Of the top 20 down-regulated genes, miR-646 host gene and homeobox A10 (HOXA10) are of great relevance given their emerging tumor suppressive functions. Expression of miR-646 has been shown to directly regulate CDK6 and FOXK1 expression in gastric cancer, suggesting its utility as a potential therapeutic target [47,48]. A lack of HOXA10 in breast cancer has been shown to decrease apoptosis and promote metastasis, and thus the role of HOXA10 in the context of palbociclib resistance warrants further investigation [49]. A limitation of our studies is the lack of validation of gene expression changes by real-time PCR; however, we believe that our initial profiling will help guide further efforts to better understand the molecular mechanisms driving drug resistance.

Metabolic reprograming is a well-established oncogenic driver that allows cells to support the increased bioenergetic and anabolic demands [50]. Importantly, CDK4/6 are key regulators of metabolic pathways, and therefore we anticipated that metabolic rewiring will be observed upon the development of resistance to palbociclib. While a previous study described an increase in glucose dependence in ER+/Her2- palbociclib-sensitive compared to palbociclib-resistant cells [51], to date little is known about global metabolic changes driving resistance to CDK4/6 inhibition. Our unbiased analysis of DEGs and metabolic pathways began to define metabolic hubs linked to palbociclib resistance. Specifically, we observed alterations in nucleotide metabolism in MCF7/pS vs. MCF7/pR cells. Importantly, these results are in line with previous reports indicating that increased expression of thymidine kinase-1 (TK1), an enzyme of the pyrimidine salvage pathway, correlates with poor prognosis in breast cancer patients treated with palbociclib [52–55].

Our findings indicate that inositol metabolism was altered in ER+ palbociclib-resistant cells. Given the role of inositols as essential membrane components crucial for the generation of secondary messengers, our results are of high biological significance and provide a direct link between signal transduction and metabolic alterations contributing to resistance. Future metabolomic profiling will be needed to confirm our initial findings and provide further evidence as to how inositol alteration contributes to resistance to palbociclib.

Collectively, our RNA-seq analysis uncovered key canonical and metabolic pathways altered in ER+ palbociclib-resistant cells and provided new insights into the molecular mechanisms and potential therapeutic targets underlying resistance to CDK4/6 inhibition.

Supplementary Materials: The following are available online at http://www.mdpi.com/2073-4425/11/4/467/s1, File S1: List of all DEG identified.

Author Contributions: Conceptualization, Y.I.-F.; methodology, L.L., C.O., N.L., J.G.G.-G., S.W.; software, E.R., X.L. and S.W.; validation, L.L., E.R., X.L. and Y.I.-F.; formal analysis, E.R., X.L., S.W. and Y.I.-F.; investigation, L.L., N.L. and Y.I.-F.; resources, E.R., X.L., S.W., J.C. and Y.I.-F.; data curation, E.R., X.L. and Y.I.-F.; writing—original draft preparation, Y.I.-F.; writing—review and editing, N.L., X.L., S.W., J.G.G.-G. and Y.I.-F.; visualization, J.G.G.-G. and Y.I.-F.; supervision, Y.I.-F.; project administration, Y.I.-F.; funding acquisition, J.C. and Y.I.-F. All authors have read and agreed to the published version of the manuscript.

Funding: This research was funded by Twisted Pink (Yoannis Imbert-Fernandez, PI). Sequencing and bioinformatics support for this work was provided by the National Institutes of Health grants P20GM103436 (Nigel Cooper, PI) and P20GM106396 (Donald Miller, PI).

Acknowledgments: We want to thank the University of Louisville's Genomics Core facility and Nigel Cooper and Donald Miller for providing support for the Next-Gen RNA-seq studies.

Conflicts of Interest: The authors declare no conflicts of interest.

References

1. Rossi, L.; McCartney, A.; Risi, E.; De Santo, I.; Migliaccio, I.; Malorni, L.; Biganzoli, L.; Di Leo, A. Cyclin-Dependent Kinase 4/6 Inhibitors in Neoadjuvant Endocrine Therapy of Hormone Receptor-Positive Breast Cancer. *Clin. Breast Cancer* **2019**, *19*, 392–398. [CrossRef]
2. Sobhani, N.; D'Angelo, A.; Pittacolo, M.; Roviello, G.; Miccoli, A.; Corona, S.P.; Bernocchi, O.; Generali, D.; Otto, T. Updates on the CDK4/6 Inhibitory Strategy and Combinations in Breast Cancer. *Cells* **2019**, *8*, 321. [CrossRef] [PubMed]
3. Zhang, J.; Wang, Q.; Wang, Q.; Cao, J.; Sun, J.; Zhu, Z. Mechanisms of resistance to estrogen receptor modulators in ER+/HER2- advanced breast cancer. *Cell. Mol. Life Sci.* **2020**, *77*, 559–572. [CrossRef] [PubMed]

4. Herrera-Abreu, M.T.; Palafox, M.; Asghar, U.; Rivas, M.A.; Cutts, R.J.; Garcia-Murillas, I.; Pearson, A.; Guzman, M.; Rodriguez, O.; Grueso, J.; et al. Early Adaptation and Acquired Resistance to CDK4/6 Inhibition in Estrogen Receptor-Positive Breast Cancer. *Cancer Res.* **2016**, *76*, 2301–2313. [CrossRef] [PubMed]
5. Yang, C.; Li, Z.; Bhatt, T.; Dickler, M.; Giri, D.; Scaltriti, M.; Baselga, J.; Rosen, N.; Chandarlapaty, S. Acquired CDK6 amplification promotes breast cancer resistance to CDK4/6 inhibitors and loss of ER signaling and dependence. *Oncogene* **2017**, *36*, 2255–2264. [CrossRef]
6. Lypova, N.; Lanceta, L.; Gibson, A.; Vega, S.; Garza-Morales, R.; McMasters, K.M.; Chesney, J.; Gomez-Gutierrez, J.G.; Imbert-Fernandez, Y. Targeting Palbociclib-Resistant Estrogen Receptor-Positive Breast Cancer Cells via Oncolytic Virotherapy. *Cancers (Basel)* **2019**, *11*, 684. [CrossRef]
7. Condorelli, R.; Spring, L.; O'Shaughnessy, J.; Lacroix, L.; Bailleux, C.; Scott, V.; Dubois, J.; Nagy, R.J.; Lanman, R.B.; Iafrate, A.J.; et al. Polyclonal RB1 mutations and acquired resistance to CDK 4/6 inhibitors in patients with metastatic breast cancer. *Ann. Oncol.* **2018**, *29*, 640–645. [CrossRef]
8. O'Leary, B.; Hrebien, S.; Morden, J.P.; Beaney, M.; Fribbens, C.; Huang, X.; Liu, Y.; Bartlett, C.H.; Koehler, M.; Cristofanilli, M.; et al. Early circulating tumor DNA dynamics and clonal selection with palbociclib and fulvestrant for breast cancer. *Nat. Commun.* **2018**, *9*, 896. [CrossRef]
9. O'Leary, B.; Cutts, R.J.; Liu, Y.; Hrebien, S.; Huang, X.; Fenwick, K.; Andre, F.; Loibl, S.; Loi, S.; Garcia-Murillas, I.; et al. The Genetic Landscape and Clonal Evolution of Breast Cancer Resistance to Palbociclib plus Fulvestrant in the PALOMA-3 Trial. *Cancer Discov.* **2018**, *8*, 1390–1403. [CrossRef]
10. Formisano, L.; Lu, Y.; Servetto, A.; Hanker, A.B.; Jansen, V.M.; Bauer, J.A.; Sudhan, D.R.; Guerrero-Zotano, A.L.; Croessmann, S.; Guo, Y.; et al. Aberrant FGFR signaling mediates resistance to CDK4/6 inhibitors in ER+ breast cancer. *Nat. Commun.* **2019**, *10*, 1373. [CrossRef]
11. Jansen, V.M.; Bhola, N.E.; Bauer, J.A.; Formisano, L.; Lee, K.M.; Hutchinson, K.E.; Witkiewicz, A.K.; Moore, P.D.; Estrada, M.V.; Sanchez, V.; et al. Kinome-Wide RNA Interference Screen Reveals a Role for PDK1 in Acquired Resistance to CDK4/6 Inhibition in ER-Positive Breast Cancer. *Cancer Res.* **2017**, *77*, 2488–2499. [CrossRef] [PubMed]
12. Gong, X.; Du, J.; Parsons, S.H.; Merzoug, F.F.; Webster, Y.; Iversen, P.W.; Chio, L.C.; Van Horn, R.D.; Lin, X.; Blosser, W.; et al. Aurora A Kinase Inhibition Is Synthetic Lethal with Loss of the RB1 Tumor Suppressor Gene. *Cancer Discov.* **2019**, *9*, 248–263. [CrossRef] [PubMed]
13. Kettner, N.M.; Vijayaraghavan, S.; Durak, M.G.; Bui, T.; Kohansal, M.; Ha, M.J.; Liu, B.; Rao, X.; Wang, J.; Yi, M.; et al. Combined Inhibition of STAT3 and DNA Repair in Palbociclib-Resistant ER-Positive Breast Cancer. *Clin. Cancer Res.* **2019**, *25*, 3996–4013. [CrossRef] [PubMed]
14. de Leeuw, R.; McNair, C.; Schiewer, M.J.; Neupane, N.P.; Brand, L.J.; Augello, M.A.; Li, Z.; Cheng, L.C.; Yoshida, A.; Courtney, S.M.; et al. MAPK Reliance via Acquired CDK4/6 Inhibitor Resistance in Cancer. *Clin. Cancer Res.* **2018**, *24*, 4201–4214. [CrossRef]
15. Illumina, I. *BaseSpace User Guide*; Part # 15044182 Rev. E; Ilumina: San Diego, CA, USA, 2014; pp. 1–98.
16. Flight, R.M.; Harrison, B.J.; Mohammad, F.; Bunge, M.B.; Moon, L.D.; Petruska, J.C.; Rouchka, E.C. categoryCompare, an analytical tool based on feature annotations. *Front. Genet.* **2014**, *5*, 98. [CrossRef]
17. Kabeya, Y.; Mizushima, N.; Ueno, T.; Yamamoto, A.; Kirisako, T.; Noda, T.; Kominami, E.; Ohsumi, Y.; Yoshimori, T. LC3, a mammalian homologue of yeast Apg8p, is localized in autophagosome membranes after processing. *EMBO J.* **2000**, *19*, 5720–5728. [CrossRef]
18. Shui, W.; Sheu, L.; Liu, J.; Smart, B.; Petzold, C.J.; Hsieh, T.Y.; Pitcher, A.; Keasling, J.D.; Bertozzi, C.R. Membrane proteomics of phagosomes suggests a connection to autophagy. *Proc. Natl. Acad. Sci. USA* **2008**, *105*, 16952–16957. [CrossRef]
19. Vijayaraghavan, S.; Karakas, C.; Doostan, I.; Chen, X.; Bui, T.; Yi, M.; Raghavendra, A.S.; Zhao, Y.; Bashour, S.I.; Ibrahim, N.K.; et al. CDK4/6 and autophagy inhibitors synergistically induce senescence in Rb positive cytoplasmic cyclin E negative cancers. *Nat. Commun.* **2017**, *8*, 15916. [CrossRef]
20. Chittaranjan, S.; Bortnik, S.; Dragowska, W.H.; Xu, J.; Abeysundara, N.; Leung, A.; Go, N.E.; DeVorkin, L.; Weppler, S.A.; Gelmon, K.; et al. Autophagy inhibition augments the anticancer effects of epirubicin treatment in anthracycline-sensitive and -resistant triple-negative breast cancer. *Clin. Cancer Res.* **2014**, *20*, 3159–3173. [CrossRef]
21. Chen, S.; Jiang, Y.Z.; Huang, L.; Zhou, R.J.; Yu, K.D.; Liu, Y.; Shao, Z.M. The residual tumor autophagy marker LC3B serves as a prognostic marker in local advanced breast cancer after neoadjuvant chemotherapy. *Clin. Cancer Res.* **2013**, *19*, 6853–6862. [CrossRef]

22. Lefort, S.; Joffre, C.; Kieffer, Y.; Givel, A.M.; Bourachot, B.; Zago, G.; Bieche, I.; Dubois, T.; Meseure, D.; Vincent-Salomon, A.; et al. Inhibition of autophagy as a new means of improving chemotherapy efficiency in high-LC3B triple-negative breast cancers. *Autophagy* **2014**, *10*, 2122–2142. [CrossRef] [PubMed]
23. Wang, H.; Nicolay, B.N.; Chick, J.M.; Gao, X.; Geng, Y.; Ren, H.; Gao, H.; Yang, G.; Williams, J.A.; Suski, J.M.; et al. The metabolic function of cyclin D3-CDK6 kinase in cancer cell survival. *Nature* **2017**, *546*, 426–430. [CrossRef] [PubMed]
24. Franco, J.; Balaji, U.; Freinkman, E.; Witkiewicz, A.K.; Knudsen, E.S. Metabolic Reprogramming of Pancreatic Cancer Mediated by CDK4/6 Inhibition Elicits Unique Vulnerabilities. *Cell Rep.* **2016**, *14*, 979–990. [CrossRef] [PubMed]
25. Rozeboom, B.; Dey, N.; De, P. ER+ metastatic breast cancer: Past, present, and a prescription for an apoptosis-targeted future. *Am. J. Cancer Res.* **2019**, *9*, 2821–2831.
26. McCartney, A.; Migliaccio, I.; Bonechi, M.; Biagioni, C.; Romagnoli, D.; De Luca, F.; Galardi, F.; Risi, E.; De Santo, I.; Benelli, M.; et al. Mechanisms of Resistance to CDK4/6 Inhibitors: Potential Implications and Biomarkers for Clinical Practice. *Front. Oncol.* **2019**, *9*, 666. [CrossRef]
27. Fry, D.W.; Harvey, P.J.; Keller, P.R.; Elliott, W.L.; Meade, M.; Trachet, E.; Albassam, M.; Zheng, X.; Leopold, W.R.; Pryer, N.K.; et al. Specific inhibition of cyclin-dependent kinase 4/6 by PD 0332991 and associated antitumor activity in human tumor xenografts. *Mol. Cancer Ther.* **2004**, *3*, 1427–1438.
28. Finn, R.S.; Dering, J.; Conklin, D.; Kalous, O.; Cohen, D.J.; Desai, A.J.; Ginther, C.; Atefi, M.; Chen, I.; Fowst, C.; et al. PD 0332991, a selective cyclin D kinase 4/6 inhibitor, preferentially inhibits proliferation of luminal estrogen receptor-positive human breast cancer cell lines in vitro. *Breast Cancer Res.* **2009**, *11*, R77. [CrossRef]
29. Verde, G.; De Llobet, L.I.; Wright, R.H.G.; Quilez, J.; Peiro, S.; Le Dily, F.; Beato, M. Unliganded Progesterone Receptor Governs Estrogen Receptor Gene Expression by Regulating DNA Methylation in Breast Cancer Cells. *Cancers (Basel)* **2018**, *10*, 371. [CrossRef]
30. Barr, F.A.; Sillje, H.H.; Nigg, E.A. Polo-like kinases and the orchestration of cell division. *Nat. Rev. Mol. Cell Biol.* **2004**, *5*, 429–440. [CrossRef]
31. Song, B.; Liu, X.S.; Liu, X. Polo-like kinase 1 (Plk1): An Unexpected Player in DNA Replication. *Cell Div.* **2012**, *7*, 3. [CrossRef]
32. Jeong, S.B.; Im, J.H.; Yoon, J.H.; Bui, Q.T.; Lim, S.C.; Song, J.M.; Shim, Y.; Yun, J.; Hong, J.; Kang, K.W. Essential Role of Polo-like Kinase 1 (Plk1) Oncogene in Tumor Growth and Metastasis of Tamoxifen-Resistant Breast Cancer. *Mol. Cancer Ther.* **2018**, *17*, 825–837. [CrossRef] [PubMed]
33. Shafer, D.; Grant, S. Update on rational targeted therapy in AML. *Blood Rev.* **2016**, *30*, 275–283. [CrossRef] [PubMed]
34. Rudolph, D.; Steegmaier, M.; Hoffmann, M.; Grauert, M.; Baum, A.; Quant, J.; Haslinger, C.; Garin-Chesa, P.; Adolf, G.R. BI 6727, a Polo-like kinase inhibitor with improved pharmacokinetic profile and broad antitumor activity. *Clin. Cancer Res.* **2009**, *15*, 3094–3102. [CrossRef] [PubMed]
35. Afratis, N.; Gialeli, C.; Nikitovic, D.; Tsegenidis, T.; Karousou, E.; Theocharis, A.D.; Pavao, M.S.; Tzanakakis, G.N.; Karamanos, N.K. Glycosaminoglycans: Key players in cancer cell biology and treatment. *FEBS J.* **2012**, *279*, 1177–1197. [CrossRef]
36. Nikitovic, D.; Kouvidi, K.; Voudouri, K.; Berdiaki, A.; Karousou, E.; Passi, A.; Tzanakakis, G.N. The motile breast cancer phenotype roles of proteoglycans/glycosaminoglycans. *Biomed. Res. Int.* **2014**, *2014*, 124321. [CrossRef]
37. Kasamatsu, A.; Uzawa, K.; Minakawa, Y.; Ishige, S.; Kasama, H.; Endo-Sakamoto, Y.; Ogawara, K.; Shiiba, M.; Takiguchi, Y.; Tanzawa, H. Decorin in human oral cancer: A promising predictive biomarker of S-1 neoadjuvant chemosensitivity. *Biochem. Biophys. Res. Commun.* **2015**, *457*, 71–76. [CrossRef]
38. Pope, W.B.; Mirsadraei, L.; Lai, A.; Eskin, A.; Qiao, J.; Kim, H.J.; Ellingson, B.; Nghiemphu, P.L.; Kharbanda, S.; Soriano, R.H.; et al. Differential gene expression in glioblastoma defined by ADC histogram analysis: Relationship to extracellular matrix molecules and survival. *AJNR Am. J. Neuroradiol.* **2012**, *33*, 1059–1064. [CrossRef]
39. Yang, Z.X.; Lu, C.Y.; Yang, Y.L.; Dou, K.F.; Tao, K.S. Lumican expression in pancreatic ductal adenocarcinoma. *Hepatogastroenterology* **2013**, *60*, 349–353. [CrossRef]
40. Nikitovic, D.; Chalkiadaki, G.; Berdiaki, A.; Aggelidakis, J.; Katonis, P.; Karamanos, N.K.; Tzanakakis, G.N. Lumican regulates osteosarcoma cell adhesion by modulating TGFbeta2 activity. *Int. J. Biochem. Cell Biol.* **2011**, *43*, 928–935. [CrossRef]

41. Yamano, Y.; Uzawa, K.; Saito, K.; Nakashima, D.; Kasamatsu, A.; Koike, H.; Kouzu, Y.; Shinozuka, K.; Nakatani, K.; Negoro, K.; et al. Identification of cisplatin-resistance related genes in head and neck squamous cell carcinoma. *Int. J. Cancer* **2010**, *126*, 437–449. [CrossRef]
42. Kwon, W.S.; Kim, T.S.; Nahm, C.H.; Moon, Y.; Kim, J.J. Aberrant cystatin-C expression in blood from patients with breast cancer is a suitable marker for monitoring tumor burden. *Oncol. Lett.* **2018**, *16*, 5583–5590. [CrossRef] [PubMed]
43. Decock, J.; Obermajer, N.; Vozelj, S.; Hendrickx, W.; Paridaens, R.; Kos, J. Cathepsin B, cathepsin H, cathepsin X and cystatin C in sera of patients with early-stage and inflammatory breast cancer. *Int. J. Biol. Markers* **2008**, *23*, 161–168. [CrossRef] [PubMed]
44. Moyano, J.V.; Evans, J.R.; Chen, F.; Lu, M.; Werner, M.E.; Yehiely, F.; Diaz, L.K.; Turbin, D.; Karaca, G.; Wiley, E.; et al. AlphaB-crystallin is a novel oncoprotein that predicts poor clinical outcome in breast cancer. *J. Clin. Investig.* **2006**, *116*, 261–270. [CrossRef] [PubMed]
45. Chelouche-Lev, D.; Kluger, H.M.; Berger, A.J.; Rimm, D.L.; Price, J.E. alphaB-crystallin as a marker of lymph node involvement in breast carcinoma. *Cancer* **2004**, *100*, 2543–2548. [CrossRef]
46. Ivanov, O.; Chen, F.; Wiley, E.L.; Keswani, A.; Diaz, L.K.; Memmel, H.C.; Rademaker, A.; Gradishar, W.J.; Morrow, M.; Khan, S.A.; et al. alphaB-crystallin is a novel predictor of resistance to neoadjuvant chemotherapy in breast cancer. *Breast Cancer Res. Treat.* **2008**, *111*, 411–417. [CrossRef]
47. Zhang, P.; Tang, W.M.; Zhang, H.; Li, Y.Q.; Peng, Y.; Wang, J.; Liu, G.N.; Huang, X.T.; Zhao, J.J.; Li, G.; et al. MiR-646 inhibited cell proliferation and EMT-induced metastasis by targeting FOXK1 in gastric cancer. *Br. J. Cancer* **2017**, *117*, 525–534. [CrossRef]
48. Xue, M.; Li, G.; Fang, X.; Wang, L.; Jin, Y.; Zhou, Q. hsa_circ_0081143 promotes cisplatin resistance in gastric cancer by targeting miR-646/CDK6 pathway. *Cancer Cell Int.* **2019**, *19*, 25. [CrossRef]
49. Mustafa, M.; Lee, J.Y.; Kim, M.H. CTCF negatively regulates HOXA10 expression in breast cancer cells. *Biochem. Biophys. Res. Commun.* **2015**, *467*, 828–834. [CrossRef]
50. Ogrodzinski, M.P.; Bernard, J.J.; Lunt, S.Y. Deciphering metabolic rewiring in breast cancer subtypes. *Transl. Res.* **2017**, *189*, 105–122. [CrossRef]
51. Lorito, N.; Bacci, M.; Smiriglia, A.; Mannelli, M.; Parri, M.; Comito, G.; Ippolito, L.; Giannoni, E.; Bonechi, M.; Benelli, M.; et al. Glucose Metabolic Reprogramming of ER Breast Cancer in Acquired Resistance to the CDK4/6 Inhibitor Palbociclib+. *Cells* **2020**, *9*, 668. [CrossRef]
52. McCartney, A.; Bonechi, M.; De Luca, F.; Biagioni, C.; Curigliano, G.; Moretti, E.; Minisini, A.M.; Bergqvist, M.; Benelli, M.; Migliaccio, I.; et al. Plasma thymidine kinase activity as a biomarker in patients with luminal metastatic breast cancer treated with palbociclib within the TREnd trial. *Clin. Cancer Res.* **2020**. [CrossRef] [PubMed]
53. Romain, S.; Spyratos, F.; Descotes, F.; Daver, A.; Rostaing-Puissant, B.; Bougnoux, P.; Colonna, M.; Bolla, M.; Martin, P.M. Prognostic of DNA-synthesizing enzyme activities (thymidine kinase and thymidylate synthase) in 908 T1-T2, N0-N1, M0 breast cancers: A retrospective multicenter study. *Int. J. Cancer* **2000**, *87*, 860–868. [CrossRef]
54. Broet, P.; Romain, S.; Daver, A.; Ricolleau, G.; Quillien, V.; Rallet, A.; Asselain, B.; Martin, P.M.; Spyratos, F. Thymidine kinase as a proliferative marker: Clinical relevance in 1,692 primary breast cancer patients. *J. Clin. Oncol.* **2001**, *19*, 2778–2787. [CrossRef] [PubMed]
55. Ma, C.X.; Gao, F.; Luo, J.; Northfelt, D.W.; Goetz, M.; Forero, A.; Hoog, J.; Naughton, M.; Ademuyiwa, F.; Suresh, R.; et al. NeoPalAna: Neoadjuvant Palbociclib, a Cyclin-Dependent Kinase 4/6 Inhibitor, and Anastrozole for Clinical Stage 2 or 3 Estrogen Receptor-Positive Breast Cancer. *Clin. Cancer Res.* **2017**, *23*, 4055–4065. [CrossRef] [PubMed]

genes

MDPI

Review

Splicing Genomics Events in Cervical Cancer: Insights for Phenotypic Stratification and Biomarker Potency

Flavia Zita Francies [1], Sheynaz Bassa [2], Aristotelis Chatziioannou [1,3], Andreas Martin Kaufmann [1,4] and Zodwa Dlamini [1,*]

1 SAMRC/UP Precision Prevention & Novel Drug Targets for HIV-Associated Cancers (PPNDTHAC) Extramural Unit, Pan African Cancer Research Institute (PACRI), University of Pretoria, Hatfield 0028, South Africa; flavia.francies@up.ac.za (F.Z.F.); achatzi@bioacademy.gr (A.C.); andreas.kaufmann@charite.de (A.M.K.)

2 Department of Radiation Oncology, Steve Biko Academic Hospital (SBAH), Faculty of Health Sciences, University of Pretoria, Hatfield 0028, South Africa; sheynaz.bassa@up.ac.za

3 Center of Systems Biology, Biomedical Research Foundation Academy of Athens, 4 Soranou Ephessiou str, 115 27 Athens, Greece

4 Clinic for Gynaecology, Laboratory for Gynaecologic Tumor Immunology, Institute of Health, Charité-Universitätsmedizin, Freie Universität Berlin, Humboldt-Universität zu Berlin, Augustenburgerplatz 1, 13353 Berlin, Germany

* Correspondence: zodwa.dlamini@up.ac.za; Tel.: +27-76-147-4878

Abstract: Gynaecological cancers are attributed to the second most diagnosed cancers in women after breast cancer. On a global scale, cervical cancer is the fourth most common cancer and the most common cancer in developing countries with rapidly increasing mortality rates. Human papillomavirus (HPV) infection is a major contributor to the disease. HPV infections cause prominent cellular changes including alternative splicing to drive malignant transformation. A fundamental characteristic attributed to cancer is the dysregulation of cellular transcription. Alternative splicing is regulated by several splicing factors and molecular changes in these factors lead to cancer mechanisms such as tumour development and progression and drug resistance. The serine/arginine-rich (SR) proteins and heterogeneous ribonucleoproteins (hnRNPs) have prominent roles in modulating alternative splicing. Evidence shows molecular alteration and expression levels in these splicing factors in cervical cancer. Furthermore, aberrant splicing events in cancer-related genes lead to chemo- and radioresistance. Identifying clinically relevant modifications in alternative splicing events and splicing variants, in cervical cancer, as potential biomarkers for their role in cancer progression and therapy resistance is scrutinised. This review will focus on the molecular mechanisms underlying the aberrant splicing events in cervical cancer that may serve as potential biomarkers for diagnosis, prognosis, and novel drug targets.

Keywords: cervical cancer; alternative splicing; biomarkers; SR proteins; hnRNP; drug resistance

Citation: Francies, F.Z.; Bassa, S.; Chatziioannou, A.; Kaufmann, A.M.; Dlamini, Z. Splicing Genomics Events in Cervical Cancer: Insights for Phenotypic Stratification and Biomarker Potency. *Genes* **2021**, *12*, 130. https://doi.org/10.3390/genes12020130

Received: 19 November 2020
Accepted: 12 January 2021
Published: 20 January 2021

Publisher's Note: MDPI stays neutral with regard to jurisdictional clai-ms in published maps and institutio-nal affiliations.

1. Introduction

Cervical cancer, also known as cervix uteri cancer, is the fourth most frequently diagnosed cancer globally and the most common malignancy in developing countries [1]. It is the most frequently diagnosed cancer in women in Sub-Saharan Africa (SSA) and the leading cause of cancer-related mortality in this region (Figure 1) [2,3]. An estimated 90% of cervical cancer-related mortality occurs in low- and middle-income countries [4]. Cervical cancer is predominantly categorised into two main histopathological subtypes—squamous cell carcinoma and adenocarcinoma [5]. Over 75–80% of all cervical cancers are squamous cell carcinomas [6,7]. Cervical cancer is attributed to a number of risk factors such as sexually transmitted infections including human immunodeficiency virus (HIV) infection, human papillomavirus (HPV) infection, socioeconomic factors, obesity, smoking [8], alcohol consumption [9], unprotected sex and multiple sexual partners, prolonged usage of oral contraceptives, and family history of cervical cancer [10]. HPV infection is the major contributor of cervical cancer [11]. HPV is a circular double-stranded DNA virus with capsid proteins with more than 200 subtypes identified and categorised as high and low risk. Of these, about 40 subtypes have an affinity for genital mucosa and are sexually transmitted. The low-risk subtypes are generally associated with genital warts, whereas high risk subtypes cause invasive cervical cancer. The most prominent high-risk HPV genotypes are HPV16 and 18. Persistent infection with these high-risk subtypes contributes to over 99% of cervical cancers [11]. HPV infections can be prevented by vaccination that confers protection against HPV 6, 11, 16, and 18 subtypes, and depending on the vaccine, subtypes 31,33,45,52,58 can also be prevented. The vaccinations are available as quadrivalent vaccine to target all four subtypes or as bivalent to target only the high-risk subtypes [12] or a combined 9-valent vaccine that targets nine subtypes [13].

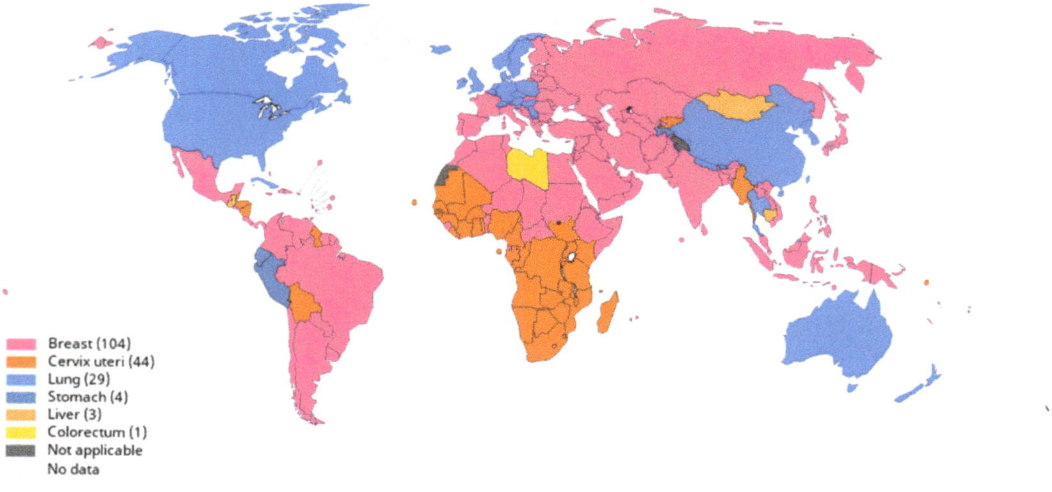

Figure 1. Global cancer mortality. The age standardised rates (ASRs) of various leading cancers worldwide in 2018. Cervical cancer is a major burden in most parts of Africa. Reprinted from [2].

In addition to HPV infections, dysregulated pathways are a fundamental feature in cervical cancer development and progression. For this reason, research in elucidating modifications in cancer-related pathways and alternative splicing is rapidly emerging. Several studies show aberrant alternative splicing and the dysregulation of gene expression in cervical cancer [14–17]. The related molecular signatures offer potential therapeutic targets for novel drug development and improved strategies in cervical cancer management, particularly for advanced disease in developing countries where HPV infections are the major contributor of cervical cancer.

The burden of cervical cancer mortality due to HPV infections is felt prominently in developing nations. Novel therapeutic targets are warranted to address this issue. Moreover, prevention strategies such as HPV vaccinations and pap smears play a significant role in cervical cancer prevention. Modifications in cervical tissue are detected through pap smears and HPV tests, and early diagnosis allows effective management of the disease [1,10]. This review will focus on the molecular mechanisms underlying the aberrant splicing events in cervical cancer that may serve as potential biomarkers for diagnosis and prognosis and as novel drug targets for their therapeutic properties.

2. Alternative Splicing and Its Implications in Cervical Cancer

Alternative splicing is an important process in gene expression and proteome diversity. In this cellular process, introns are spliced to join exons for the production of proteins through several mechanisms (Figure 2). Alternative splicing maintains cellular diversity and regulates the synthesis of multiple protein isoforms from the same gene. These protein isoforms perform several biological functions that are necessary for normal cellular functionality. Alternative splicing is an intricate process that is closely regulated by numerous spliceosome factors that aid in recognition of intron and splice sites such as small nuclear ribonucleoproteins (snRNP) particles and the serine/arginine-rich (SR) proteins [18]. In this process, proteins are synthesised, when introns are spliced and functional exons are joined together. The negative regulation of alternative splicing is achieved by heterogeneous ribonucleoproteins (hnRNPs) that block the intron and exon boundaries [19]. These two protein families—the SR proteins and the hnRNPs—are important trans-acting regulatory factors in splicing and are known to be altered in cervical cancer [20–23]. Enhanced levels of SR lead to splicing induction, whereas splicing is inhibited when hnRNPs are overly expressed. Aberrant alternative splicing, resulting from DNA damage, mutations and expression alterations in splicing factors, miRNA disruptions, and unregulated gene expression, are implicated in cancer mechanisms, such as sustained cell proliferation, apoptotic evasion, tumour suppressor inhibition, angiogenesis, metastasis, and drug resistance [19,24–26]. Evidence suggest that aberrant alternative splicing plays an important role in the development of cervical cancer. In cervical cancer, alternative splicing is primarily HPV-mediated. Next generation sequencing (NGS) offers a platform to identify potential disease-causing splice variants and genomic changes in splicing regulatory factors/proteins. Elucidating the functions of these splice variants may provide underlying information on malignant transformation and be beneficial in developing novel strategies for therapeutic interventions [27]. For these reasons, modifications in alternative splicing are becoming a significant biomarker with diagnostic and therapeutic potential.

Alternative splicing of key genes may facilitate the development and progression of cervical malignancy. For instance, the 5′ alternative splicing of the *KLHDC7B* gene is closely associated with cellular differentiation and tumour size in 67.5% of squamous cell carcinoma [28]. Similarly, 35% of exon skipping in the *SYCP2* gene was reported in cervical squamous cell carcinoma and associated with invasion and metastases [28]. Evidence also shows the association of cervical cancer and the aberrant alternative splicing of the *IL1RAP* gene. SRSF10 regulates the splicing of *IL1RAP* gene and promotes the production of its oncogenic isoform, MIL1RAP. This in turn facilitates the malignant cell evasion of phagocytosis by macrophages. Therefore, aberrant alternative splicing of *IL1RAP* gene promotes immune evasion and promotes cervical cancer [29]. A recent study by Ouyang et al. (2020) provides evidence that supports the notion that aberrant splicing events are closely associated with cervical cancer development, and the identification of these splicing biomarkers may provide useful prognostic and therapeutic tools [30]. The authors identified 2860 alternative splicing events. Of these, SNRPA and CCDC12 were associated with the tumour suppressor gene, p53, and were identified as hub genes in cervical cancer [30]. These results highlight the need to screen candidate biomarkers associated with cervical cancer that may have a clinical utility in diagnosis, prognosis, and therapy. Biomarkers related to cervical cancer are shown in Table 1.

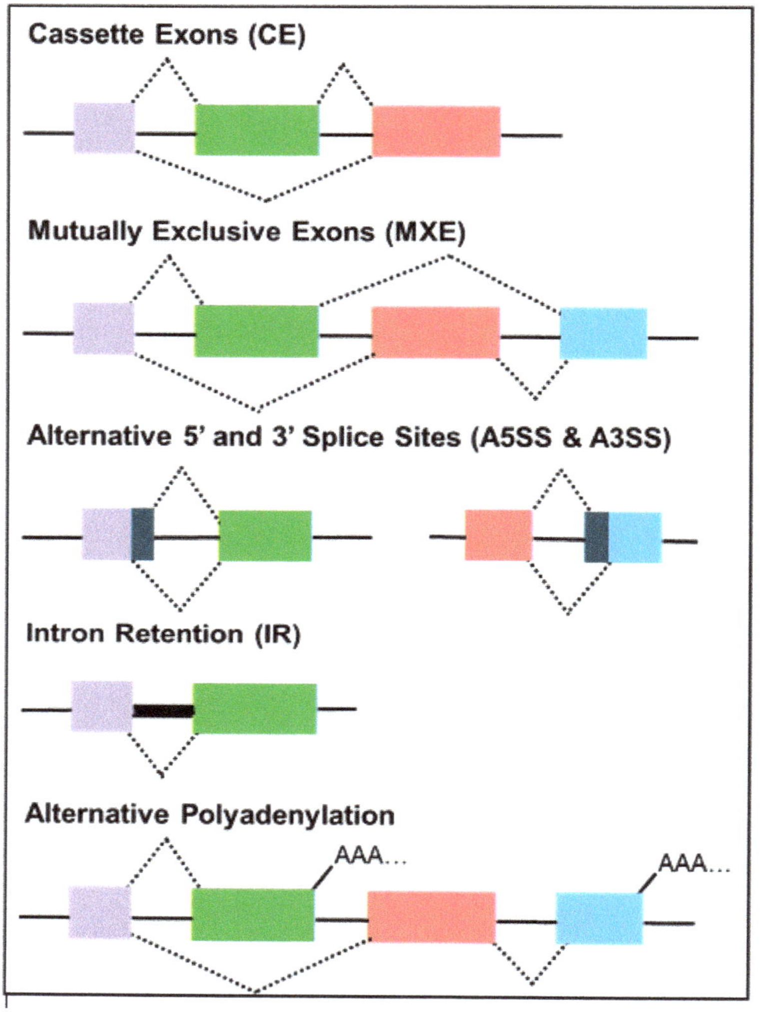

Figure 2. Frequent types of alternative splicing mechanisms. Alternative spliced mRNA produces mature transcripts, namely, cassette exons (CEs), mutually exclusive exons (MXEs), alternative 5′ or 3′ splice sites (A5SS and A3SS), intron retention (IR), and alternative polyadenylation (AP). Coloured boxes: exons; black lines: introns [31–35].

HPV contributes to the development and progression of cervical cancer by disrupting alternative splicing and other cellular functions. The HPV genome is double-stranded and circular; it is divided into three regions, namely, the long control region (LCR) and early and late region. Each region produces proteins that have different functions in the life cycle of HPV and in cancer development [36]. Persistent HPV infection gives rise to malignancy by producing viral proteins necessary to maintain virus replication and oncoproteins. Viral oncoproteins facilitate disease development and progression by abrogating normal cellular functions such as G1 arrest, cell proliferation, apoptosis, DNA repair, and chromosomal instability [37]. In addition, HPV oncoproteins bind to splicing factors and induce aberrant alternative splicing events. Moreover, HPV-related cervical cancer has a number of genes and splicing factors that are significantly upregulated. These include genes with vital functions such as immune surveillance, inflammatory response, and tumour suppressors [29,37]. Collectively, the interference of HPV in alternative splicing and cellular function promotes transformation, leading to carcinogenesis.

Table 1. Overview of biomarkers associated with cervical cancer.

Biomarkers	Biological Function	Modifications in Cervical Cancer	Clinical Utility	Ref.
HPV E6	p53 degradation	Overexpressed in cervical cancer cells	Screening and prevention	[38,39]
HPV E7	pRb degradation	Overexpressed in cervical cancer cells	Screening and prevention	[38,39]
AURKA	Genomic stability	Overexpressed in precancerous and cancerous cervical cells	Early detection	[40–42]
DTL	Checkpoint regulation	Overexpressed in cervical cancer cells	Early detection	[41]
HMGB3	Maintain balance in stem cell population	Overexpressed in cervical cancer cells	Early detection	[41]
KIF2C	Cell proliferation	Overexpressed in cervical cancer cells	Early detection	[41]
NEK2	Mitotic and cell cycle regulation	Overexpressed in cervical cancer cells	Early detection	[41]
RFC4	DNA replication	Overexpressed in cervical cancer cells	Early detection	[41]
p16ink4a	Tumour suppressor	Overexpressed in precancerous and cancerous cervical cells	Screening and diagnosis	[43–45]
Ki-67	Cell proliferation	Increased expression in proliferating epithelial cervical lesions	Screening and diagnosis	[43,46]
MCM2/TOP2A	DNA synthesis	Overexpressed in cervical dysplasia	Diagnosis	[38,47]
MSI1	RNA binding protein	Overexpressed in cervical cancer cells	Diagnostic and therapeutic	[48–51]
miR-21, miR-127 and miR-199a	-	Increased expression in cervical cancer cells	Prognosis	[39]
miR-143, miR214, miR-218 and miR-34a	-	Decreased expression in cervical cancer cells	Prognosis	[39]
ALDH1	Cellular differentiation and proliferation	Overexpressed in cervical cancer cells	Prognosis and predictive	[48,51,52]
EGFR	Transmembrane protein	Overexpressed in cervical cancer cells	Prognosis and predictive	[53,54]
Oct3/4	Transcription factor	Overexpressed in cervical cancer cells	Prognosis and predictive	[48,52,55]
Sox2	Transcription factor	Overexpressed in cervical cancer cells	Prognosis and predictive	[51,55]
CD49f	Stem cell marker	Overexpressed in cervical cancer cells	Prognosis and predictive	[48,51,56]
CD133	Cell surface antigen	Overexpressed in cervical cancer cells	Prognosis and predictive	[48,57]
CD44	Cellular differentiation and proliferation	Overexpressed in cervical cancer cells	Predictive	[48,58]
KAT2B	Mitotic and cell cycle regulation	Downregulated in cervical cancer cells	Predictive	[59]

2.1. HPV-Mediated Disruptions in Serine/Arginine-Rich (SR) Proteins

The spliceosome is crucial in regulating alternative splicing. In addition, other regulatory factors that are short DNA sequences, known as exonic splicing enhancers (ESEs), exonic splicing silencers (ESSs), intronic splicing enhancers (ISEs), and intronic splicing silencers (ISSs), ensure accurate splicing [60]. The splicing regulators have either a positive or negative effect on alternative splicing. ESE and ISE are cis-acting elements that are capable of binding the SR protein family to facilitate recognition of exons and initiate assembly of the spliceosome prior to alternative splicing (Figure 3) [61]. SR proteins are also known as SR splicing factors (SRSF) with SRSF1–12 as the major proteins in this family that have been identified as splicing regulators [62]. SR proteins have other vital cellular functions that are hallmarks of cancer, namely, cell cycle regulation, apoptosis, genome stability, cell adhesion and metastasis [27,63], and angiogenesis [64].

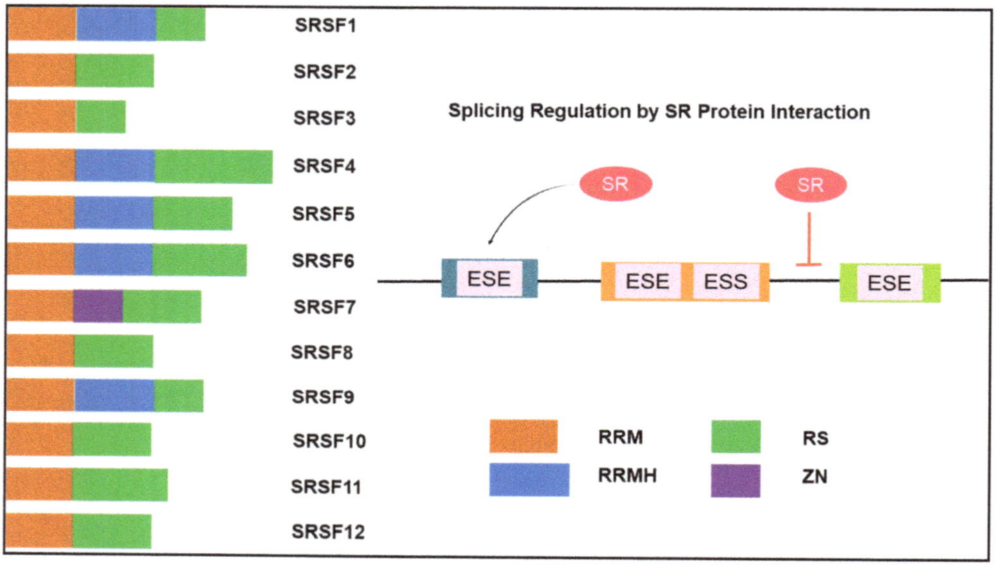

Figure 3. Alternative splicing regulation by SR binding. The domain structure of the 12 serine-rich (SR) proteins indicating the RRM (RNA recognition motif), RRMH (RRM homology), RS (arginine/serine-rich domain), and Zn (Zink knuckle). SR proteins bind to exonic splicing enhancers (ESEs), facilitating splice site recognition and stimulating the activation of splicing. In comparison, splicing is inhibited by the binding of the SR to introns [64–66].

The SRSF regulates splicing by determining the cycle of phosphorylation of SR proteins. CDC-like kinases (Clks), SR protein-specific kinases (SRPKs), and Topoisomerase 1 modulate the activation of SRSFs through a cycle of phosphorylation and dephosphorylation [18]. In the event of dephosphorylation, SRSFs begin to accumulate in the cytoplasm [64]. In comparison, phosphorylated SRSFs are transported to the nucleus to stimulate splicing. The SRPKs are capable of splicing regulation by the binding action to Clks in the nucleus and the cytoplasm [21]. In addition to splicing regulation, evidence suggests that SRPKs are able to modulate viral genomic material such as the HPV [18,21,31]. Evidence shows the binding of HPV E4 protein to SRPK1 [18]. This binding action impedes activation of SR protein by inhibiting the phosphorylation of SRSF1, SRSF3, SRSF4, and SRSF7 and impedes the pre-mRNA processing (Figure 4) [67]. This leads to aberrant cellular splicing that results in oncoprotein production and cervical cancer [18].

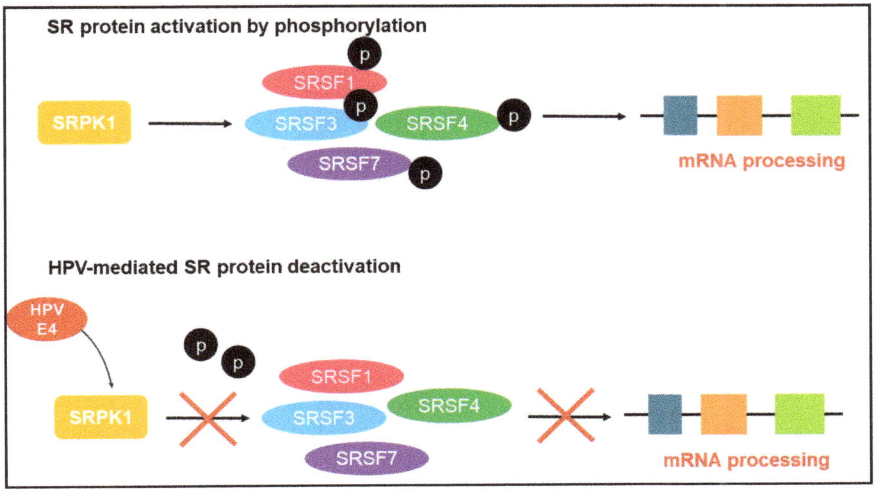

Figure 4. Human papillomavirus (HPV)-mediated disruption of accurate mRNA processing. The binding of HPV E4 protein leads to the deactivation of SR splicing factors (SRSFs) 1, 3, 4, and 7 by the loss of phosphorylation. This concomitantly inhibits pre-mRNA splicing, leading to inaccurate splicing and production of oncoproteins that give rise to malignancy [18,67]. SRPK: SR protein-specific kinase.

Evidence sheds light on the oncogenic role of SRSF1 [68] and a recent report shows its involvement in cervical malignancy [21]. Mole et al. (2020) showed enhanced levels of SRSF1 in cervical cancer cells. The authors showed the trans-activation of the SRSF1 gene promoter by the high-risk HPV16 E2 protein, with differing levels in the nucleus and cytoplasm [21]. Modifications of SRSF1 abrogate alternative splicing and facilitate genomic instability and cervical malignancy. Henceforth, the results suggest that the increased cytoplasmic levels of SRSF1 are associated with early tumour progression [21]. Other evidence shows the interaction of SRSF1 binding to long non-coding RNAs (lnRNA) to regulate expression levels of keratin 17. Cervical cancer cells display enhanced levels of keratin 17. Dong et al. (2019) showed the interplay between SRSF1 and lnRNA to modulate expression of keratin 17 through alternative splicing in cervical cancer [22]. In addition to SRSF1, SRSF3 regulates the expression of a number of genes and the overexpression of SRSF3 has been shown to modulate cell proliferation by inducing G2/M cell cycle arrest and apoptosis [69,70]. SRSF3 induces production of interleukin enhancer binding factor 3 (ILF3) isoform 1 and 2 through aberrant alternative splicing. These isoforms are involved in malignant transformation [71]. Furthermore, SRSF3 regulates expression of p300, a tumour suppressor, and induces cell proliferation [70]. In HPV-infected cervical cells, SRSF3 plays a significant part in the E6* splicing that is vital for E7 production [72] and in E1^E4 for viral replication [73]. Silencing SRSF3 in HPV-infected cells shows downregulation of viral E6 and E7 [72] and suppresses the E1^E4 splicing [73]. These results highlight the oncogenic potential of SRSF3 that may lead to cellular transformation and may contribute to cervical cancer [69].

DNA damage response plays a vital role in maintaining genomic stability and preventing carcinogenesis. Several important genes are involved in DNA damage pathways such as RAD51, ATM, p53, and ERCC1 [74]. Detecting modifications in DNA repair genes could be beneficial as biomarkers for diagnosis, prognosis, and targets for therapy. For instance, evidence shows the upregulated RAD51 mRNA levels in cervical cancer, which are associated with poor prognosis [75]. In addition to somatic mutations, HPV induces DNA damage in cervical cancer cells [76] and the resulting DNA damage response gene expression serves as prognostic biomarkers [77]. New evidence shows the association between SRSF6 and DNA damage genes. Yang et al. (2020) evaluated the function of SRSF6 in cervical cancer cells and showed that overexpressed SRSF6 influenced the alternative splicing of DNA damage genes [78]. SRSF6-induced aberrant alternative splicing of DNA damage genes is associated with the hallmarks of cancer such as cell proliferation, tumour progression, and apoptosis [78]. Elucidating the functional impact of SRSF6 in alternative splicing of DNA damage genes could offer a target for cervical cancer therapy.

2.2. HPV-Mediated Disruptions in Heterogeneous Ribonucleoproteins (hnRNPs)

The ESS and ISS act as negative regulators to repress alternative splicing and bind to the hnRNP family of proteins. Similar to SR proteins, the hnRNPs can either positively or negatively regulate splicing by binding to ESS and ISS, negatively prompting exon definition (Figure 5). There are currently at least 20 hnRNPs identified with several important cellular functions including alternative splicing [31]. Loss of regulation in hnRNPs leads to modified gene expression of tumour suppressors and other cancer-related genes [27,79]. Henceforth, hnRNPs are implicated in malignant transformation and could be scrutinised as potential cancer-related biomarkers.

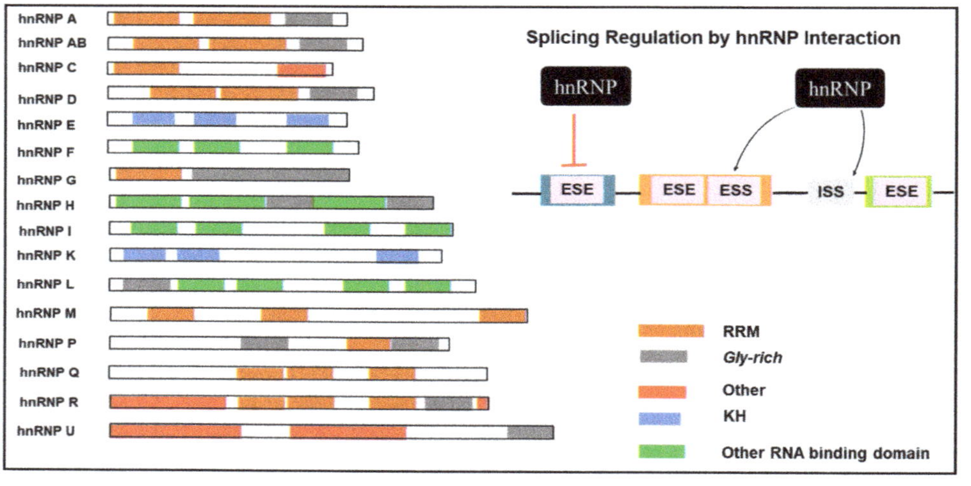

Figure 5. Alternative splicing regulation and the structural domains of hnRNP family. The domain structure of the hnRNP showing the RRM (RNA recognition motif), KH (K homology domain), and other RNA binding domain that is structurally different from RRM. hnRNP negatively regulates this process by binding to either exonic splicing silencers (ESSs) or intronic splicing silencers (ISSs). In addition, hnRNP blocks the activity of exonic splicing enhancers (ESEs) by binding to it [34,65,80].

Alternative splicing events are frequent in cervical cancer and are significantly associated with diagnosis and prognosis. Major splicing factors promote cervical malignancy by facilitating the production of HPV mRNAs and oncoproteins required. In addition, cellular oncogenic protein production is favoured to enhance the development of cervical cancer (Table 2). Cervical cancer cells have elevated expression of hnRNPs. For instance, hnRNPA1 is highly expressed in cervical cancer cells and can disrupt cancer-related genes. The alternative splicing of pyruvate kinase mRNA is induced by hnRNPA1 and favours aerobic glycolysis, resulting in uncontrolled cell proliferation. In the event where hnRNPA1 is downregulated, cancer-specific apoptosis is induced. hnRNPA1 is thus a good biomarker for cervical cancer diagnosis [23]. Another recent study investigated prognostic biomarkers of alternative splicing in cervical cancer and revealed hnRNPA1, ubiquitin C, and RNA polymerase II subunit L as effective prognostic biomarkers [81]. As a crucial component in alternative splicing, scrutinising the aberrant splicing induced by hnRNPA1 in cervical cancer is critical. Additionally, during the HPV infection-related differentiation of cervical epithelial cells, hnRNPA1 is further upregulated and enables oncoviral protein transcription. Deleterious mutations in hnRNPA1 have been identified and may alter expression levels contributing to aberrant alternative splicing, mRNA processing, and translation [82].

Table 2. The role of major splicing factors, the human papillomavirus (HPV) binding region, and function of transcripts in cancer progression.

Splicing Factor	HPV Binding Region	HPV16 mRNA	Cancer Promoting Function
SRSF1	E4	Production of E6/E7 mRNA	Apoptotic regulation
SRSF3	E4	Production of E6/E7 mRNA	Increased cell proliferation
SRSF9	E4	Production of late mRNAs	Increased cell proliferation and suppressed apoptosis
hnRNPA1	L1	Production of the isoform E6*I/E7	Apoptotic regulation
hnRNPA2/B1	E4	Production of the isoform E6*I/E7	Apoptotic regulation
hnRNPC	Early 3′-UTR	Production of L1 mRNA	-
hnRNPD	E4	Production of late mRNAs	-
hnRNPE1/E2	L2	Inhibition of L2 mRNA	-
hnRNPG	E4	Production of late mRNAs	-
hnRNPH	L2	Inhibition of late mRNAs	-
hnRNPI	Early 3′-UTR	Inhibition of late mRNAs	Cell proliferation and cell invasion
hnRNPK	L2	Inhibition of L2 mRNA	Cell cycle regulation
hnRNPL	E4 and L1	Inhibition of late mRNAs	-

UTR: Untranslated region. Reviewed in [31].

Prolonged HPV infections influence cellular and viral alternative splicing to enhance viral oncogene production, leading to malignant transformation of the cervix. Malignant transformation is initiated and sustained by the high-risk HPV16 E6 and E7 proteins that interact with tumour suppressor genes p53 and retinoblastoma protein (pRb), respectively. The interaction of E6 with p53 results in apoptosis, whereas E7 steers cell proliferation by interacting with pRb [83,84]. Moreover, E6 and E7 are essential in viral replication [85]. Zheng et al. (2020) showed splicing regulation of E6 and E7 by cellular hnRNPA1 and hnRNPA2 [20]. This study revealed the direct interaction of hnRNPA1 and hnRNPA2 with high-risk HPV16 splice site SA742 and SA409. The authors showed the inhibition of SA409 when hnRNPA1 is overexpressed and favouring viral E6 mRNA production. In comparison, when hnRNPA2 is upregulated, the viral E6Ê7, E1, and E4 mRNA transcripts are favoured [20]. Adequate amounts of both E6 and E7 transcripts are required for the development of cervix carcinoma. Furthermore, evidence also shows that HPV interacts with hnRNPA1 and the silencing of hnRNPA1 suppresses E6 intron retention [73]. Hence, targeting hnRNPA1 and hnRNPA2 to modulate viral E6 and E7 mRNA transcripts may provide novel therapeutic strategies.

3. Alternative Splicing and Therapy Resistance

Drug resistance is a considerable hurdle in cancer treatment and management. Aberrant alternative splicing events are a common theme in cancer drug resistance and, therefore, strategies targeted to silence variants that promote drug resistance are highly warranted. Aberrant splice variants can promote resistance to chemotherapy and radiotherapy [24,86–88] by mechanisms that include apoptotic regulation, modified drug metabolism, response to DNA damage, and regulation of cell proliferation (Figure 6) [89]. Radiotherapy is an important therapeutic modality for the management of advanced cervical cancer and radioresistance may be detrimental. In cervical cancer, a splice variant of nucleophosmin

(NPM) protein resulting from alternative splicing causes radioresistance [86]. NPM functions in mRNA processing, genome stability, and apoptotic regulation [90]. Enhanced expression of the NPM2 variant is correlated with a radio-protective function. Evidence shows that silencing the NPM2 splice variant decreases radioresistance in cervical cancer cells in a dose-dependent manner [86]. Similarly, enhanced levels of ΔNp73, a splice variant of p73, have anti-apoptotic functions and display radioresistance in cervical cancer cells [91]. p73 (i) is a p53 homologue that expresses the oncogenic isoform ΔNp73 [92]; (ii) functions in DNA damage repair, cell cycle regulation, and apoptosis with p73 polymorphism closely associated with cervical cancer [93]; and (iii) is a prognostic biomarker for cervical cancer [94]. Cervical cancer cells exposed to high-LET radiation degrade ΔNp73 to exhibit enhanced apoptosis and cell cycle arrest at the G2/M phase when compared with low-LET radiation [91]. In addition, ΔNp73 promotes malignant transformation by interacting with RAS and inducing drug resistance to chemotherapy and radiotherapy [87]. Furthermore, the HPV oncoprotein, E6, suppresses the activity of p53 expression and alters sensitivity to radiotherapy. The overexpression of the splice variant, p73α, in p53 deficient cervical cancer cells, enhances sensitivity to radiotherapy [95]. These results highlight the importance of targeting aberrant splice variants to reverse radioresistance in cervical cancer, which is significantly relevant in treating advanced metastatic disease.

Cervical cancer is often managed with chemotherapy and radiotherapy concurrently. An estimated 50% of patients do not attain a complete response to therapy due to resistance. Alterations in molecular pathways that promote drug resistance are potential drug targets to counteract resistance [96]. For instance, the CRK-like (CRKL) adapter protein is overexpressed in approximately 50% of cervical cancers. Moreover, evidence shows that CRKL significantly regulates alternative splicing of pre-mRNA in cancer-related genes in cervical carcinoma to promote malignant transformation, metastases, and chemoresistance by binding to BCR-ABL and activating the Src and Akt signally pathway through phosphorylation [97,98]. Additionally, recent evidence shows the role of AKT3 mRNA in inducing cisplatin resistance [99]. By blocking the activity of Src and Akt through pharmacological inhibitors such as dasatinib [97] and fucoxanthin [100], respectively, aberrant splicing events that facilitate chemoresistance in cervical cancer can be reversed and promote a complete therapy response in advanced metastatic disease.

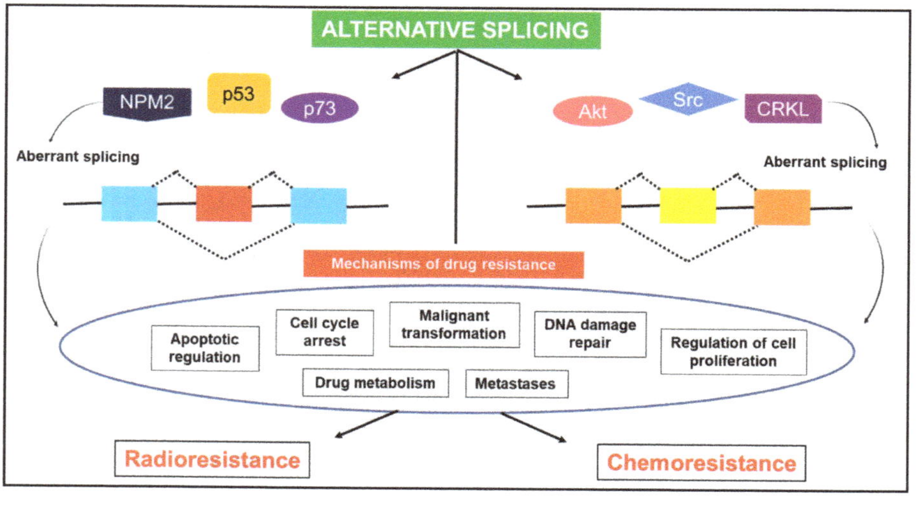

Figure 6. Alternative splicing-induced drug resistance. Aberrant splicing events of vital genes in cervical cancer cells promote drug resistance through several mechanisms by regulating apoptosis, cell cycle arrest, cell proliferation, and DNA damage response. In addition, splice variants may also alter drug targets that effect drug metabolism and lead to chemoresistance and alter the sensitivity to radiotherapy [86,87,89,95,97–99].

4. Clinical Utility of Biomarkers in Cervical Cancer

Altered expression of splicing regulators, deleterious mutations in splicing regulators and splicing regulatory sequences, and suppressed activity of splicing regulators can cause aberrant alternative splicing, which may result in tumourigenesis and therapy resistance (Figure 7). However, alternative splicing biomarkers have been studied extensively as potential targets of novel therapy [24,27]. The current diagnostic and prognostic indicators of cervical cancer are largely clinicopathology and HPV screening intensive. With the introduction of NGS, large-scale RNA sequencing can be clinically utilised to identify tissue-specific molecular biomarkers. Subsequent to the identification of onco-biomarkers, functional biological assays are imperative to characterise the properties of effective and clinically significant biomarkers for novel clinical utility in diagnosis, prognosis, and therapeutic interventions [27].

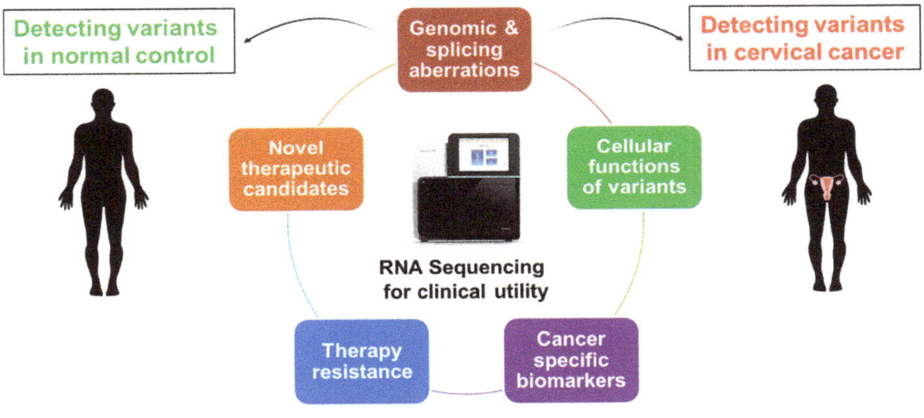

Figure 7. Overview of clinical biomarker identification. Aberrant alternative splice variants are often expressed in significantly higher levels compared with normal splice variants that can be identified through next generation sequencing (NGS). These aberrations can contribute to the development of tumourigenesis, therapy resistance, and poor prognosis. The effects of aberrant alternative splicing can be addressed by identifying cervical cancer-specific genomic and splicing aberrations that are clinically relevant for novel diagnostic, prognostic, and therapeutic purposes [24,81].

Reversing aberrant alternative splicing or silencing oncogenic variants could offer therapeutic strategies in managing cervical cancer. Pharmacological agents are frequently evaluated for their splicing inhibitory or silencing effects in cancer cells. The current alternative splicing modulators studied are small molecule splicing inhibitors, transsplicing, antisense oligonucleotides, and gene therapy. These modulators can regulate alternative splicing by controlling the functioning of spliceosomal activity [27]. For instance, caffeine suppresses the expression of SRSF2/3 and p53α, while upregulating the alternative spliced variant of p53β. Caffeine regulates cellular functions such as cell cycle arrest, DNA damage, and apoptosis by modulating the SRSF3 [101,102]. Cervical cancer cells treated with caffeine showed tumour suppression through the modulation of splicing factors. In addition, the recent evidence shows that pladienolide B inhibits the splicing factor SF3b1, which is a subunit of the spliceosome, to induce the G2/M cell cycle arrest, apoptosis, and p73 splicing in cervical cancer cells [103]. Other small molecules evaluated in cervical cancer include RI-1, a RAD51 inhibitor [104]. Modified gene expression is a central characteristic of cancer cells such as the altered expression of RAD51 mRNA in cervical cancer cells compared with healthy cells [75]. RI-1 promotes cell cycle arrest from G0/G1 to S phase and inhibits the RAD51-induced cell proliferation in cervical cancer cells [104]. These results indicate the potential of pharmacological agents to regulate alternative splicing in cervical cancer and their therapeutic potential.

Inhibiting splicing factors can evoke a tumour suppressive function. For instance, blocking the function of SRSF1 may contribute to apoptotic activity. Cervical cancer cells treated with an AURKA kinase inhibitor, such as the pharmacological agent VX-680, downregulate the post-transcriptional expression levels of SRSF1 [105]. AURKA kinases, part of the aurora family of proteins, are cell division regulators. Dysregulation of these proteins leads to uncontrolled cell division and proliferation, resulting in malignancy [42]. Cervical cancer cells treated with VX-680 promote aberrant alternative splicing of apoptotic regulating genes, Bcl-x and Mcl-1, and inhibit the anti-apoptotic function of SRSF1, leading to apoptosis [105]. Silencing of SRSF1, therefore, signifies a novel therapeutic target for cervical cancer.

5. Conclusions

The mortality associated with cervical cancer is increasing at an alarming rate. The development of cervical cancer is largely influenced by HPV infections in low- and middle-income countries that add to this encumbrance. Vaccination programs addressing HPV have been successful in lowering HPV infections in high-risk women. Moreover, screening and prevention programs are useful in early detection and treatment. In addition to HPV infections, molecular alterations at the RNA level contribute to cervical carcinoma. These include modifications in cellular alternative splicing induced by HPV. RBPs like SRs and hnRNPs are essential in maintaining the stability and packing of mRNAs, as well as transport to the cytoplasm for further processing. These processes are intricately balanced by several splicing factors and proteins to ensure accurate alternative splicing. Despite the stringent regulation, SR proteins and hnRNPs are often dysregulated in cervical cancer and lead to aberrant alternative splicing of many important cancer-related genes, including therapy resistance. For these reasons, SR proteins and hnRNPs are ideal candidates for drug targets. Hence, identifying biomarkers crucial to the development of cervical malignancy, its pathogenesis, and splice variants that are highly expressed in cervical cancer will be beneficial in developing novel therapeutic targets, especially in low- and middle-income countries where the burden of cervical cancer is rapidly increasing.

Author Contributions: F.Z.F. was involved in writing original draft preparation and editing; S.B. was involved in review and editing; A.C. was involved in final review and editing; A.M.K. was involved in final review and editing; Z.D. conceived the idea, funding acquisition, supervision, and involved in review and editing. All authors have read and agreed to the published version of the manuscript.

Funding: This research was funded by the Medical Research Council of South Africa, grant number: 23108.

Institutional Review Board Statement: Not applicable.

Informed Consent Statement: Not applicable.

Data Availability Statement: Not applicable.

Acknowledgments: We would like to thank the Medical Research Council of South Africa (SAMRC) and the University of Pretoria for funding this research.

Conflicts of Interest: The authors declare no conflict of interest. The funders had no role in the design of the study; in the collection, analyses, or interpretation of data; in the writing of the manuscript; or in the decision to publish the results.

References

1. Arbyn, M.; Weiderpass, E.; Bruni, L.; de Sanjosé, S.; Saraiya, M.; Ferlay, J.; Bray, F. Estimates of incidence and mortality of cervical cancer in 2018: A worldwide analysis. *Lancet Glob. Health* **2020**, *8*, e191–e203. [CrossRef]
2. IARC. Cancer Today. Available online: https://gco.iarc.fr/today/online-analysis-map?v=2018&mode=cancer&mode_population=continents&population=900&populations=900&key=asr&sex=2&cancer=39&type=1&statistic=5&prevalence=0&population_group=0&ages_group%5B%5D=0&ages_group%5B%5D=17&nb_items=10&group_cancer=1&include_nmsc=1&include_nmsc_other=1&projection=natural-earth&color_palette=default&map_scale=quantile&map_nb_colors=5&continent=0&rotate=%255B10%252C0%255D (accessed on 20 October 2020).

3. Bray, F.; Ferlay, J.; Soerjomataram, I.; Siegel, R.L.; Torre, L.A.; Jemal, A. Global cancer statistics 2018: GLOBOCAN estimates of incidence and mortality worldwide for 36 cancers in 185 countries. *CA Cancer J. Clin.* **2018**, *68*, 394–424. [CrossRef] [PubMed]
4. Ferlay, J.; Soerjomataram, I.; Dikshit, R.; Eser, S.; Mathers, C.; Rebelo, M.; Parkin, D.M.; Forman, D.; Bray, F. Cancer incidence and mortality worldwide: Sources, methods and major patterns in GLOBOCAN 2012. *Int. J. Cancer* **2015**, *136*, E359–E386. [CrossRef] [PubMed]
5. Lax, S. Histopathology of cervical precursor lesions and cancer. *Acta Dermatovenerol. Alp. Pannon. Adriat.* **2011**, *20*, 125–133.
6. Chiang, Y.-C.; Chen, Y.-Y.; Hsieh, S.-F.; Chiang, C.-J.; You, S.-L.; Cheng, W.-F.; Lai, M.-S.; Chen, C.-A. Screening frequency and histologic type influence the efficacy of cervical cancer screening: A nationwide cohort study. *Taiwan. J. Obstet. Gynecol.* **2017**, *56*, 442–448. [CrossRef] [PubMed]
7. Safaeian, M.; Solomon, D.; Castle, P.E. Cervical cancer prevention—Cervical screening: Science in evolution. *Obstet. Gynecol. Clin. N. Am.* **2007**, *34*, 739–760. [CrossRef]
8. Momenimovahed, Z.; Salehiniya, H. Incidence, mortality and risk factors of cervical cancer in the world. *Biomed. Res. Ther.* **2017**, *4*, 1795–1811. [CrossRef]
9. Kabekkodu, S.P.; Chakrabarty, S.; Ghosh, S.; Brand, A.; Satyamoorthy, K. Epigenomics, Pharmacoepigenomics, and Personalized Medicine in Cervical Cancer. *Public Health Genom.* **2017**, *20*, 100–115. [CrossRef]
10. Akinlotan, M.; Bolin, J.N.; Helduser, J.; Ojinnaka, C.; Lichorad, A.; McClellan, D. Cervical Cancer Screening Barriers and Risk Factor Knowledge Among Uninsured Women. *J. Community Health* **2017**, *42*, 770–778. [CrossRef]
11. Chan, C.K.; Aimagambetova, G.; Ukybassova, T.; Kongrtay, K.; Azizan, A. Human Papillomavirus Infection and Cervical Cancer: Epidemiology, Screening, and Vaccination—Review of Current Perspectives. *J. Oncol.* **2019**, *2019*, 3257939. [CrossRef]
12. Sankaranarayanan, R. HPV vaccination: The most pragmatic cervical cancer primary prevention strategy. *Int. J. Gynaecol. Obstet.* **2015**, *131* (Suppl. 1), S33–S35. [CrossRef] [PubMed]
13. Joura, E.A.; Giuliano, A.R.; Iversen, O.E.; Bouchard, C.; Mao, C.; Mehlsen, J.; Moreira, E.D., Jr.; Ngan, Y.; Petersen, L.K.; Lazcano-Ponce, E.; et al. A 9-valent HPV vaccine against infection and intraepithelial neoplasia in women. *N. Engl. J. Med.* **2015**, *372*, 711–723. [CrossRef] [PubMed]
14. Wu, K.; Yi, Y.; Liu, F.; Wu, W.; Chen, Y.; Zhang, W. Identification of key pathways and genes in the progression of cervical cancer using bioinformatics analysis. *Oncol. Lett.* **2018**, *16*, 1003–1009. [CrossRef] [PubMed]
15. Sharma, G.; Dua, P.; Agarwal, S.M. A Comprehensive Review of Dysregulated miRNAs Involved in Cervical Cancer. *Curr. Genom.* **2014**, *15*, 310–323. [CrossRef]
16. Campos-Parra, A.D.; Padua-Bracho, A.; Pedroza-Torres, A.; Figueroa-González, G.; Fernández-Retana, J.; Millan-Catalan, O.; Peralta-Zaragoza, O.; Cantú de León, D.; Herrera, L.A.; Pérez-Plasencia, C. Comprehensive transcriptome analysis identifies pathways with therapeutic potential in locally advanced cervical cancer. *Gynecol. Oncol.* **2016**, *143*, 406–413. [CrossRef]
17. Lin, M.; Ye, M.; Zhou, J.; Wang, Z.P.; Zhu, X. Recent Advances on the Molecular Mechanism of Cervical Carcinogenesis Based on Systems Biology Technologies. *Comput. Struct. Biotechnol. J.* **2019**, *17*, 241–250. [CrossRef]
18. Graham, S.V.; Faizo, A.A.A. Control of human papillomavirus gene expression by alternative splicing. *Virus Res.* **2017**, *231*, 83–95. [CrossRef]
19. Qi, F.; Li, Y.; Yang, X.; Wu, Y.-P.; Lin, L.-J.; Liu, X.-M. Significance of alternative splicing in cancer cells. *Chin. Med. J.* **2020**, *133*, 221–228. [CrossRef]
20. Zheng, Y.; Jönsson, J.; Hao, C.; Shoja Chaghervand, S.; Cui, X.; Kajitani, N.; Gong, L.; Wu, C.; Schwartz, S. Heterogeneous Nuclear Ribonucleoprotein A1 (hnRNP A1) and hnRNP A2 Inhibit Splicing to Human Papillomavirus 16 Splice Site SA409 through a UAG-Containing Sequence in the E7 Coding Region. *J. Virol.* **2020**, *94*, e01509–e01520. [CrossRef]
21. Mole, S.; Faizo, A.A.A.; Hernandez-Lopez, H.; Griffiths, M.; Stevenson, A.; Roberts, S.; Graham, S.V. Human papillomavirus type 16 infection activates the host serine arginine protein kinase 1 (SRPK1)—Splicing factor axis. *J. Gen. Virol.* **2020**, *101*, 523–532. [CrossRef]
22. Dong, M.; Dong, Z.; Zhu, X.; Zhang, Y.; Song, L. Long non-coding RNA MIR205HG regulates KRT17 and tumor processes in cervical cancer via interaction with SRSF1. *Exp. Mol. Pathol.* **2019**, *111*, 104322. [CrossRef] [PubMed]
23. Kim, Y.-J.; Kim, B.-R.; Ryu, J.-S.; Lee, G.-O.; Kim, H.-R.; Choi, K.-H.; Ryu, J.-W.; Na, K.-S.; Park, M.-C.; So, H.-S.; et al. HNRNPA1, a Splicing Regulator, Is an Effective Target Protein for Cervical Cancer Detection: Comparison With Conventional Tumor Markers. *Int. J. Gynecol. Cancer* **2017**, *27*, 326. [CrossRef] [PubMed]
24. Singh, B.; Eyras, E. The role of alternative splicing in cancer. *Transcription* **2017**, *8*, 91–98. [CrossRef] [PubMed]
25. Mthembu, N.N.; Mbita, Z.; Hull, R.; Dlamini, Z. Abnormalities in alternative splicing of angiogenesis-related genes and their role in HIV-related cancers. *HIV AIDS (Auckl)* **2017**, *9*, 77–93. [CrossRef]
26. Urbanski, L.M.; Leclair, N.; Anczuków, O. Alternative-splicing defects in cancer: Splicing regulators and their downstream targets, guiding the way to novel cancer therapeutics. *J. Virol.* **2018**, *9*, e1476. [CrossRef]
27. Bergsma, A.J.; van der Wal, E.; Broeders, M.; van der Ploeg, A.T.; Pim Pijnappel, W.W.M. Alternative Splicing in Genetic Diseases: Improved Diagnosis and Novel Treatment Options. In *International Review of Cell and Molecular Biology*; Loos, F., Ed.; Academic Press: Cambridge, MA, USA, 2018; Volume 335, pp. 85–141.
28. Guo, P.; Wang, D.; Wu, J.; Yang, J.; Ren, T.; Zhu, B.; Xiang, Y. The landscape of alternative splicing in cervical squamous cell carcinoma. *Oncotargets Ther.* **2014**, *8*, 73–79. [CrossRef]

29. Liu, F.; Dai, M.; Xu, Q.; Zhu, X.; Zhou, Y.; Jiang, S.; Wang, Y.; Ai, Z.; Ma, L.; Zhang, Y.; et al. SRSF10-mediated IL1RAP alternative splicing regulates cervical cancer oncogenesis via mIL1RAP-NF-κB-CD47 axis. *Oncogene* **2018**, *37*, 2394–2409. [CrossRef]
30. Ouyang, D.; Yang, P.; Cai, J.; Sun, S.; Wang, Z. Comprehensive analysis of prognostic alternative splicing signature in cervical cancer. *Cancer Cell Int.* **2020**, *20*, 221. [CrossRef]
31. Cerasuolo, A.; Buonaguro, L.; Buonaguro, F.M.; Tornesello, M.L. The Role of RNA Splicing Factors in Cancer: Regulation of Viral and Human Gene Expression in Human Papillomavirus-Related Cervical Cancer. *Front. Cell Dev. Biol.* **2020**, *8*. [CrossRef]
32. Chen, J.; Weiss, W.A. Alternative splicing in cancer: Implications for biology and therapy. *Oncogene* **2015**, *34*, 1–14. [CrossRef]
33. Iñiguez, L.P.; Hernández, G. The Evolutionary Relationship between Alternative Splicing and Gene Duplication. *Front. Genet.* **2017**, *8*, 8. [CrossRef] [PubMed]
34. Shkreta, L.; Cloutier, A.; Toutant, J.; Rendos, H.V.; Chabot, B. Regulation of alternative splicing and the case of Bcl-x. *Pak. J. Biochem. Mol. Biol.* **2015**, *48*, 27–38.
35. Makhafola, T.J.; Mbele, M.; Yacqub-Usman, K.; Hendren, A.; Haigh, D.B.; Blackley, Z.; Meyer, M.; Mongan, N.P.; Bates, D.O.; Dlamini, Z. Apoptosis in Cancer Cells Is Induced by Alternative Splicing of hnRNPA2/B1 Through Splicing of Bcl-x, a Mechanism that Can Be Stimulated by an Extract of the South African Medicinal Plant, Cotyledon orbiculata. *Front. Oncol.* **2020**, *10*, 10. [CrossRef] [PubMed]
36. Li, W.; Qi, Y.; Cui, X.; Huo, Q.; Zhu, L.; Zhang, A.; Tan, M.; Hong, Q.; Yang, Y.; Zhang, H.; et al. Characteristic of HPV Integration in the Genome and Transcriptome of Cervical Cancer Tissues. *BioMed Res. Int.* **2018**, *2018*, 6242173. [CrossRef] [PubMed]
37. Burd, E.M. Human papillomavirus and cervical cancer. *Clin. Microbiol. Rev.* **2003**, *16*, 1–17. [CrossRef]
38. Yim, E.-K.; Park, J.-S. Biomarkers in cervical cancer. *Biomark Insights* **2007**, *1*, 215–225. [CrossRef]
39. Sahasrabuddhe, V.V.; Luhn, P.; Wentzensen, N. Human papillomavirus and cervical cancer: Biomarkers for improved prevention efforts. *Future Microbiol.* **2011**, *6*, 1083–1098. [CrossRef]
40. Twu, N.F.; Yuan, C.C.; Yen, M.S.; Lai, C.R.; Chao, K.C.; Wang, P.H.; Wu, H.H.; Chen, Y.J. Expression of Aurora kinase A and B in normal and malignant cervical tissue: High Aurora A kinase expression in squamous cervical cancer. *Eur. J. Obstet. Gynecol. Reprod. Biol.* **2009**, *142*, 57–63. [CrossRef]
41. van Dam, P.A.; Rolfo, C.; Ruiz, R.; Pauwels, P.; Van Berckelaer, C.; Trinh, X.B.; Ferri Gandia, J.; Bogers, J.P.; Van Laere, S. Potential new biomarkers for squamous carcinoma of the uterine cervix. *J. ESMO Open* **2018**, *3*, e000352. [CrossRef]
42. Suman, S.; Mishra, A. Network analysis revealed aurora kinase dysregulation in five gynecological types of cancer. *Oncol. Lett.* **2018**, *15*, 1125–1132. [CrossRef]
43. Ahmed, S.; Obaseki, D.; Mayun, A.; Mohammed, A.; Rafindadi, A.; Abdul, M. The role of biomarkers ($p16^{INK4a}$ and Ki-67) in cervical cancer screening: An appraisal. *Ann. Trop. Pathol.* **2017**, *8*, 1–4. [CrossRef]
44. Cheah, P.L.; Looi, L.M.; Mun, K.S.; Abdoul Rahman, N.; Teoh, K.H. Implications of continued upregulation of p16(INK4a) through the evolution from high-grade squamous intraepithelial lesion to invasive squamous carcinoma of the cervix. *Malays. J. Pathol.* **2011**, *33*, 83–87. [PubMed]
45. von Knebel Doeberitz, M.; Reuschenbach, M.; Schmidt, D.; Bergeron, C. Biomarkers for cervical cancer screening: The role of p16(INK4a) to highlight transforming HPV infections. *Expert Rev. Proteom.* **2012**, *9*, 149–163. [CrossRef] [PubMed]
46. Shi, Q.; Xu, L.; Yang, R.; Meng, Y.; Qiu, L. Ki-67 and P16 proteins in cervical cancer and precancerous lesions of young women and the diagnostic value for cervical cancer and precancerous lesions. *Oncol. Lett.* **2019**, *18*, 1351–1355. [CrossRef] [PubMed]
47. Tornesello, M.L.; Buonaguro, L.; Giorgi-Rossi, P.; Buonaguro, F.M. Viral and Cellular Biomarkers in the Diagnosis of Cervical Intraepithelial Neoplasia and Cancer. *BioMed Res. Int.* **2013**, *2013*, 519619. [CrossRef]
48. Organista-Nava, J.; Gómez-Gómez, Y.; Garibay-Cerdenares, O.L.; Leyva-Vázquez, M.A.; Illades-Aguiar, B. Cervical cancer stem cell-associated genes: Prognostic implications in cervical cancer (Review). *Oncol. Lett.* **2019**, *18*, 7–14. [CrossRef]
49. Gong, P.; Wang, Y.; Gao, Y.; Gao, M.; Liu, L.; Qu, P.; Jin, X.; Gao, Q. Msi1 promotes tumor progression by epithelial-to-mesenchymal transition in cervical cancer. *Hum. Pathol.* **2017**, *65*, 53–61. [CrossRef]
50. Liu, X.; Yang, W.T.; Zheng, P.S. Msi1 promotes tumor growth and cell proliferation by targeting cell cycle checkpoint proteins p21, p27 and p53 in cervical carcinomas. *Oncotarget* **2014**, *5*, 10870–10885. [CrossRef]
51. Hou, T.; Zhang, W.; Tong, C.; Kazobinka, G.; Huang, X.; Huang, Y.; Zhang, Y. Putative stem cell markers in cervical squamous cell carcinoma are correlated with poor clinical outcome. *BMC Cancer* **2015**, *15*, 785. [CrossRef]
52. Tulake, W.; Yuemaier, R.; Sheng, L.; Ru, M.; Lidifu, D.; Abudula, A. Upregulation of stem cell markers ALDH1A1 and OCT4 as potential biomarkers for the early detection of cervical carcinoma. *Oncol. Lett.* **2018**, *16*, 5525–5534. [CrossRef]
53. Yadav, S.; Verma, A.; Sarin, N.; Singh, S. Expression of epidermal growth factor receptor in squamous cell carcinoma of uterine cervix. *Clin. Cancer Investig. J.* **2019**, *8*, 227–231. [CrossRef]
54. Chen, Q.; Huang, Y.; Shao, L.; Han-Zhang, H.; Yang, F.; Wang, Y.; Liu, J.; Gan, J. An EGFR-Amplified Cervical Squamous Cell Carcinoma Patient with Pulmonary Metastasis Benefits from Afatinib: A Case Report. *Oncotargets Ther.* **2020**, *13*, 1845–1849. [CrossRef] [PubMed]
55. Kim, B.W.; Cho, H.; Choi, C.H.; Ylaya, K.; Chung, J.-Y.; Kim, J.-H.; Hewitt, S.M. Clinical significance of OCT4 and SOX2 protein expression in cervical cancer. *BMC Cancer* **2015**, *15*, 1015. [CrossRef] [PubMed]
56. Ammothumkandy, A.; Maliekal, T.T.; Bose, M.V.; Rajkumar, T.; Shirley, S.; Thejaswini, B.; Giri, V.G.; Krishna, S. CD66 and CD49f expressing cells are associated with distinct neoplastic phenotypes and progression in human cervical cancer. *Eur. J. Cancer* **2016**, *60*, 166–178. [CrossRef]

57. Javed, S.; Sharma, B.K.; Sood, S.; Sharma, S.; Bagga, R.; Bhattacharyya, S.; Rayat, C.S.; Dhaliwal, L.; Srinivasan, R. Significance of CD133 positive cells in four novel HPV-16 positive cervical cancer-derived cell lines and biopsies of invasive cervical cancer. *BMC Cancer* **2018**, *18*, 357. [CrossRef]
58. Gutiérrez-Hoya, A.; Zerecero-Carreón, O.; Valle-Mendiola, A.; Moreno-Lafont, M.; López-Santiago, R.; Weiss-Steider, B.; Soto-Cruz, I. Cervical Cancer Cells Express Markers Associated with Immunosurveillance. *J. Immunol. Res.* **2019**, *2019*, 1242979. [CrossRef]
59. Kori, M.; Yalcin Arga, K. Potential biomarkers and therapeutic targets in cervical cancer: Insights from the meta-analysis of transcriptomics data within network biomedicine perspective. *PLoS ONE* **2018**, *13*, e0200717. [CrossRef]
60. Wang, Z.; Burge, C.B. Splicing regulation: From a parts list of regulatory elements to an integrated splicing code. *RNA* **2008**, *14*, 802–813. [CrossRef]
61. Lam, B.J.; Hertel, K.J. A general role for splicing enhancers in exon definition. *RNA* **2002**, *8*, 1233–1241. [CrossRef]
62. Long, J.C.; Caceres, J.F. The SR protein family of splicing factors: Master regulators of gene expression. *Biochem. J.* **2008**, *417*, 15–27. [CrossRef]
63. Howard, J.M.; Sanford, J.R. The RNAissance family: SR proteins as multifaceted regulators of gene expression. *Wiley Interdiscip. Rev. RNA* **2015**, *6*, 93–110. [CrossRef] [PubMed]
64. Zhou, Z.; Fu, X.-D. Regulation of splicing by SR proteins and SR protein-specific kinases. *Chromosoma* **2013**, *122*, 191–207. [CrossRef] [PubMed]
65. Somberg, M. Cellular and Viral Factors that Control Human Papillomavirus Type 16 Late Gene Expression. Ph.D. Thesis, Comprehensive Summary. Acta Universitatis Upsaliensis, Uppsala, Sweden, 2011.
66. Twyffels, L.; Gueydan, C.; Kruys, V. Shuttling SR proteins: More than splicing factors. *FEBS J.* **2011**, *278*, 3246–3255. [CrossRef] [PubMed]
67. Prescott, E.L.; Brimacombe, C.L.; Hartley, M.; Bell, I.; Graham, S.; Roberts, S. Human papillomavirus type 1 E1^E4 protein is a potent inhibitor of the serine-arginine (SR) protein kinase SRPK1 and inhibits phosphorylation of host SR proteins and of the viral transcription and replication regulator E2. *J. Virol.* **2014**, *88*, 12599–12611. [CrossRef] [PubMed]
68. Karni, R.; de Stanchina, E.; Lowe, S.W.; Sinha, R.; Mu, D.; Krainer, A.R. The gene encoding the splicing factor SF2/ASF is a proto-oncogene. *Nat. Struct. Mol. Biol.* **2007**, *14*, 185–193. [CrossRef] [PubMed]
69. Jia, R.; Li, C.; McCoy, J.P.; Deng, C.-X.; Zheng, Z.-M. SRp20 is a proto-oncogene critical for cell proliferation and tumor induction and maintenance. *Int. J. Biol. Sci.* **2010**, *6*, 806–826. [CrossRef] [PubMed]
70. Ajiro, M.; Jia, R.; Yang, Y.; Zhu, J.; Zheng, Z.-M. A genome landscape of SRSF3-regulated splicing events and gene expression in human osteosarcoma U2OS cells. *Nucleic Acids Res.* **2016**, *44*, 1854–1870. [CrossRef]
71. Jia, R.; Ajiro, M.; Yu, L.; McCoy, P., Jr.; Zheng, Z.M. Oncogenic splicing factor SRSF3 regulates ILF3 alternative splicing to promote cancer cell proliferation and transformation. *RNA* **2019**, *25*, 630–644. [CrossRef]
72. Jia, R.; Liu, X.; Tao, M.; Kruhlak, M.; Guo, M.; Meyers, C.; Baker, C.C.; Zheng, Z.-M. Control of the Papillomavirus Early-to-Late Switch by Differentially Expressed SRp20. *J. Virol.* **2009**, *83*, 167–180. [CrossRef]
73. Ajiro, M.; Tang, S.; Doorbar, J.; Zheng, Z.-M. Serine/Arginine-Rich Splicing Factor 3 and Heterogeneous Nuclear Ribonucleoprotein A1 Regulate Alternative RNA Splicing and Gene Expression of Human Papillomavirus 18 through Two Functionally Distinguishable cis Elements. *J. Virol.* **2016**, *90*, 9138–9152. [CrossRef]
74. Wieringa, H.W.; van der Zee, A.G.J.; de Vries, E.G.E.; van Vugt, M.A.T.M. Breaking the DNA damage response to improve cervical cancer treatment. *Cancer Treat. Rev.* **2016**, *42*, 30–40. [CrossRef] [PubMed]
75. Leonardi, S.; Buttarelli, M.; De Stefano, I.; Ferrandina, G.; Petrillo, M.; Babini, G.; Scambia, G.; Marino, C.; Mancuso, M.; Gallo, D. The relevance of prelamin A and RAD51 as molecular biomarkers in cervical cancer. *Oncotarget* **2017**, *8*, 94247–94258. [CrossRef] [PubMed]
76. Banerjee, N.S.; Moore, D.; Parker, C.J.; Broker, T.R.; Chow, L.T. Targeting DNA Damage Response as a Strategy to Treat HPV Infections. *Int. J. Mol. Sci.* **2019**, *20*, 5455. [CrossRef] [PubMed]
77. Sample, K.M. DNA repair gene expression is associated with differential prognosis between HPV16 and HPV18 positive cervical cancer patients following radiation therapy. *Sci. Rep.* **2020**, *10*, 2774. [CrossRef]
78. Yang, X.; Zhan, P.; Feng, S.; Ji, H.; Tian, W.; Wang, M.; Cheng, C.; Song, B. SRSF6 regulates alternative splicing of genes involved in DNA damage response and DNA repair in HeLa cells. *Oncol. Rep.* **2020**, *44*, 1851–1862. [CrossRef]
79. Geuens, T.; Bouhy, D.; Timmerman, V. The hnRNP family: Insights into their role in health and disease. *Hum. Genet.* **2016**, *135*, 851–867. [CrossRef]
80. Chen, Q.; Jin, M.; Zhu, J.; Xiao, Q.; Zhang, L. Functions of Heterogeneous Nuclear Ribonucleoproteins in Stem Cell Potency and Differentiation. *BioMed Res. Int.* **2013**, *2013*, 623978. [CrossRef]
81. Shao, X.-Y.; Dong, J.; Zhang, H.; Wu, Y.-S.; Zheng, L. Prognostic Value and Potential Role of Alternative mRNA Splicing Events in Cervical Cancer. *Front. Genet.* **2020**, *11*, 11. [CrossRef]
82. Erdem, M.; Özgül, I.; Erson-Bensan, A.E. HNRNPA1 (Heterogeneous Nuclear Ribonucleoprotein A1). *Atlas Genet. Cytogenet. Oncol. Haematol.* **2019**, *23*, 137–142. [CrossRef]
83. Roman, A.; Munger, K. The papillomavirus E7 proteins. *Virology* **2013**, *445*, 138–168. [CrossRef]
84. Vande Pol, S.B.; Klingelhutz, A.J. Papillomavirus E6 oncoproteins. *Virology* **2013**, *445*, 115–137. [CrossRef] [PubMed]

85. Narisawa-Saito, M.; Kiyono, T. Basic mechanisms of high-risk human papillomavirus-induced carcinogenesis: Roles of E6 and E7 proteins. *Cancer Sci.* **2007**, *98*, 1505–1511. [CrossRef] [PubMed]
86. Dalenc, F.; Drouet, J.; Ader, I.; Delmas, C.; Rochaix, P.; Favre, G.; Cohen-Jonathan, E.; Toulas, C. Increased expression of a COOH-truncated nucleophosmin resulting from alternative splicing is associated with cellular resistance to ionizing radiation in HeLa cells. *Int. J. Cancer* **2002**, *100*, 662–668. [CrossRef]
87. Zhu, W.; Pan, X.; Yang, Z.; Xing, P.; Zhang, Y.; Li, F.; Lu, X. Expression and prognostic significance of TAp73 and ΔNp73 in FIGO stage I-II cervical squamous cell carcinoma. *Oncol. Lett.* **2015**, *9*, 2090–2094. [CrossRef] [PubMed]
88. Siegfried, Z.; Karni, R. The role of alternative splicing in cancer drug resistance. *Curr. Opin. Genet. Dev.* **2018**, *48*, 16–21. [CrossRef]
89. Sciarrillo, R.; Wojtuszkiewicz, A.; Assaraf, Y.G.; Jansen, G.; Kaspers, G.J.L.; Giovannetti, E.; Cloos, J. The role of alternative splicing in cancer: From oncogenesis to drug resistance. *Drug Resist. Updates* **2020**, *53*, 100728. [CrossRef]
90. Box, J.K.; Paquet, N.; Adams, M.N.; Boucher, D.; Bolderson, E.; O'Byrne, K.J.; Richard, D.J. Nucleophosmin: From structure and function to disease development. *BMC Mol. Biol.* **2016**, *17*, 19. [CrossRef]
91. Di, C.X.; Yang, L.N.; Zhang, H.; An, L.Z.; Zhang, X.; Ma, X.F.; Sun, C.; Wang, X.H.; Yang, R.; Wu, Z.H.; et al. Effects of carbon-ion beam or X-ray irradiation on anti-apoptosis ΔNp73 expression in HeLa cells. *Gene* **2013**, *515*, 208–213. [CrossRef]
92. Ishimoto, O.; Kawahara, C.; Enjo, K.; Obinata, M.; Nukiwa, T.; Ikawa, S. Possible Oncogenic Potential of ΔNp73: A newly identified isoform of human p73. *Cancer Res.* **2002**, *62*, 636–641.
93. Liang, X.; Chen, B.; Zhong, J. Association of P73 polymorphisms with susceptibilities of cervical carcinoma: A meta-analysis. *Oncotarget* **2017**, *8*, 57409–57413. [CrossRef]
94. Ye, H.; Guo, X. TP73 is a credible biomarker for predicting clinical progression and prognosis in cervical cancer patients. *Biosci. Rep.* **2019**, *39*, BSR20190095. [CrossRef] [PubMed]
95. Liu, S.S.; Chan, K.Y.-K.; Leung, R.C.-Y.; Law, H.K.-W.; Leung, T.-W.; Ngan, H.Y.-S. Enhancement of the radiosensitivity of cervical cancer cells by overexpressing p73α. *J. Mol. Cancer Ther.* **2006**, *5*, 1209–1215. [CrossRef] [PubMed]
96. Kumar, L.; Harish, P.; Malik, P.S.; Khurana, S. Chemotherapy and targeted therapy in the management of cervical cancer. *Curr. Probl. Cancer* **2018**, *42*, 120–128. [CrossRef] [PubMed]
97. Ji, H.; Li, B.; Zhang, S.; He, Z.; Zhou, Y.; Ouyang, L. Crk-like adapter protein is overexpressed in cervical carcinoma, facilitates proliferation, invasion and chemoresistance, and regulates Src and Akt signaling. *Oncol. Lett.* **2016**, *12*, 3811–3817. [CrossRef] [PubMed]
98. Song, Q.; Yi, F.; Zhang, Y.; Jun Li, D.K.; Wei, Y.; Yu, H.; Zhang, Y. CRKL regulates alternative splicing of cancer-related genes in cervical cancer samples and HeLa cell. *BMC Cancer* **2019**, *19*, 499. [CrossRef] [PubMed]
99. Wang, Y.; Liu, L.; Chen, Z. Transcriptome profiling of cervical cancer cells acquired resistance to cisplatin by deep sequencing. *Artif. Cells Nanomed. Biotechnol.* **2019**, *47*, 2820–2829. [CrossRef]
100. Ye, G.; Lu, Q.; Zhao, W.; Du, D.; Jin, L.; Liu, Y. Fucoxanthin induces apoptosis in human cervical cancer cell line HeLa via PI3K/Akt pathway. *Tumor Biol.* **2014**, *35*, 11261–11267. [CrossRef]
101. Lu, G.Y.; Huang, S.M.; Liu, S.T.; Liu, P.Y.; Chou, W.Y.; Lin, W.S. Caffeine induces tumor cytotoxicity via the regulation of alternative splicing in subsets of cancer-associated genes. *Int. J. Biochem. Cell Biol.* **2014**, *47*, 83–92. [CrossRef]
102. Di, C.; Syafrizayanti; Zhang, Q.; Chen, Y.; Wang, Y.; Zhang, X.; Liu, Y.; Sun, C.; Zhang, H.; Hoheisel, J.D. Function, clinical application, and strategies of Pre-mRNA splicing in cancer. *Cell Death Differ.* **2019**, *26*, 1181–1194. [CrossRef]
103. Zhang, Q.; Di, C.; Yan, J.; Wang, F.; Qu, T.; Wang, Y.; Chen, Y.; Zhang, X.; Liu, Y.; Yang, H.; et al. Inhibition of SF3b1 by pladienolide B evokes cycle arrest, apoptosis induction and p73 splicing in human cervical carcinoma cells. *Artif. Cells Nanomed. Biotechnol.* **2019**, *47*, 1273–1280. [CrossRef]
104. Chen, Q.; Cai, D.; Li, M.; Wu, X. The homologous recombination protein RAD51 is a promising therapeutic target for cervical carcinoma. *Oncol. Rep.* **2017**, *38*, 767–774. [CrossRef] [PubMed]
105. Kędzierska, H.; Piekiełko-Witkowska, A. Splicing factors of SR and hnRNP families as regulators of apoptosis in cancer. *Cancer Lett.* **2017**, *396*, 53–65. [CrossRef] [PubMed]

genes

MDPI

Article

Quantitative Proteomics of Urinary Bladder Cancer Cell Lines Identify UAP1 as a Potential Therapeutic Target

Vinuth N. Puttamallesh [1,2], Barnali Deb [1,3], Kirti Gondkar [1,2], Ankit Jain [1], Bipin Nair [2], Akhilesh Pandey [1,3,4,5,6], Aditi Chatterjee [1,2,3], Harsha Gowda [1,2,3] and Prashant Kumar [1,2,3,*]

1 Institute of Bioinformatics, International Technology Park, Bangalore 560066, India; vinuth@ibioinformatics.org (V.N.P.); barnali@ibioinformatics.org (B.D.); kirti@ibioinformatics.org (K.G.); ankit@ibioinformatics.org (A.J.); pandey.akhilesh@mayo.edu (A.P.); aditi@ibioinformatics.org (A.C.); harsha@ibioinformatics.org (H.G.)

2 Amrita School of Biotechnology, Amrita Vishwa Vidyapeetham, Kollam 690525, India; bipin@am.amrita.edu

3 Manipal Academy of Higher Education, Madhav Nagar, Manipal 576104, India

4 Department of Laboratory Medicine and Pathology, Mayo Clinic, Rochester, MN 55905, USA

5 Center for Individualized Medicine, Mayo Clinic, Rochester, MN 55905, USA

6 Center for Molecular Medicine, National Institute of Mental Health and Neurosciences, Bangalore 560029, India

* Correspondence: prashant@ibioinformatics.org; Tel.: +91-80-28416140; Fax: +91-80-28416132

Received: 4 May 2020; Accepted: 28 May 2020; Published: 8 July 2020

Abstract: Bladder carcinoma (BC) incidence and mortality rates are increasing worldwide. The development of novel therapeutic strategies is required to improve clinical management of this cancer. Aberrant protein expression may lead to cancer initiation and progression. Therefore, the identification of these potential protein targets and limiting their expression levels would provide alternative treatment options. In this study, we utilized a liquid-chromatography tandem mass spectrometry-based global proteomics approach to identify differentially expressed proteins in bladder cancer cell lines. A total of 3913 proteins were identified in this study, of which 479 proteins were overexpressed and 141 proteins were downregulated in 4 out of 6 BC cell lines when compared with normal human urothelial cell line (TERT-NHUC). We evaluated the role of UDP-N-acetylhexosamine pyrophosphorylase (UAP1) in bladder cancer pathogenesis. The silencing of UAP1 led to reduction in proliferation, invasion, colony formation and migration capability of bladder cancer cell lines. Thus, our study reveals UAP1 as a promising therapeutic target for bladder cancer.

Keywords: quantitative proteomics; urothelial carcinoma; molecular subtypes; therapeutic target

1. Introduction

Bladder carcinoma (BC) is the most common malignancy of the urinary tract. Majority of the patients are diagnosed as non-muscle invasive (NMIBC; 60%) with a recurrence rate of 50%–70% and about 20% of patients are diagnosed as muscle invasive bladder cancer (MIBC) [1]. The risk of progression for NMIBC to MIBC after 5 years ranges from 6% to 45% [2,3]. MIBCs are biologically aggressive, with less than 15% five-year survival rate [4]. In the past few years, the surgical advancements have improved the outcomes of the patients. However, despite the use of neoadjuvant and adjuvant therapies, the long term survival rates in BC patients has remained unchanged for over a decade [5]. Furthermore, progress is required to develop a common molecular classifier that can be used for effective clinical decisions [6].

Several FDA approved immune checkpoint inhibitors, such as atezoliumab, nivoluma, pembrolizmab, durlumab and avelumab, were used to treat bladder cancer, with response rates ranging from 15% to 25% [7,8]. Moreover, extensive efforts have been made to identify therapeutic targets for BC, by targeting proteins belonging to various biological functions. This includes casein kinase 2 (CK2), C-X-C motif chemokine ligand 1(CXCL1), eukaryotic translation initiation factor 3 subunit D (EIF3D), adenylate kinase 4 and yes-associated protein 1 (YAP1) [9–12]. Alternatively, several proteins which are studied in other cancer types, such as are androgen receptor (AR), aurora kinase A (AURKA), epidermal growth factor receptor (EGFR), focal adhesion kinase (FAK), fibroblast growth factor receptor (FGFR) and transforming growth factor-β-induced (TGFBI) have also been explored in BC for their potential as therapeutic targets [13–16]. Though all these efforts provided multiple avenues for reducing the tumor burden in BC, they fell short in successful translation as treatment options.

Proteomics is a high-throughput technology that can be used to identify differentially expressed proteins that potentially play an important role in pathogenesis. A rapidly evolving technology platform is known to have the potential to identify novel proteins involved in key biological processes in the cell that may serve as potential drug targets. Therefore, an unbiased investigation of the proteomic alterations in BC using high resolution mass spectrometry-based approach will aid in identifying alternative therapeutic targets.

In this study, we investigated six BC cell lines (SW780, RT112, VMCUB1, T24, J82 and UMUC3), which were previously characterized as luminal, basal and non-type subtype [17]. Previously, it was shown that the EMT score of non-type molecular subtype is more "mesenchymal-like," whereas the luminal/basal subtypes are "epithelial-like." The non-type subtype cell lines show an increased migratory and invasive phenotype, reflecting typical characteristics of a mesenchymal-like phenotype [18]. We performed quantitative proteomic analysis to identify global proteomic changes in these BC cell lines. Our study identified and quantified 3913 proteins. We explored the role of UAP1 for the first time in bladder carcinoma, which was overexpressed in basal and non-type subtype cell lines. UAP1 plays an important role in energy metabolism and has been reported to be involved in prostate cancer pathogenesis [19]. Furthermore, the functional characterization of UAP1 using siRNA-based silencing in non-type subtype BC cell lines resulted in a decrease in the cell proliferation, colony formation, invasion and migration properties of highly aggressive bladder cancer cell lines.

2. Materials and Methods

2.1. Cell Lines and Culture Conditions

BC cell lines SW780, RT112, VMCUB1, T24, J82 and UMUC3 and normal human urothelial cell line TERT-NHUC (JTERT) were received from Prof. Jean Paul Thiery (Department of Biochemistry, National University of Singapore, Singapore). The non-type (UMUC3, J82, T24), luminal (RT-112, SW780) and basal (VMCUB-1) BC cell lines were cultured using DMEM medium supplemented with 10% Fetal Bovine Serum (FBS) and 1% Penicillin/Streptomycin. TERT-NHUC was grown in KGM gold™ keratinocyte growth medium containing supplements (bovine pituitary extract, human epidermal growth factor, insulin, hydrocortisone, epinephrine, transferrin, gentamicin and amphotericin-B). All the cell lines were maintained at 37 °C in a humidified 5% CO_2 incubator.

2.2. Cell lysis and Protein Digestion

Cells were lysed in lysis buffer (50 mM triethyl ammonium bicarbonate (TEABC) pH 8.0, 2% SDS, 1 mM sodium orthovanadate, 2.5 mM sodium pyrophosphate, 1 mM β-glycerophosphate, 1 mM sodium fluoride), sonicated and centrifuged at 16,000× *g* for 20 min. Protein estimation of the clarified lysate was done using bicinchoninic acid assay (BCA) (Pierce, Waltham, MA, USA) according to manufacturer's instructions. Equal amount of protein from all the cell lines were reduced, using 10 mM Dithiotheritol (DTT) for 30 min at 60 °C, followed by alkylation using 20 mM Iodoacetamide (IAA) in dark for 10 min at room temperature. Reduced and alkylated protein lysate was subjected to acetone precipitation, using 7 volumes of pre-chilled acetone to remove SDS from the solution. Protein digestion was performed using sequencing grade trypsin (Promega, Madison, WI, USA) at 1:20

enzyme:substrateratio at 37 °C for 12–14 h. The tryptic peptides were vacuum dried and stored until further use.

2.3. TMT Labeling and Basic pH Reverse Phase Liquid Chromatography (bRPLC)

TMT labeling was done according to the manufacturer's instructions, with minor modifications. Briefly, the TMT labels were reconstituted in 41 µL of anhydrous acetonitrile (ACN) and trypsin digested peptide samples were reconstituted in 100 µL of 50 mM TEABC (pH 8.0). Tandem mass tag (TMT) labels 128N, 128C, 129C, 130N, 130C and 131 were used for labeling bladder cancer cell line samples, and TMT label 126 was used for TERT-NHUC control cell line. TMT labels were mixed with respective samples and the reaction was incubated for 1 h at room temperature. After incubation, the reaction was quenched with 8 µL of 5% hydroxylamine. The labeled peptides were lyophilized and subjected to basic pH reverse phase chromatography (bRPLC). Lyophilized samples were reconstituted in bRPLC solvent A (10 mM TAEBC, pH 9) and were separated on XBridge BEH C18 Column (Waters, UK), using solvent B (10 mM TEABC with 90% acetonitrile, pH 9). The column was equilibrated at 5% Solvent A from 0-to-5 min; the solvent B percentage was gradually increased from 5% to 55% in the following 60 min and then increased from 55–90% for the following 10 min, and then maintained at 90% Solvent B for 10 min before being decreased to 5% for 2 min on an Agilent 1100 Liquid Chromatography (LC) system, with a flow rate of 1 mL/min. A total of 96 fractions were collected over a period of 60 min gradient and later concatenated into 12 fractions. Fractionated samples were lyophilized before LC-MS/MS analysis.

2.4. LC-MS/MS Analysis

LC-MS/MS analysis was done on an Orbitrap Fusion mass spectrometer (Thermo Electron, Bremen, Germany), interfaced with Easy-nLC1000 nanoflow LC system (Thermo Scientific, Odense, Denmark). The peptides were reconstituted in 0.1% formic acid and loaded onto a trap column (nanoviper 2 cm, 3 µ magic C18Aq, Thermo Scientific). Peptides were resolved on an analytical column (nanoviper 25 cm (75 µm silica capillary, 3 µm magic C18, Thermo Scientific)), at a flow rate of 300 nL/min, using a linear gradient of 7–35% solvent B (0.1% formic acid in 100% ACN) over 100 min. The total run time including sample loading and column reconditioning was 120 min. Data-dependent acquisition with full scans in 350–1700 m/z range was carried out using an Orbitrap mass analyzer at a mass resolution of 120,000 at 400 m/z. Most intense precursor ions from a survey scan were selected for MS/MS fragmentation using higher energy collision dissociation (HCD) fragmentation, with 35% normalized collision energy and detected at a mass resolution of 30,000 at 400 m/z. AGC target value was set to 50,000, with maximum ion injection time of 150 ms. For MS3 analysis, synchronous precursor selection was enabled and 10 precursor ions were selected for fragmentation with 55% HCD collision energy.

2.5. Data Analysis

The mass spectrometry data were searched for using MASCOT (version 2.2.0) and SEQUEST search algorithms against Human RefSeq database (version 89), using Proteome Discoverer (version 2.2 (Thermo Fisher Scientific, Bremen, Germany). The search parameters for both algorithms included carbamidomethylation of cysteine residues (57.02 Da), TMT modification at peptide N-terminus and lysine side chain as a fixed modification, oxidation of methionine (15.99 Da) as dynamic modification. MS/MS spectra were searched with a precursor mass tolerance of 10 ppm and fragment mass tolerance of 0.1 Da. Trypsin was specified as the protease, and a maximum of two missed cleavages were allowed. The data was searched against target decoy database and the false discovery rate was set to 1% at the peptide level. The TMT ratio for each peptide-spectrum match was calculated by the quantitation node.

2.6. Protein-Protein Interaction Networks

Interaction network was analyzed using the STRING functional protein association network (https://string-db.org; version: 11.0; University of Zurich, Zurich, Switzerland) [20]. The input consisted of the 122 proteins that were overexpressed in all the BC cell lines and was set to highest confidence (0.90) of active interaction. The disconnected nodes were hidden, and K-means clustering was conducted to identify three clusters in the dataset.

2.7. Western Blotting

Cell lines were cultured up to 70% confluency and cells were harvested using RIPA lysis buffer (10 mM Tris pH 7.4, 150 mMNaCl, 5mM EDTA, 1% Triton-X-100, 0.1% SDS containing protease and phosphatase inhibitor cocktails) and sonicated to extract proteins. Western blot analysis was performed as previously described, using 30 μg protein lysates [21]. Nitrocellulose membranes were hybridized with primary antibodies and developed using Luminol reagent (Santa Cruz Biotechnology, Dallas, TX, USA), as per the manufacturer's instructions. Anti-UAP1 (HPA014659) antibody and β-actin antibody were obtained from Sigma (St. Louis, MO, USA). Anti-GAPDH antibody was obtained from Abcam (Cambridge, MA, USA).

2.8. siRNA Transfection

ON-TARGETplusSMARTpool control siRNA and UAP1siRNA (catalog ID: L-017160-01-0005) were procured from Dharmacon (Lafayette, CO, USA), and BC cell lines (UMUC3, J82 and T24) were transfected with controlandUAP1siRNA using RNAiMAX reagent (Invitrogen, Carlsbad, CA, USA), according to the manufacturer's instructions. Cells were subjected to invasion assays and colony formation assays 48 h post-transfection, unless otherwise stated.

2.9. Cell Proliferation Assay

Cell proliferation assays were carried out as described previously [22]. Cells from UMUC3, J82 and T24 cell lines were seeded at a density of 4000 cells/well into a 96-well plate and cells were counted subsequently after every 48 h. Cell proliferation was determined using MTT (3-(4, 5-dimethylthiazol-2yl)-2, 5-diphenyl tetrazolium bromide) assays, as described. Absorbance was measured at 570 nm and 650 nm over a period of 4 days and growth kinetics were plotted. All the experiments were carried out in triplicates.

2.10. Colony Formation Assay

Colony formation assays were carried out as described previously, with minor modifications [21]. UMUC3, J82 and T24 cells were transfected with UAP1siRNA or scramble siRNA. 48 h post transfection, 1000 cells/well were seeded in 6-well plates with complete media and allowed to grow for 7 days. The resulting colonies were fixed with methanol, and stained with Giemsa (Sigma, St. Louis, MO, USA). The number of colonies per dish was counted. All experiments were performed in triplicate and standard deviation was calculated.

2.11. Wound Healing Assay

The wound healing assays were performed as described previously [23]. Briefly, 100,000 cells were seeded in 6 well plates in triplicates for each condition. UMUC3, J82 and T24 cells were treated with UAP1siRNA or control siRNA. The cells were allowed to grow until full confluency was achieved. Uniform size wound was introduced from end to end in a 6 well plate for each condition and a photograph was taken at 0 h. Cells were then observed for wound healing periodically and photomicrograph at 8 h was taken under microscope. All experiments were performed in triplicate, unless otherwise indicated.

2.12. Invasion Assay

Invasion assays were performed in a transwell system (BD Biosciences), with Matrigel-coated filters and cellular invasion was evaluated after 48 h, as described previously [24]. Briefly, invasiveness of the cells was assayed in the membrane invasion culture system using polyethylene terephthalate (PET) membrane (8-μm pore size), in the upper compartment of a transwell coated with Matrigel (BioCoat Matrigel Invasion Chamber; BD Biosciences). The cells were transfected with either control or UAP1siRNA and seeded at 2.0×10^4 cells per 500 μL of media on the Matrigel-coated PET membrane in the upper compartment. The lower compartment was filled with complete growth media and the plates were maintained at 37 °C for 48 h. At the end of the incubation time, the upper surface of the membrane was wiped with a cotton-tip applicator, to remove non migratory cells. Cells that migrated to the lower side of membrane were fixed and stained using 4% methylene blue. The membranes were cut out from the transwell and mounted on glass slide using Dibutylphthalate Polystyrene Xylene (DPX) and covered with a microscope cover slip. The number of cells that penetrated was counted for six randomly selected viewing fields at 10x magnification.

2.13. Statistical Analysis

Statistical analyses were carried out using GraphPad Prism version 6 (GraphPad Software, La Jolla, CA, USA). For cell culture-based assays (proliferation, invasion, colony formation and wound healing) non-parametric test (Mann–Whitney U test) was used to assess statistical significance. For proteomics data, a statistical analysis was done using one-way ANOVA for individual proteins.

3. Results

3.1. Identification of Proteins with Altered Expression in Urinary Bladder Cancer Cell Lines through Quantitative Proteomics

We studied the protein expression of BC cell lines (SW780, RT112, VMCUB1, T24, J82 and UMUC3) and normal human urothelial carcinoma cell line (TERT-NHUC) using tandem mass tag (TMT)-based labeling technology, coupled with high-resolution mass spectrometry to identify differentially expressed proteins. The experimental workflow used for differential proteomic expression analysis in this study is depicted in Figure 1. We identified and quantified 3913 proteins across all the cell lines in triplicate mass spectrometry analysis and calculated their fold-change values based on reporter ion intensities (Supplementary Table S1; Supplementary Figure S1).

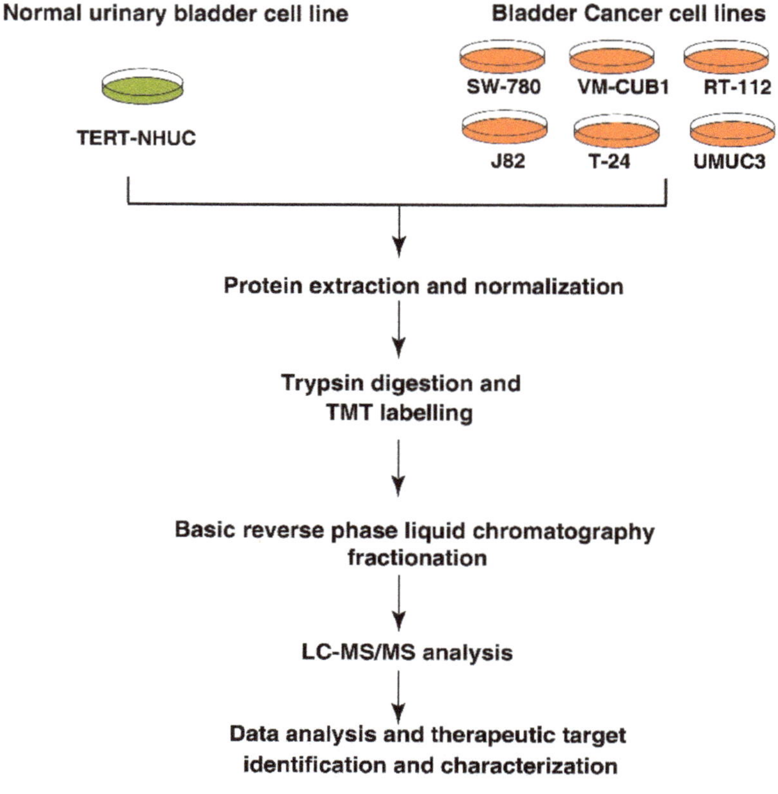

Figure 1. Workflow for quantitative proteomic analysis of urinary bladder cancer cell lines. Cultured bladder cancer (BC) cell lines and normal human urothelial carcinoma cell line (TERT-NHUC) were harvested in cell lysis buffer. Protein extraction was done using probe sonicator, followed by protein estimation, normalization, trypsin digestion and tandem mass tag (TMT) labeling. TMT labeled peptides were pooled and subjected to fractionation and LC-MS/MS analysis. Proteomics data were analyzed and candidate molecule was validated.

3.2. Distinct Protein Expression Pattern of Non-Type/Basal as Compared to Luminal Bladder Carcinoma Cell Lines

We compared the protein expression pattern of the non-type/basal and luminal subtypes of bladder carcinoma cell lines. Unsupervised clustering based on both rows and columns was conducted with one minus Kendall's correlation with average linkage and we observed that the luminal cell lines showed a distinct expression pattern (Figure 2A). We compared differentially expressed proteins (>2 fold) from all the non-type/basal and luminal cell lines. We identified 135 and 53 proteins, which were overexpressed and downregulated exclusively in non-type/basal cell lines (Figure 2B). Principle component analysis revealed that the luminal cell lines show distinct expression pattern as compared to the non-type/basal cell lines (Figure 2C).

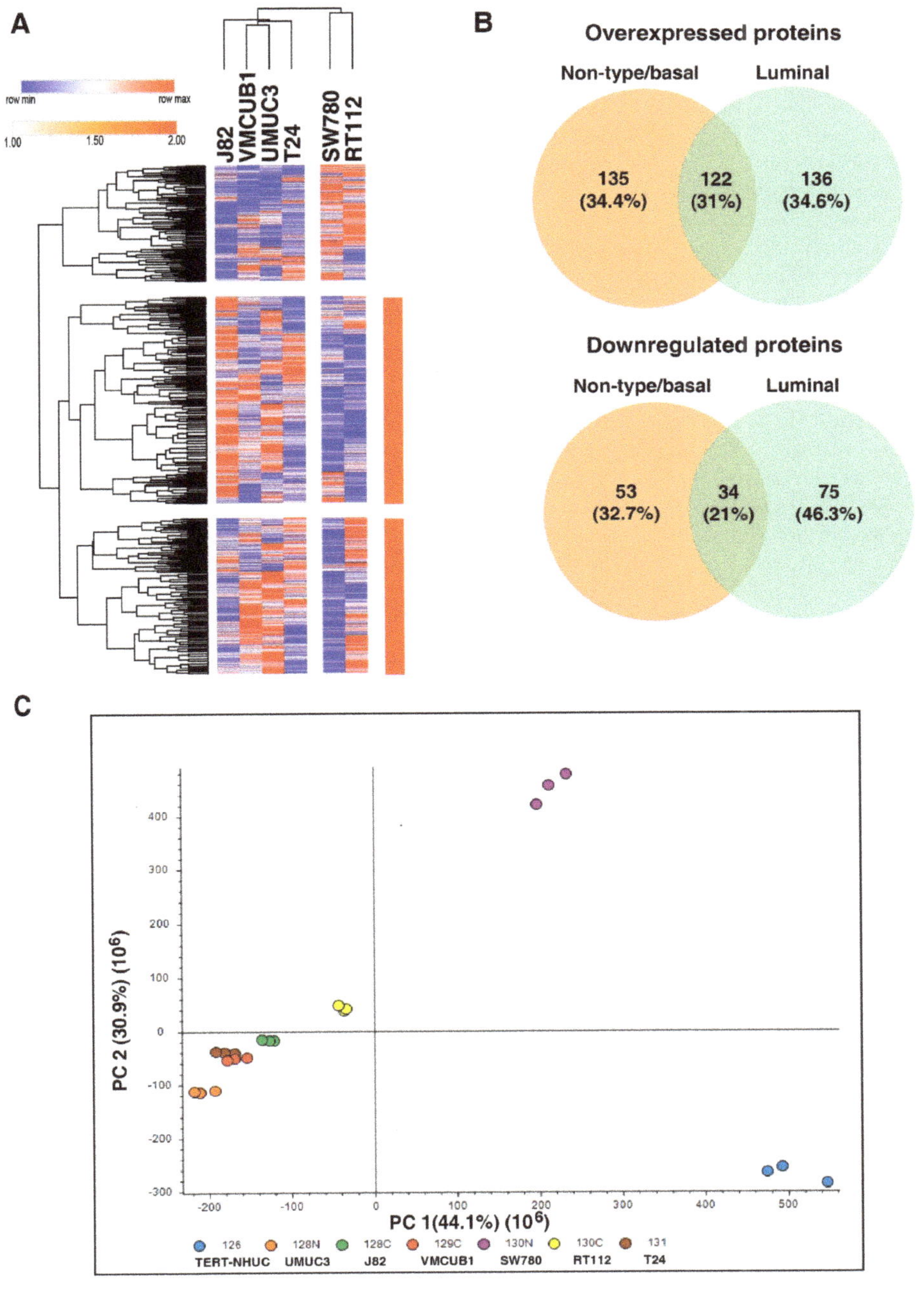

Figure 2. Comparison of the protein expression of non-type/basal and luminal subtype. (**A**) Unsupervised clustering showing the distinct expression pattern of non-type/basal and luminal cell lines. (**B**) Overexpressed and downregulated proteins in non-type/basal and luminal cell lines. (**C**) Principle component analysis of the protein expression of the bladder carcinoma cell lines.

3.3. DNA Replication Process and Cell Cycle Regulation and Progression Were Dysregulated across All Bladder Carcinoma Cell Lines

We identified 122 proteins to be overexpressed and 34 proteins to be downregulated across all the 6 BC cell lines, compared to the TERT-NHUC (control) cell line (Figure 3A). 479 proteins were overexpressed, and 141 proteins were downregulated ≥2 fold in four or more BC cell lines when compared with the TERT-NHUC (control) cell line. We checked the interaction between the proteins that were overexpressed across all BC cell lines. We observed that the proteins formed two major clusters; one cluster involved the MCM proteins; namely MCM2, MCM3, MCM4, MCM5, MCM6 and MCM7, which closely interact with each other and are involved in DNA replication, while the other cluster involved proteins involved in cell cycle regulation and progression, such as AURKB, AURKA, CDC20, RRM2, TOP2A, RACGAP KIF23, KIF4A, TPX2, CEP55, NUSAP1, CKAP2, CENPF, and so on (Figure 3B).

3.4. UAP1 Was Overexpressed in Bladder Carcinoma

We sought to identify the proteins that show higher expression in the basal and non-type subtype of bladder carcinoma, which might be attributed to the aggressive phenotype in these cells. We identified 39 proteins which were overexpressed in non-type and basal subtype (Figure 3C; Supplementary Table S2). We selected UDP-N-acetylhexosamine pyrophosphorylase (UAP1) as a candidate molecule to study the functional implication in BC, based on its function and association with cancer pathogenesis. Expression pattern of UAP1 in BC cell lines is represented in the box plot, as shown in Supplementary Figure S2.

3.5. Silencing of UAP1 Decreases Cellular Proliferation in Urinary Bladder Cancer Cells

Western blotting analysis confirmed the overexpression of UAP1 in most of the BC cell lines compared with the TERT-NHUC cell line (Figure 4A). We selected the 3 non-type BC cell lines (UMUC3, J82, T24), to study the role of UAP1 in cellular proliferation. Endogenous expression of UAP1 was silenced-using siRNA mediated knockdown and Western blot analysis confirmed the successful knockdown of UAP1 in BC cell lines (Figure 4B). We then assessed the effect of UAP1 knockdown on the cellular proliferation of BC cell lines. UAP1 siRNA and control siRNA transfected BC cell lines were grown in triplicate for 96 h, and cell viability was assessed at every 24 h intervals using MTT assay. We observed significant reduction in cellular proliferation of UAP1 knockdown BC cell lines when compared with the control cell line (Figure 4C). These results indicate that UAP1 plays a role in cellular proliferation in non-type BC cells.

3.6. Silencing of UAP1 Decreases Colony Forming Ability in Urinary Bladder Cancer Cells

After demonstrating the role of UAP1 in cellular proliferation, we continued to study the role of UAP1 in the colony forming ability of the BC cells. UAP1 siRNA transfected cells and control siRNA transfected cells were seeded onto 6 well plates at 1000 cells per well density and allowed to grow for 7 days. The colonies formed were fixed, stained, counted under microscope and photographed for representation. Silencing of UAP1 led to reduction in the number of colonies formed and reduced the size of colonies formed in UAP1 knockdown non-type BC cells (Figure 5A,B). Our results show the importance of UAP1 in the colony formation of BC cells.

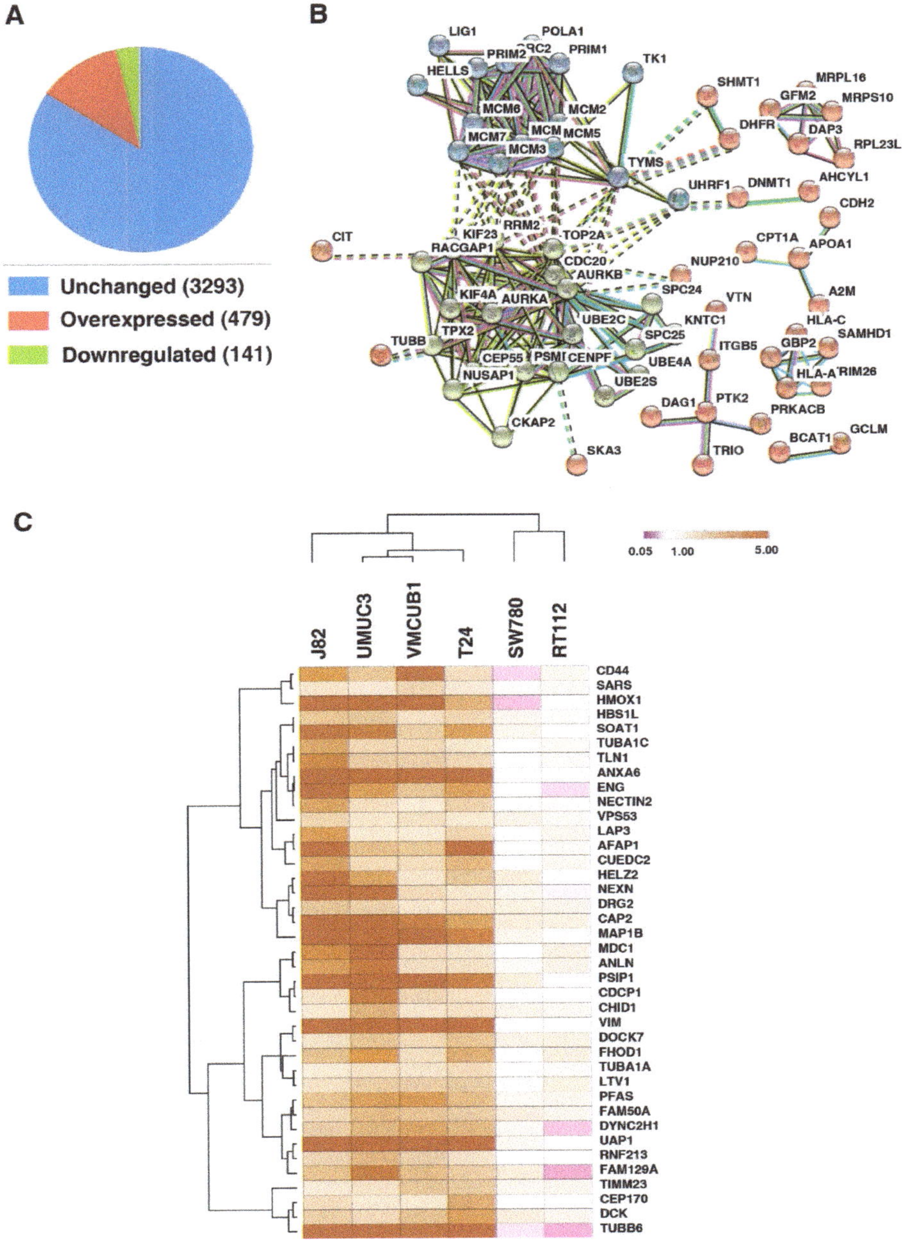

Figure 3. Global proteomic profiling of bladder carcinoma cell lines. (**A**) Total dysregulated proteins identified in the study in four or more bladder carcinoma cell lines, with 479 proteins overexpressed and 141 proteins downregulated. (**B**) Protein-protein interaction network showing the clusters of proteins with highest confidence (0.90) of interaction. (**C**) Heatmap showing 39 proteins that were overexpressed in the basal and non-type bladder carcinoma cell lines.

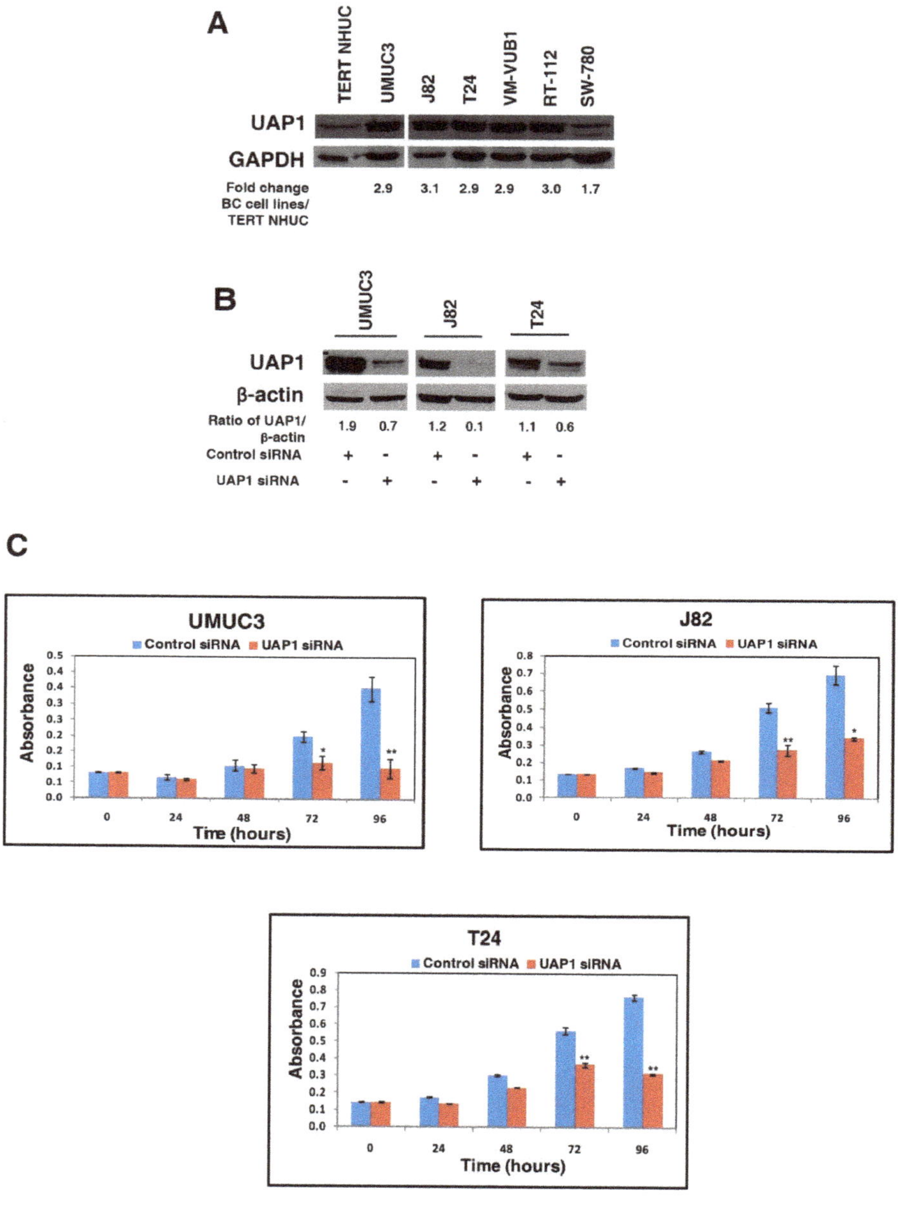

Figure 4. UAP1 silencing reduces cell proliferation of urinary bladder cancer cells. (**A**) Western blot analysis of UAP1 expression in BC cell lines and TERT-NHUC cell line validates the high expression of UAP1 in BC cell lines, as discovered from mass spectrometry analysis. (**B**) Western blot analysis of UAP1 silencing using siRNA mediated knockdown confirms the reduced expression of UAP1 after silencing in BC cell lines; β actin was used as a loading control. (**C**) Inhibition of UAP1 reduces cellular proliferation of BC cell lines (* $p < 0.05$; ** $p < 0.01$).

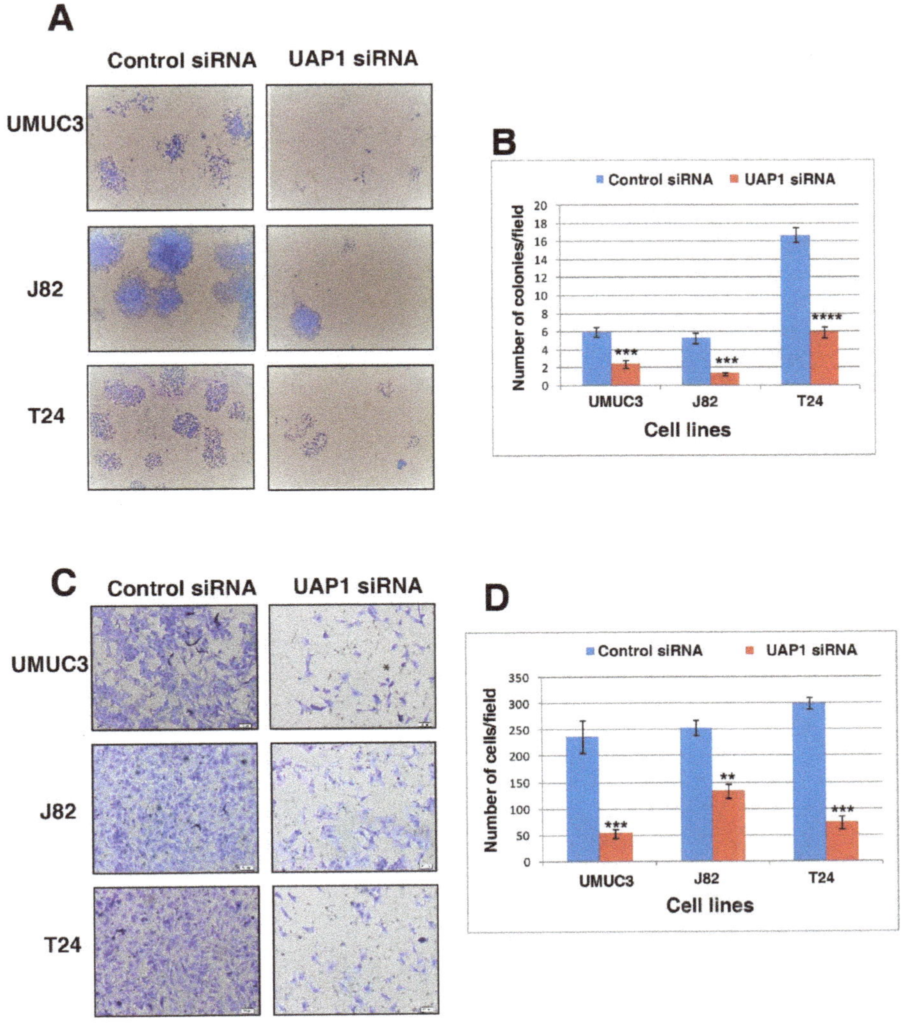

Figure 5. Inhibition of UAP1 affects the colony forming ability and reduces the invasive ability of urinary bladder cancer cell lines. (**A**) Colony formation assay was performed after siRNA mediated knockdown of UAP1 or control siRNA in BC cell lines. Colonies formed were fixed and stained using methylene blue. Images were captured and colonies were counted; reduced colony forming ability of BC cells after siRNA silencing was observed. (**B**) Graphical representation of reduction in colony forming ability of BC cell lines after UAP1 silencing (*** $p < 0.001$; **** $p < 0.0001$). (**C**) BC cell lines were transfected with UAP1 siRNA or control siRNA and invasion assay was performed. The assay was done in transwell system using Matrigel-coated filters and the cells that migrated to the lower chamber fixed, stained and images captured for counting and representation. (**D**) Graphical representation of reduction in invasive ability of BC cell lines after UAP1 silencing (** $p < 0.001$; *** $p < 0.001$).

3.7. Silencing of UAP1 Decreases Invasive Property in Urinary Bladder Cancer Cells

We further examined whether UAP1 has a role in BC invasiveness. Invasion assays were performed after silencing the endogenous expression of UAP1 using siRNA. BC cells transfected with UAP1 siRNA or control siRNA were transferred to the matrigel-coated PET membrane in the upper compartment and the plates were incubated at 37 °C for 48 h. We fixed and stained the migrated cells and counted the cells under a microscope. siRNA mediated silencing of UAP1 showed reduction in invasive capability of all the non-type BC cell lines (Figure 5C,D). Our results indicate a clear role of UAP1 in regulating invasive capability, which is a critical process during metastasis.

3.8. Silencing of UAP1 Decreases Cell Motility in Urinary Bladder Cancer Cells

Since the silencing of UAP1 reduces both the cell proliferation and colony forming ability of non-type subtype BC cells, we decided to explore whether UAP1 has any role in tumor cell motility. We performed scratch wound assays using non-type subtype BC cells with or without UAP1 silencing. Cells were grown till full confluence in 6 well plates and wounds were made in uniform size. After incubation for 8 h, control BC cells showed increased migration compared to UAP1 silenced BC cells. We took the images of wounds at 0 h and 8 h and calculated the distance covered by cells using Image J software I (Figure 6A,B). Our results from the cell motility assay clearly showed that UAP1 influences cell motility in BC cells.

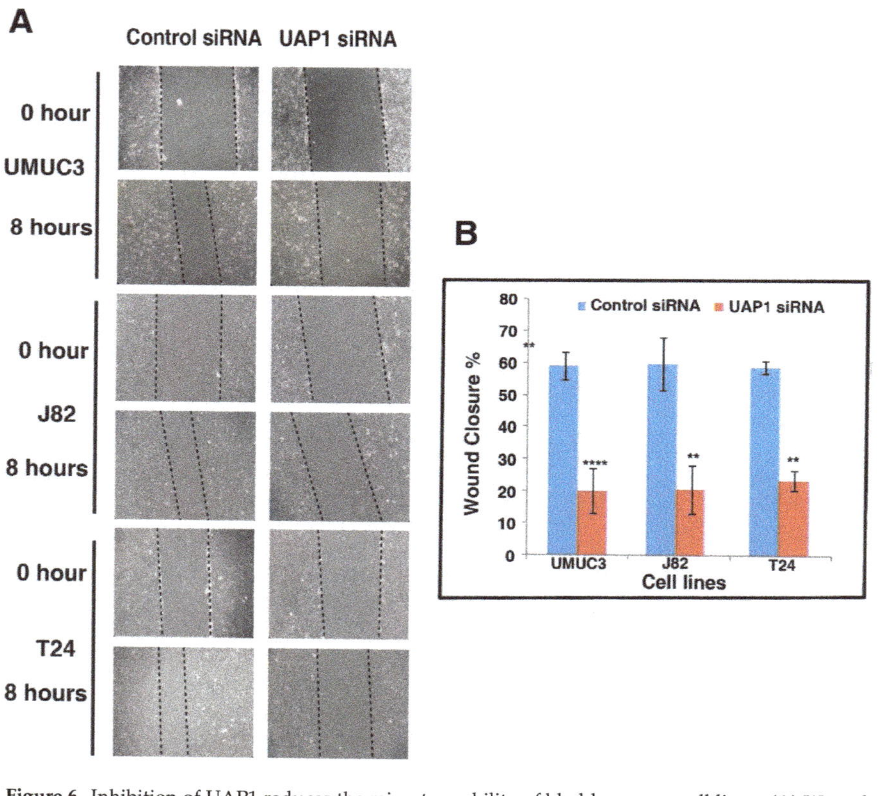

Figure 6. Inhibition of UAP1 reduces the migratory ability of bladder cancer cell lines. (**A**) Wound healing assay was carried out after transfection of BC cell lines using either UAP1 siRNA or control siRNA; scratches were made after cell confluence and monitored for 8 h till wound healing. (**B**) Representative graph shows the distance migrated by BC cell lines (* $p < 0.05$; ** $p < 0.01$; *** $p < 0.001$; **** $p < 0.0001$).

4. Discussion

BC management is challenging due to its presentation, histological subtypes and high recurrence rates. Surgical intervention and systemic chemotherapy are the preferred treatment options by clinicians, due to lack of bladder cancer specific targeted therapies [25]. The deeper knowledge on molecular mechanisms would help enormously to develop precision therapeutics. In addition, alternative therapeutic approaches need to be explored to improve the overall survival of patients, particularly in the case of aggressive tumors. In this study, we have conducted a global quantitative proteomics study to identify dysregulated proteins and potential targets which could lead to better treatment strategies in bladder carcinoma.

Cancer cell lines provide abundant information about cancer cell biology and alterations in protein repertoire in disease pathogenesis. In this study, we identified 122 proteins that were overexpressed across all the BC cell lines. These proteins might play an important role in cancer pathogenesis. The protein-protein interaction network revealed interaction between MCM proteins which formed the hub proteins. These proteins include a panel of minichromosome maintenance complex component (MCM) proteins; namely MCM2, MCM3, MCM4, MCM5, MCM6 and MCM7 [26,27]. MCM complex plays an important role in the DNA replication initiation. The active MCM2-7 double hexamer conducts bidirectional DNA synthesis at the S-phase of the cell cycle [28,29]. It has been reported that MCM complex subunits have been implicated in cell proliferation, invasion and metastasis in cancer [30–32]. A meta-analysis of the MCM proteins also reported that the higher expression of these proteins were related to worse prognosis in cancer patients [28]. Indeed, MCM expression is often observed in all epithelial layers and a high frequency of MCM-positive cells correlates with adverse clinical outcome in bladder carcinoma [33].

Apart from MCM family proteins, we also identified thymidine kinase (TK1), aurora kinase B (AURKB), cell division cycle protein 20 (CDC20), DNA topoisomerase 2-α (TOP2A), exportin-5 (XPO5) and several other proteins significantly overexpressed in our study [34–37]. Pharmaceutical compounds targeting Aurora A (such as Alisertib), have been extensively studied in preclinical models and observed to have synergy with several other targeted therapies that lead to tumor regression in various cancers [38]. Furthermore, TK1 has been reported as a tumor biomarker and it also exhibits potential in drug discovery and as a therapeutic target [39]. Interestingly, TK1 is not essential for normal cell growth, but it is important in the repair of DNA damage that may be caused by chemotherapy [6]. Alternatively, the overexpression of TK1 can reduce the efficacy of chemotherapy as well [39]. Cell cycle proteins are being widely studied for their potential in cancer therapeutics and further mechanistic studies would lead to their clinical utility in personalized therapy.

Our proteomics data identified 39 proteins which were specifically overexpressed more than 2-fold in basal and non-type subtype cell lines compared to the luminal subtype (table A). Among the list, only few molecules have been previously studied in bladder cancer tumorigenesis. Annexin-6 was overexpressed in pT1 grade3 in urothelial carcinoma [40]. Expression of vimentin was found in 69% cases of transitional cell carcinoma and its expression is associated with the grade of transitional cell carcinoma [41]. In accordance with our study, Wu et al. have also identified the expression of Anilin actin binding protein to be subtype-specific [42]. Similarly, the expression of Anilin actin binding protein was higher in the basal subtype compared to luminal subtype in bladder cancer cell lines.

Aberrant glycosylation has been gaining importance as one of the hallmarks of cancer. Alteration in metabolism and dysregulation in cellular energetics leads to cancer progression [43]. A tight association between the hexosamine biosynthetic pathway (HBP) and cell metabolism is widely reported [44]. Post-translational modification like glycosylation plays a key role in cell adhesion, migration, immune surveillance, cell signaling and cellular metabolism. HBP mediates glycosylation events through the final substrate UDP-GlcNAc. The conversion of GlcNAc-1P substrate to UDP-GlcNAc is mediated by UDP-N-acetylhexosamine pyrophosphorylase (UAP1) [45].

In our data, we identified that UDP-N-acetylhexosamine pyrophosphorylase (UAP1) was overexpressed more than 2-fold in all the basal and non-type subtypes of bladder cancer cell lines. Although the lack of biological replicates for proteomic analysis of each cell line is a limitation, UAP1 was chosen for further characterization, as it was overexpressed in four out of six cell lines analyzed. Functional assays to evaluate role of UAP1 in regulating proliferation, cell migration and colony formation were carried out in biological replicates across three different cell lines. UAP1 is the final enzyme in the hexosamine biosynthetic pathway (HBP), producing UDP-N-Acetylglucosamine (UDP-GlcNAc), which is a substrate for protein glycosylation. Abnormal protein glycosylation was previously shown to be associated with poor prognosis in bladder cancer patients [45]. A previous study in prostate cancer has shown the ability of UAP1 to protect cancer cells from endoplasmic reticulum stress and provide a growth advantage. High expression of UAP1 exhibited resistance against inhibitors of N-linked glycosylation (tunicamycin and 2-deoxyglucose) and targeting UAP1 blocked anchorage independent growth [19]. Earlier work from our group on N-glycoproteomic profile in bladder cancer cell lines has identified aberrant glycosylation in aggressive non-type subtype BC cell lines [46]. These results prompted us to further investigate UAP1 role in BC. Western blot analysis in our study confirmed that UAP1 was also overexpressed in the luminal cell lines, although not as high as in the basal or non-type cell lines. Further studies are required to understand its role in cancer progression, to comment on its role on the aggressive phenotype. Furthermore, knockdown of UAP1 using siRNA-based silencing in vitro showed a reduction in cell proliferation, invasion, colony formation and migration properties of BC cell lines. These results are novel to BC and demonstrate UAP1 role in tumorigenesis.

5. Conclusions

In conclusion, comprehensive quantitative proteomic analysis of urinary BC cell lines identified the overexpression of UAP1 for the first time. Knockdown of UAP1 reduced BC cell proliferation, invasion, colony formation and migration ability of these cell lines. Our data suggests that UAP1 could be a promising therapeutic target for aggressive urinary bladder cancer.

Supplementary Materials: The following are available online at http://www.mdpi.com/2073-4425/11/7/763/s1, Figure S1: Unsupervized clustering of all the proteins identified and quantified across 6 BC cell lines, Figure S2: Box plot representing expression pattern of UAP1 in BC cell lines. Table S1: Complete list of proteins identified and quantified in the study, Table S2: Proteins overexpressed in non-type and basal subtype BC cell lines.

Author Contributions: Conceptualization, P.K., A.P., and H.G.; methodology, V.N.P., AJ., and K.G.; software, V.N.P.; B.D., validation, V.N.P., A.C., and A.J.; formal analysis, V.N.P., H.G., and P.K.; investigation, V.N.P., H.G., and P.K.; writing—original draft preparation, V.N.P., B.D., K.G., and P.K.; writing—review and editing, A.C., B.N., H.G., and P.K.; visualization, V.N.P., H.G., and P.K.; supervision, H.G., and P.K.; project administration, P.K.; funding acquisition, H.G., A.P., and P.K. All authors have read and approved the final version of the manuscript.

Funding: This research was funded by Department of Science and Technology (DST), Ramanujan Fellowship, Government of India, grant number SB/S2/RJN-077/2015, and the APC was funded by SB/S2/RJN-077/2015, Department of Science and Technology (DST), Ramanujan Fellowship, Government of India.

Acknowledgments: We thank Jean Paul Thiery (Department of Biochemistry, National University of Singapore, Singapore) for providing bladder cancer cell lines and normal human urothelial cell line TERT-NHUC (JTERT). We thank the Department of Biotechnology (DBT), Government of India, for research support to the Institute of Bioinformatics (IOB), Bangalore. We thank the "Manipal Academy of Higher Education", Madhav Nagar, Manipal 576104, India, for research support to the Institute of Bioinformatics. Harsha Gowda is a Wellcome trust/DBT India Alliance Intermediate carrier Fellow. Prashant Kumar is a recipient of the Ramanujan Fellowship awarded by the Department of Science and Technology (DST), Government of India. BD is a recipient of the INSPIRE Fellowship, Department of Science and Technology, Government of India. KG is a recipient of the Senior Research Fellowship from the University Grants Commission (UGC), Government of India.

Conflicts of Interest: The authors declare no conflict of interest.

References

1. Knowles, M.A.; Hurst, C.D. Molecular biology of bladder cancer: New insights into pathogenesis and clinical diversity. *Nat. Rev. Cancer* **2014**, *15*, 25–41. [CrossRef] [PubMed]
2. Sylvester, R.J.; Van Der Meijden, A.P.; Oosterlinck, W.; Witjes, J.A.; Bouffioux, C.; Denis, L.; Newling, D.W.; Kurth, K. Predicting Recurrence and Progression in Individual Patients with Stage Ta T1 Bladder Cancer Using EORTC Risk Tables: A Combined Analysis of 2596 Patients from Seven EORTC Trials. *Eur. Urol.* **2006**, *49*, 466–477. [CrossRef] [PubMed]
3. Mandhani, A.; Mitash, N.; Agnihotri, S.; Mittal, B.; Tiwari, S. Molecular cystoscopy: Micro-RNAs could be a marker for identifying genotypic changes for transitional cell carcinoma of the urinary bladder. *Indian J. Urol.* **2016**, *32*, 149–153. [CrossRef] [PubMed]
4. Lobo, N.; Mount, C.; Omar, K.; Nair, R.; Thurairaja, R.; Khan, M.S. Landmarks in the treatment of muscle-invasive bladder cancer. *Nat. Rev. Urol.* **2017**, *14*, 565–574. [CrossRef]
5. Zehnder, P.; Studer, U.E.; Skinner, E.C.; Thalmann, G.; Miranda, G.; Roth, B.; Cai, J.; Birkhäuser, F.D.; Mitra, A.P.; Burkhard, F.C.; et al. Unaltered oncological outcomes of radical cystectomy with extended lymphadenectomy over three decades. *BJU Int.* **2013**, *112*, E51–E58. [CrossRef]
6. Gakis, G. Management of Muscle-invasive Bladder Cancer in the 2020s: Challenges and Perspectives. *Eur. Urol. Focus* **2020**. [CrossRef]
7. Bellmunt, J.; Powles, T.; Vogelzang, N.J. A review on the evolution of PD-1/PD-L1 immunotherapy for bladder cancer: The future is now. *Cancer Treat. Rev.* **2017**, *54*, 58–67. [CrossRef]
8. Gopalakrishnan, D.; Koshkin, V.S.; Ornstein, M.C.; Papatsoris, A.; Grivas, P. Immune checkpoint inhibitors in urothelial cancer: Recent updates and future outlook. *Ther. Clin. Risk Manag.* **2018**, *14*, 1019–1040. [CrossRef]
9. Zhang, X.; Yang, X.; Yang, C.; Li, P.; Yuan, W.; Deng, X.; Cheng, Y.; Li, P.; Yang, H.; Tao, J.; et al. Targeting protein kinase CK2 suppresses bladder cancer cell survival via the glucose metabolic pathway. *Oncotarget* **2016**, *7*, 87361–87372. [CrossRef]
10. Kawanishi, H.; Matsui, Y.; Ito, M.; Watanabe, J.; Takahashi, T.; Nishizawa, K.; Nishiyama, H.; Kamoto, T.; Mikami, Y.; Tanaka, Y.; et al. Secreted CXCL1 Is a Potential Mediator and Marker of the Tumor Invasion of Bladder Cancer. *Clin. Cancer Res.* **2008**, *14*, 2579–2587. [CrossRef]
11. Latosinska, A.; Mokou, M.; Makridakis, M.; Mullen, W.; Zoidakis, J.; Lygirou, V.; Frantzi, M.; Katafigiotis, I.; Stravodimos, K.; Hupe, M.C.; et al. Proteomics analysis of bladder cancer invasion: Targeting EIF3D for therapeutic intervention. *Oncotarget* **2017**, *8*, 69435–69455. [CrossRef] [PubMed]
12. Gershoni-Baruch, R.; Nachlieli, T.; Guilburd, J.N. Apert's syndrome with occipital encephalocele and absence of corpus callosum. Child's nervous system: ChNS: Official journal of the International Society for Pediatric. *Neurosurgery* **1991**, *7*, 231–232.
13. Guo, M.; Lu, S.; Huang, H.; Wang, Y.; Yang, M.Q.; Yang, Y.; Fan, Z.; Jiang, B.; Deng, Y. Increased AURKA promotes cell proliferation and predicts poor prognosis in bladder cancer. *BMC Syst. Boil.* **2018**, *12*, 118. [CrossRef] [PubMed]
14. Rebouissou, S.; Bernard-Pierrot, I.; De Reyniès, A.; Lepage, M.-L.; Krucker, C.; Chapeaublanc, E.; Herault, A.; Kamoun, A.; Caillault, A.; Letouzé, E.; et al. EGFR as a potential therapeutic target for a subset of muscle-invasive bladder cancers presenting a basal-like phenotype. *Sci. Transl. Med.* **2014**, *6*, 244ra91. [CrossRef]
15. Baldia, P.H.; Maurer, A.; Heide, T.; Rose, M.; Stoehr, R.; Hartmann, A.; Williams, S.V.; Knowles, M.A.; Knuechel, R.; Gaisa, N.T. Fibroblast growth factor receptor (FGFR) alterations in squamous differentiated bladder cancer: A putative therapeutic target for a small subgroup. *Oncotarget* **2016**, *7*, 71429–71439. [CrossRef]
16. Zou, J.; Huang, R.; Li, H.; Wang, B.; Chen, Y.; Chen, S.; Ou, K.; Wang, X. Secreted TGF-β-induced protein promotes aggressive progression in bladder cancer cells. *Cancer Manag. Res.* **2019**, *11*, 6995–7006. [CrossRef]
17. Warrick, J.I.; Walter, V.; Yamashita, H.; Chung, E.; Shuman, L.; Amponsa, V.O.; Zheng, Z.; Chan, W.; Whitcomb, T.L.; Yue, F.; et al. FOXA1, GATA3 and PPARvCooperate to Drive Luminal Subtype in Bladder Cancer: A Molecular Analysis of Established Human Cell Lines. *Sci. Rep.* **2016**, *6*, 38531. [CrossRef]
18. Deb, B.; Puttamallesh, V.N.; Gondkar, K.; Thiery, J.; Gowda, H.; Kumar, P. Phosphoproteomic Profiling Identifies Aberrant Activation of Integrin Signaling in Aggressive Non-Type Bladder Carcinoma. *J. Clin. Med.* **2019**, *8*, 703. [CrossRef]

19. Itkonen, H.M.; Engedal, N.; Babaie, E.; Lühr, M.; Guldvik, I.J.; Minner, S.; Hohloch, J.; Tsourlakis, M.C.; Schlomm, T.; Mills, I.G. UAP1 is overexpressed in prostate cancer and is protective against inhibitors of N-linked glycosylation. *Oncogene* **2014**, *34*, 3744–3750. [CrossRef]
20. Franceschini, A.; Szklarczyk, D.; Frankild, S.; Kuhn, M.; Simonovic, M.; Roth, A.; Lin, J.; Minguez, P.; Bork, P.; Von Mering, C.; et al. STRING v9.1: Protein-protein interaction networks, with increased coverage and integration. *Nucleic Acids Res.* **2012**, *41*, D808–D815. [CrossRef]
21. Subbannayya, Y.; Syed, N.; Barbhuiya, M.A.; Raja, R.; Marimuthu, A.; Sahasrabuddhe, N.; Pinto, S.M.; Manda, S.S.; Renuse, S.; Manju, H.; et al. Calcium calmodulin dependent kinase kinase 2—A novel therapeutic target for gastric adenocarcinoma. *Cancer Boil. Ther.* **2015**, *16*, 336–345. [CrossRef] [PubMed]
22. Syed, N.; Chavan, S.; Sahasrabuddhe, N.A.; Renuse, S.; Sathe, G.; Nanjappa, V.; Radhakrishnan, A.; Raja, R.; Pinto, S.M.; Srinivasan, A.; et al. Silencing of high-mobility group box 2 (HMGB2) modulates cisplatin and 5-fluorouracil sensitivity in head and neck squamous cell carcinoma. *Proteomics* **2015**, *15*, 383–393. [CrossRef] [PubMed]
23. Raja, R.; Sahasrabuddhe, N.A.; Radhakrishnan, A.; Syed, N.; Solanki, H.S.; Puttamallesh, V.N.; Balaji, S.A.; Nanjappa, V.; Datta, K.K.; Babu, N.; et al. Chronic exposure to cigarette smoke leads to activation of p21 (RAC1)-activated kinase 6 (PAK6) in non-small cell lung cancer cells. *Oncotarget* **2016**, *7*, 61229–61245. [CrossRef] [PubMed]
24. Radhakrishnan, A.; Nanjappa, V.; Raja, R.; Sathe, G.; Puttamallesh, V.N.; Jain, A.; Pinto, S.M.; Balaji, S.A.; Chavan, S.; Sahasrabuddhe, N.A.; et al. A dual specificity kinase, DYRK1A, as a potential therapeutic target for head and neck squamous cell carcinoma. *Sci. Rep.* **2016**, *6*, 36132. [CrossRef]
25. Ma, G.; Yang, X.; Liang, Y.; Wang, L.; Li, D.; Chen, Y.; Liang, Z.; Wang, Y.; Niu, H. Precision medicine and bladder cancer heterogeneity. *Bull Cancer* **2018**, *105*, 925–931. [CrossRef]
26. Burger, M.; Denzinger, S.; Hartmann, A.; Wieland, W.-F.; Stoehr, R.; Obermann, E.C. Mcm2 predicts recurrence hazard in stage Ta/T1 bladder cancer more accurately than CK20, Ki67 and histological grade. *Br. J. Cancer* **2007**, *96*, 1711–1715. [CrossRef]
27. Choy, B.; LaLonde, A.; Que, J.; Wu, T.; Zhou, Z. MCM4 and MCM7, potential novel proliferation markers, significantly correlated with Ki-67, Bmi1, and cyclin E expression in esophageal adenocarcinoma, squamous cell carcinoma, and precancerous lesions. *Hum. Pathol.* **2016**, *57*, 126–135. [CrossRef]
28. Gou, K.; Liu, J.; Feng, X.; Li, H.; Yuan, Y.; Xing, C. Expression of Minichromosome Maintenance Proteins (MCM) and Cancer Prognosis: A meta-analysis. *J. Cancer* **2018**, *9*, 1518–1526. [CrossRef]
29. Hyrien, O. How MCM loading and spreading specify eukaryotic DNA replication initiation sites. *F1000 Res.* **2016**, *5*, 2063. [CrossRef]
30. Lau, K.-M.; Chan, Q.K.Y.; Pang, J.; Li, K.K.W.; Yeung, W.W.; Chung, N.Y.F.; Lui, P.C.; Tam, Y.-S.; Li, H.-M.; Zhou, L.; et al. Minichromosome maintenance proteins 2, 3 and 7 in medulloblastoma: Overexpression and involvement in regulation of cell migration and invasion. *Oncogene* **2010**, *29*, 5475–5489. [CrossRef]
31. DaFonseca, C.J.; Shu, F.; Zhang, J.J. Identification of two residues in MCM5 critical for the assembly of MCM complexes and Stat1-mediated transcription activation in response to IFN-. *Proc. Natl. Acad. Sci. USA* **2001**, *98*, 3034–3039. [CrossRef] [PubMed]
32. Zheng, T.; Chen, M.; Han, S.; Zhang, L.; Bai, Y.; Fang, X.; Ding, S.-Z.; Yang, Y. Plasma minichromosome maintenance complex component 6 is a novel biomarker for hepatocellular carcinoma patients. *Hepatol. Res.* **2014**, *44*, 1347–1356. [CrossRef] [PubMed]
33. Korkolopoulou, P.; Givalos, N.; Saetta, A.A.; Goudopoulou, A.; Gakiopoulou, H.; Thymara, I.; Thomas-Tsagli, E.; Patsouris, E. Minichromosome maintenance proteins 2 and 5 expression in muscle-invasive urothelial cancer: A multivariate survival study including proliferation markers and cell cycle regulators. *Hum. Pathol.* **2005**, *36*, 899–907. [CrossRef] [PubMed]
34. Rausch, S.; Hennenlotter, J.; Teepe, K.; Kuehs, U.; Aufderklamm, S.; Bier, S.; Mischinger, J.; Gakis, G.; Stenzl, A.; Schwentner, C.; et al. Muscle-invasive bladder cancer is characterized by overexpression of thymidine kinase 1. *Urol. Oncol. Semin. Orig. Investig.* **2015**, *33*, 426.e21–426.e29. [CrossRef]
35. Simon, R.; Atefy, R.; Wagner, U.; Forster, T.; Bruderer, J.; Wilber, K.; Mihatsch, M.J.; Gasser, T.; Sauter, G. HER-2 andTOP2A coamplification in urinary bladder cancer. *Int. J. Cancer* **2003**, *107*, 764–772. [CrossRef]
36. Choi, J.-W.; Kim, Y.; Lee, J.H.; Kim, Y.-S. High expression of spindle assembly checkpoint proteins CDC20 and MAD2 is associated with poor prognosis in urothelial bladder cancer. *Virchows Arch. A Pathol. Anat. Histopathol.* **2013**, *463*, 681–687. [CrossRef]

37. Mobley, A.; Zhang, S.; Bondaruk, J.; Wang, Y.; Majewski, T.; Caraway, N.P.; Huang, L.; Shoshan, E.; Velazquez-Torres, G.; Nitti, G.; et al. Aurora Kinase A is a Biomarker for Bladder Cancer Detection and Contributes to its Aggressive Behavior. *Sci. Rep.* **2017**, *7*, 40714. [CrossRef]
38. Otto, T.; Sicinski, P. Cell cycle proteins as promising targets in cancer therapy. *Nat. Rev. Cancer* **2017**, *17*, 93–115. [CrossRef]
39. Jagarlamudi, K.K.; Shaw, M. Thymidine kinase 1 as a tumor biomarker: Technical advances offer new potential to an old biomarker. *Biomark. Med.* **2018**, *12*, 1035–1048. [CrossRef]
40. Langbein, S.; Lehmann, J.; Harder, A.; Steidler, A.; Michel, M.S.; Alken, P.; Badawi, J.K. Protein profiling of bladder cancer using the 2D-PAGE and SELDI-TOF-MS technique. *Technol. Cancer Res. Treat.* **2006**, *5*, 67–71. [CrossRef]
41. Rahmani, A.H.; Babiker, A.Y.; AlWanian, W.M.; Elsiddig, S.A.; Faragalla, H.E.; Aly, S.M. Association of Cytokeratin and Vimentin Protein in the Genesis of Transitional Cell Carcinoma of Urinary Bladder Patients. *Dis. Markers* **2015**, *2015*, 1–5. [CrossRef] [PubMed]
42. Wu, S.; Nitschke, K.; Heinkele, J.; Weis, C.-A.; Worst, T.S.; Eckstein, M.; Porubsky, S.; Erben, P. ANLN and TLE2 in Muscle Invasive Bladder Cancer: A Functional and Clinical Evaluation Based on In Silico and In Vitro Data. *Cancers* **2019**, *11*, 1840. [CrossRef] [PubMed]
43. Akella, N.M.; Ciraku, L.; Reginato, M.J. Fueling the fire: Emerging role of the hexosamine biosynthetic pathway in cancer. *BMC Boil.* **2019**, *17*, 52. [CrossRef]
44. Chiaradonna, F.; Ricciardiello, F.; Palorini, R. The Nutrient-Sensing Hexosamine Biosynthetic Pathway as the Hub of Cancer Metabolic Rewiring. *Cells* **2018**, *7*, 53. [CrossRef]
45. De Queiroz, R.M.; Oliveira, I.A.; Piva, B.; Catão, F.B.; Rodrigues, B.D.C.; Pascoal, A.D.C.; Diaz, B.; Todeschini, A.R.; Caarls, M.B.; Dias, W. Hexosamine Biosynthetic Pathway and Glycosylation Regulate Cell Migration in Melanoma Cells. *Front. Oncol.* **2019**, *9*, 116. [CrossRef]
46. Deb, B.; Patel, K.; Sathe, G.; Kumar, P. N-Glycoproteomic Profiling Reveals Alteration In Extracellular Matrix Organization In Non-Type Bladder Carcinoma. *J. Clin. Med.* **2019**, *8*, 1303. [CrossRef] [PubMed]

genes

MDPI

Review

Genetic Drivers of Head and Neck Squamous Cell Carcinoma: Aberrant Splicing Events, Mutational Burden, HPV Infection and Future Targets

Zodwa Dlamini [1,*], Mohammed Alaouna [2], Sikhumbuzo Mbatha [3], Ahmed Bhayat [4], Mzubanzi Mabongo [5], Aristotelis Chatziioannou [1,6,7] and Rodney Hull [1]

1 SA-MRC/UP Precision Prevention & Novel Drug Targets for HIV-Associated Cancers Extramural Unit, Pan African Cancer Research Institute, University of Pretoria, Hatfield, Pretoria 0028, South Africa; achatzi@bioacademy.gr (A.C.); Rodney.hull@up.ac.za (R.H.)
2 Department of Internal Medicine, Faculty of Health Sciences, University of the Witwatersrand, Johannesburg 2050, South Africa; 321814@students.wits.ac.za
3 Department of Surgery, Steve Biko Academic Hospital, University of Pretoria, Hatfield, Pretoria 0028, South Africa; sikhumbuzo.mbatha@up.ac.za
4 Department of Community Dentistry, School of Dentistry, University of Pretoria, Hatfield, Pretoria 0028, South Africa; ahmed.bhayat@up.ac.za
5 Department of Maxillofacial and Oral Surgery, School of Dentistry, University of Pretoria, Hatfield, Pretoria 0028, South Africa; mzubanzi.mabongo@up.ac.za
6 e-NIOS Applications PC, Kallithea, 17676 Athens, Greece
7 Center of Systems Biology, Biomedical Research Foundation of the Academy of Athens, 11527 Athens, Greece
* Correspondence: Zodwa.Dlamini@up.ac.za; Tel.: +27-76-147-4878

Citation: Dlamini, Z.; Alaouna, M.; Mbatha, S.; Bhayat, A.; Mabongo, M.; Chatziioannou, A.; Hull, R. Genetic Drivers of Head and Neck Squamous Cell Carcinoma: Aberrant Splicing Events, Mutational Burden, HPV Infection and Future Targets. *Genes* **2021**, *12*, 422. https://doi.org/10.3390/genes12030422

Academic Editor: Selvarangan Ponnazhagan

Received: 27 January 2021
Accepted: 11 March 2021
Published: 15 March 2021

Abstract: Head and neck cancers include cancers that originate from a variety of locations. These include the mouth, nasal cavity, throat, sinuses, and salivary glands. These cancers are the sixth most diagnosed cancers worldwide. Due to the tissues they arise from, they are collectively named head and neck squamous cell carcinomas (HNSCC). The most important risk factors for head and neck cancers are infection with human papillomavirus (HPV), tobacco use and alcohol consumption. The genetic basis behind the development and progression of HNSCC includes aberrant non-coding RNA levels. However, one of the most important differences between healthy tissue and HNSCC tissue is changes in the alternative splicing of genes that play a vital role in processes that can be described as the hallmarks of cancer. These changes in the expression profile of alternately spliced mRNA give rise to various protein isoforms. These protein isoforms, alternate methylation of proteins, and changes in the transcription of non-coding RNAs (ncRNA) can be used as diagnostic or prognostic markers and as targets for the development of new therapeutic agents. This review aims to describe changes in alternative splicing and ncRNA patterns that contribute to the development and progression of HNSCC. It will also review the use of the changes in gene expression as biomarkers or as the basis for the development of new therapies.

Keywords: head and neck squamous cell carcinoma (HNSCC); aberrant splicing events; human papillomavirus (HPV) infection; non-coding RNA (ncRNA); methylation; mutational burden

1. Introduction

The term head and neck cancers are used to describe a variety of tumors that arise in the mouth, nose, throat, sinuses or salivary glands [1]. Head and neck cancers are the sixth most common form of malignancy, with a total of 600,000 reported cases around the globe each year [2]. Over 90% of these cases are squamous carcinoma of the head and neck, head and neck squamous cell carcinoma (HNSCC) [3]. More than two-thirds of HNSCC incidents are diagnosed in developing countries [4]. The estimated average age of patients is 60 years, and the incident rate is highest in males [5]. The first indications that a patient is suffering from these types of cancers include changes in the sound of the voice, a persistent sore throat that will not heal, difficulty in swallowing and most notably, the development of lumps or lesions in the throat [6]. Even with major advances in diagnosis, radiation therapy and immunotherapy, the 5-year survival rate for HNSCC patients has not improved in recent decades [7,8]. Additionally, due to the lack of appropriate biomarkers for the early diagnosis of HNSCC, in many patients, the cancer is only detected at the later stages of the disease, leading to a poor prognosis [4,9].

The primary risk factors for HNSCC involve smoking and heavy alcohol use [10]. Human papillomavirus (HPV) is classified as a distinct risk factor, giving rise to tumors that are distinct from those caused by other risk factors [11]. Genome-wide systematic sequencing of mRNAs, microRNAs (miRNAs), long non-coding RNAs (lncRNAs), and circular RNAs have led to the identification of probable methylation sites, single nucleotide polymorphisms (SNPs), mutations and variations in copy number in a variety of different genres. This has led to the identification of numerous potential biomarkers for HNSCC [12–16]. In addition to these genomic and epigenetic changes, alternative splicing events have also been implicated in the initiation and progression of head and neck cancer [17].

2. The Role Played by HPV Infection in HNSCC Development and Progression

Another important factor for the changes in gene expression that occur in head and neck cancer is infection with HPV. This is an independent etiological factor in the development of HNSCC and has been the target of interest for a large amount of recent research [11,18,19]. Human papillomaviruses are epitheliotropic DNA viruses with an average genome size of 8 kb [20]. The virus generates two oncoproteins encoded by the E6 and E7 genes that effectively inhibit the proteins p53 and pRb. This leads to the initiation of the cell cycle and DNA synthesis, which is required for viral replication [20]. Numerous studies have shown that HPV positive (HPV+) and HPV negative (HPV-) HNSCCs are separate entities with distinct etiologies, clinical behaviors, treatment outcomes, pathological toxicity, and molecular profiles [21–24]. HPV type 16 is identified as the causative agent in more than 90% of HNSCC cases [25].

Tumors expressing HPV genes (particularly HPV16) displayed no TP53 mutations and low losses of segments of chromosomes 3p, 9p and 17p [19]. Traditionally HPV+ HNSCC is more sensitive to treatment; however, resistance to these treatments, such as chemotherapy, radiation, and surgery, are on the rise [26]. Most HPV+ HNSCC tumors were found to be infected with HPV16 alone and showed expression of the HPV16-E6 oncogene [26]. A distinct class of 123 member genes was specifically deregulated in HPV16 positive HNSCC. These genes were deregulated in both smokers and non-smokers [26]. The symptoms of HPV+ and HPV- HNSCC are very different from each other, which causes confusion about whether these cancers are considered distinct tumors [27]. HPV + oral cancers show changes in the expression of genes regulating the cell cycle or a decrease in the levels of tumor suppressor proteins, such as pRb and cyclin D1. These proteins are usually overexpressed in oral HPV-e tumors [27,28].

HPV-Infected HNSCC Expression Profiles

HPV + HNSCC tumors overexpress retinoblastoma-binding protein factor-C replication gene, and transcription factor partner E2F-dimerization protein (TFDP2). A large group of genes that play a role in the defense against viral infection and immune response have been shown to be ineffective against HPV, including interleukin and interferon-induced proteins [26].

A study has demonstrated that there is a difference in the expression pattern of host genes in HPV + tumors from smokers, ex-smokers and non–smokers. HPV16-+ tumors from smokers could be monitored through the expression of p53 or E2F-transcription factors, such as insulin growth factor (IGF), protein transcription factor-C4 (RFC4), cell division cycle (CDC7), cytochrome P450 (CYP4V2), mini-chromosome maintenance protein complex 2 (MCM 2) and mitotic checkpoint complex protein (MCC) [29,30]. It was proposed that the expression of some of these genes might be linked to tobacco use [26]. Cyclin-dependent kinase inhibitor 2C (CDKN2C) and retinoblastoma (RB) genes were among the genes whose expression is consistently and greatly altered in HPV16-positive tumors from non–smokers. Neither of these genes has been conclusively linked with HPV16 in smokers despite their upregulation [26]. CDKN2C encodes the enzyme p18, a tumor suppressor and cyclin-dependent kinase receptor. It binds protein kinases and acts in conjunction with the retinoblastoma tumor protein (pRb) to inhibit cell cycle progression and regulate growth [26]. Enhanced CDKN2C and RB expression; suggests the lack of the negative feedback loop, a situation that is observed when the expression of HPV16-E7 is repressed [22]. Studies found that the expression of CDKN2 was able to regulate the growth of cell lines derived from HPV16-+ HNSCC tumors [31]. Cancer development and progression can also be inhibited in HPV-infected cells through the ability of pRb to suppresses the function of the E2F transcription factors in cells infected with HPV [32,33].

In head and neck cancer, p16 is inactivated by gene mutation or methylation, which triggers the functional inactivation of pRb [26,34]. HPV-E6/E7-related overexpression of p16 protein occurs in oral lesions [35]. In HPV16-+ HNSCC, elevated levels of the expression of the cell cycle control genes (CDC7, MCM2) were also recorded [22,31]. Regulation of interferon-inducible protein (IFN-inducible) and interleukin-1 receptor antagonist (IL-1RA) was recorded in-HPV16 expressing immortalized head and neck tumor cell lines [36]. Spontaneous degradation of the early HPV protein E2 led to increased transcription of viral DNA. This is accompanied by the rise of antiviral gene expression in the form of IFN and an increase in viral E6/E7 oncoprotein production [37]. Other studies also demonstrated that transfection of malignant cells with oncogene E7 would render them more vulnerable to IFN-alpha-induced apoptosis [38]. This suggests that active chronic oncogenic HPV infection can impact the vulnerability of cells to IFN-induced apoptosis in tumor tissue and may similarly affect IFN-based HPV therapy in associated diseases, such as HNSCC [34].

3. Alternative Splicing in HNSCC

A study published in 2019 reported the occurrence of alternative splicing (AS) events in 519 HNSCC patients. It was found that in these 519 samples, there were 4626 AS-related survival events in 3280 genes. These changes in AS signatures resulted in multiple, cumulative survival outcomes [39]. A study in 2020 identified 4068 splicing events associated with changes in the survival of HNSCC patients, using records from The Cancer Genome Atlas (TCGA). These results imply that a patient's AS signature can be used as a prognostic biomarker [17]. The top five AS events that correlated with survival were exon skipping, use of alternate promoter sites, use of alternate terminator sites, use of alternate acceptor sites, and use of alternate donor sites (Figure 1). These results imply that AS events are capable of being not only diagnostic and prognostic biomarkers but also therapeutic targets for the treatment of HNSCC patients. GO and KEGG analysis indicates that most genes whose splicing is altered in HNSCC are implicated in playing a role in functions, such as apoptosis, DNA repair, mRNA splicing and metabolism [39].

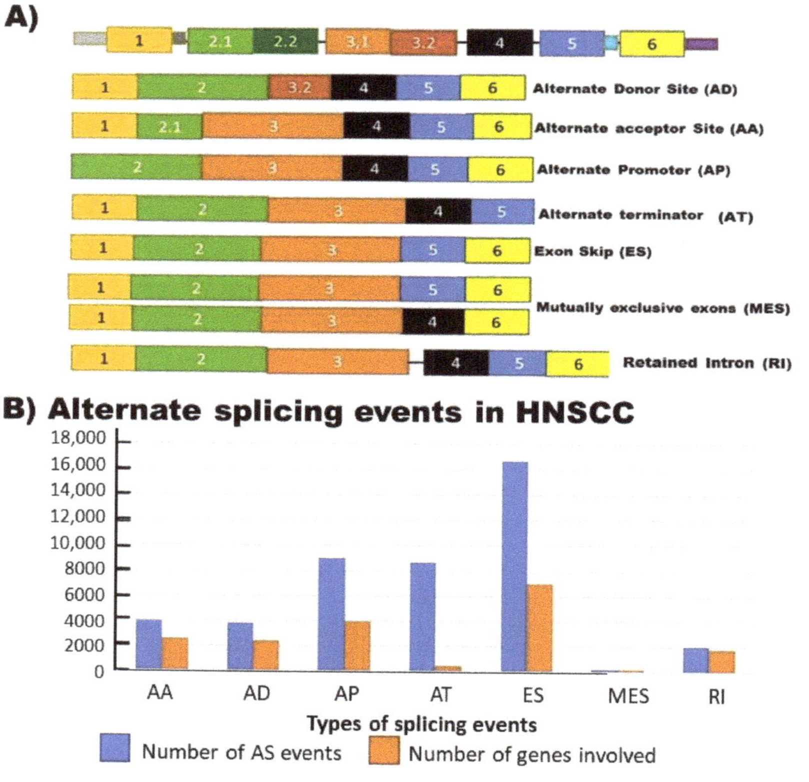

Figure 1. Analysis of alternative splicing (AS) events in head and neck squamous cell carcinomas (HNSCC). (**A**) Seven types of AS events were identified in HNSCC patient samples. These include changes in the location of alternate acceptor sites (AA), donor sites (AD), alternate promoter sites (AP) and alternate terminator sites (AT). Other AS events involve changes in the incorporation of exons and exclusion of introns due to exon skipping (ES), the use of mutually exclusive exons (ME) and retained introns (RI) [40]. (**B**) The number of each type of AS event taking place in 519 HNSCC patients. AA, alternative accepter; AD, alternative donor; AP, alternative promoter. AT, alternative terminator site; ES, exon skip; ME, mutually exclusive exons; RI, intron preserved/retained intron [39].

Further analysis indicated that HNSCC patient survival was associated with AS of five specific genes. These five genes are C5orf30, eEF1A lysine and N-terminal methyltransferase (*METTL13*), Ras homolog gene family member T1 (*RHOT1*), ATP-binding cassette sub-family C member 5 (*ABCC5*), and Myelin protein zero-like protein 1 (*MPZL1*). The role played by AS in these five genes is currently not fully known. METTL13 di-methylates eukaryotic elongation factor 1A (eEF1A), leading to increased translation and protein expression and can promote cancer formation and progression [41]. METL13 is alternately spliced to give rise to 5 isoforms. The full-length protein contains two methyltransferase domains. The second, of which is missing in at least two of the isoforms. These isoforms would then be less efficient at methylating targets, implying that these isoforms could help prevent cancer formation and progression (Figure 2A). RHOT1 is a membrane receptor that promotes proliferation and cancer [42]. It is spliced to give rise to six isoforms. Some of these isoforms lack the transmembrane receptor, implying that these isoforms can block signaling and could, therefore, prevent cell migration. (Figure 2B) MPZL1 activates Src kinases, which results in increased cancer cell proliferation and migration [43]. The full-length variant has two transmembrane domains. Two of the splice variants lack one of these domains. This may interfere with the recognition of ligands by this receptor (Figure 2C). In addition to these genes, there are a variety of genes that have been found to be alternately spliced in HNSCC. Examples of these genes are discussed in more detail below.

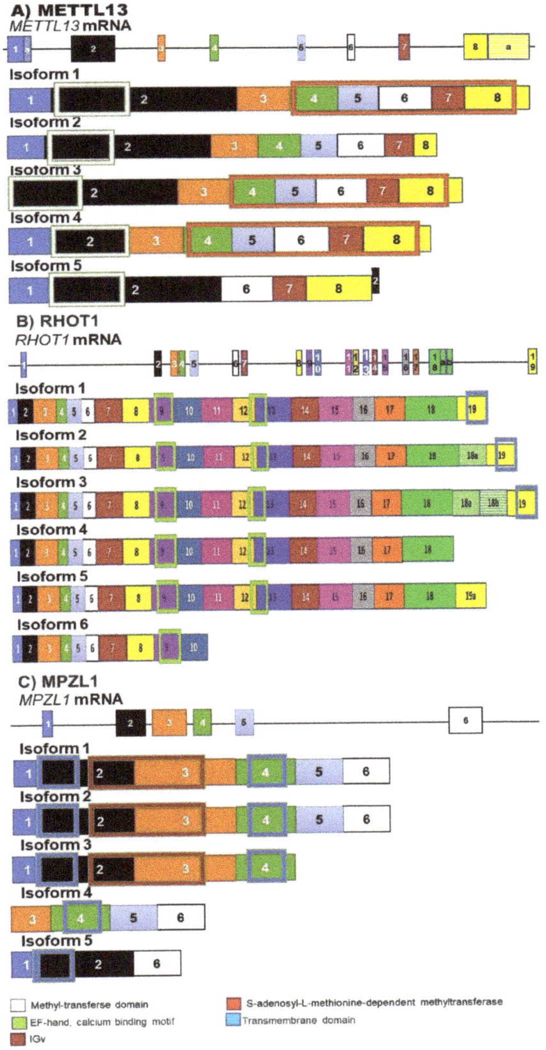

Figure 2. Isoforms of some of the proteins where alternative splicing had a strong correlation with survival. (**A**) eEF1A lysine and N-terminal methyltransferase (METTL13) are responsible for the activation of the eukaryotic elongation factor 1A (eEF1A) through methylation. There are 5 known isoforms of this protein. Some of these isoforms are missing one of the two methyltransferase domains. (**B**) Ras homolog gene family member T1 (RHOT1) is the gene that codes for the Mitochondrial Rho GTPase protein, a membrane receptor that is spliced to give rise to at least 6 isoforms. Some of these isoforms lack the transmembrane receptor, implying that these isoforms can block signaling. (**C**) Myelin protein zero-like protein 1 (MPZL1) is spliced to give rise to 5 known isoforms, some of which lack one of the transmembrane domains. Numbered boxes indicate exons, while colored boxes show the position of the corresponding domain. The blocks in this figure represent the different exons making up each isoform. The size of the boxes indicates the relative size of the exons.

3.1. DOCK5

Dedicator of cytokinesis 5 (DOCK5) is an intracellular signaling protein that is alternately spliced to give rise to at least two isoforms. By analyzing the expression changes of different isoforms of dedicator of cytokinesis 5 (DOCK5) and comparing this to clinical parameters, a link was discovered between the expression of certain DOCK5 variants and the patient's tobacco usage. This indicates that smoking decreases the overall survival of a patient through the alteration of the expression of DOCK5 variants [44]. AS of the DOCK5 mRNA gives rise to two splice variants. One variant contains an exon with an alternate terminator site, resulting in a truncated variant of DOCK5. This variant of DOCK5 enabled HNSCC cell proliferation, migration, and invasion of HPV-negative HNSCC [44]. The DOCK family of proteins are members of the guanine nucleotide exchange factor (GEF) group. These contain two DOCK homology (DHR) domains, DHR-1, and DHR-2, where DHR-2 is the GEF catalytic element [45]. AS events in DOCK5 observed in HNSCC results in the loss of this catalytic domain [44]. The expression of a truncated variant lacking the catalytic domain can promote the development and progression of HPV-negative HNSCC [44]. This also implies that this splice variant, as well as the processes it is associated with, can serve as possible therapeutic targets.

3.2. Lysyl Oxidase (LOXL2) Facilitates the Development of HPV-Negative HNSCC

Lysyl oxidases (LOXL) are a family of copper-containing amine oxidases that catalyze the deamination of lysine residues in collagen and/or elastin. These lysines are involved in the formation of crosslinking in the extracellular matrix leading to increased fibroblast growth and adhesion. In this way, the overexpression of members of the LOXL family promotes metastasis [46]. There are four members of the LOXL family, and these are named LOXL1-4 [47,48]. One of these family members, LOXL2, contributes to the initiation and development of tumors [49]. LOXL2 overexpression seems to enhance the ability of cancer cells to invade tissue layers and promote metastasis [49,50]. LOXL2 activity can be inhibited by hypoxia and hydrogen peroxide (Figure 3A).

Several splice variants of LOXL2 have been reported in various cancers, including esophageal squamous cell carcinoma. Two isoforms—the exon 13-free form (Figure 3B) and the form containing a 72-nucleotide-deletion—result in tumor progression through a new molecular mechanism distinct from the canonical model for LOXL2 [51,52]. This alternate LOX2 mechanism involves the activation of signaling pathways, such as focal adhesion kinase (FAK) and protein kinase B (PKB), and leads to the transition from epithelial to mesenchymal tissue [53]. Overexpression of the isoform lacking exon 13 activated these pathways to a higher degree in HPV-negative HNSCC patients; this includes both the focal adhesion kinase (FAK) and protein kinase B (PKB) pathways. This is associated with modifications in p-FAK, p-AKTT308, p-AKTS473 and p-S6 (Phospho-S6 Ribosomal Protein) [54]. Therefore, it is predicted that this LOXL2 splice variant may promote the proliferation, migration, and invasion of HPV-negative HNSCC cells [54].

All these findings promote the idea that LOXL2 isoforms can be used as biomarkers or therapeutic targets. Its use as a biomarker is further promoted by the fact that it is secreted [55] and can, therefore, be used as a biomarker in non-invasive liquid biopsies.

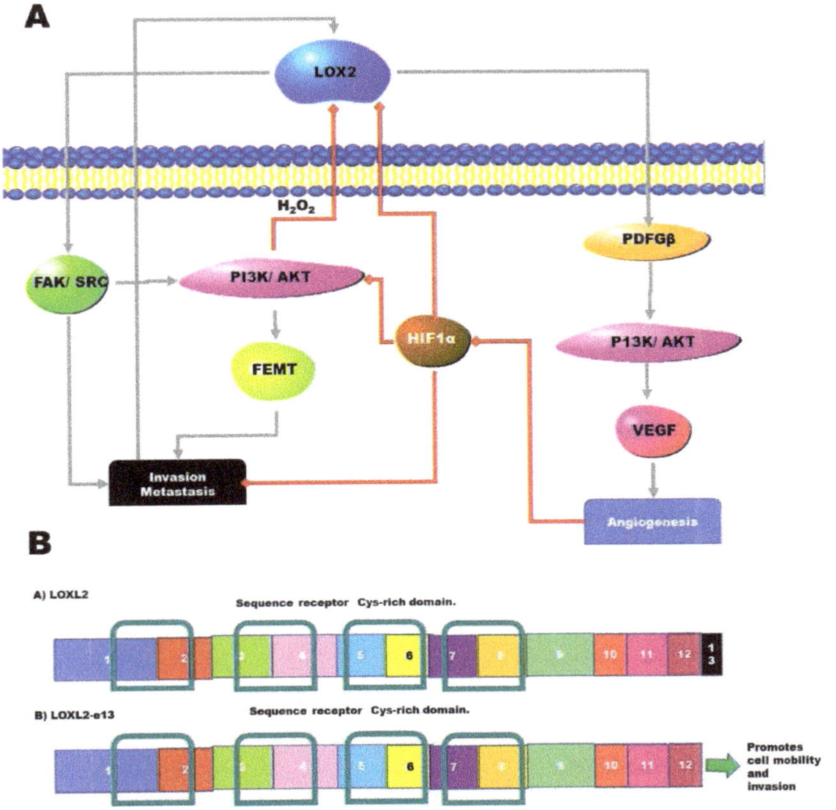

Figure 3. Alternate splicing of the LOXL2 mRNA. (**A**) The lysyl oxidase-like (LOXL2) signaling pathway stimulates invasion and angiogenesis in cancer cells. The pathway is also inhibited because of hypoxia. Green lines with green arrows indicate stimulation or induction, while red lines with red diamonds indicate inhibition. (**B**) The e13 splice variant of LOXL2 results from the exclusion of the final exon, exon 13. This variant promotes stronger signaling promoting invasion and metastasis. Both isoforms contain all four SR domains (scavenger receptor Cys-rich domain). These are shown in the figure as the blue boxes. These domains are responsible for facilitating binding to cell membranes during phagocytosis. FAK, focal adhesion kinase; PI3K/AKT, phosphatidylinositol 3-kinase/protein kinase B; HIF1α, hypoxia-inducible factor 1α; VEGF, vascular endothelial growth factor; EMT, epithelial-mesenchymal transition.

3.3. Transcription Factor Dp-2 (TFDP2)

The transcription factor Dp-2, also known as the E2F dimerization partner 2 (*TFDP2)* gene, encodes a member of a family of transcription factors that heterodimerize with E2F proteins to enhance their DNA-binding activity and promote transcription of E2F target genes. The expression of TFDP2 is upregulated in HPV16-positive HNSCC tumors from non-smokers [26]. Some of these E2F target genes function to control the transcriptional activity of numerous genes involved in the progression from the G1 to the S phase of the cell cycle. TFDP2 is alternately spliced to give rise to 8 isoforms (Figure 4). One of these isoforms lacks the E2F transcription factor dimerization partner domain. This DNA-binding domain stimulates E2F transcription. Increased expression of this isoform may inhibit cell cycle progression (Figure 4).

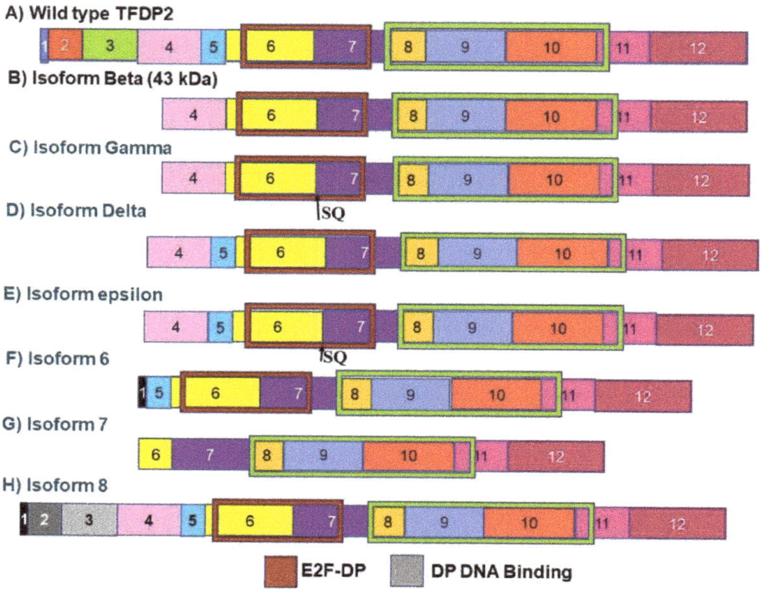

Figure 4. (**A–H**) Isoforms of TFDP2. The TFDP2 transcription factor is known to be expressed at high levels in HNSCC and is alternately spliced to give rise to 8 isoforms. Isoform 7 is missing the E2F/DP dimerization domain. This isoform can block E2F associated transcription and, therefore, inhibit cell cycle progression. Although the precise role of these isoforms in HNSCC is unknown, the high expression levels of the wild-type variant in HNSCC imply that isoform 7 may serve as a negative regulator of cancer progression. The blocks in this figure represent the different exons making up each isoform. The size of the boxes indicates the relative size of the exons.

3.4. Splicing of p53 in HNSCC

The abolition of normal p53 function is one of the most common genetic changes in human cancer. P53 mutations are assumed to contribute significantly to the development of around 40% of HNSCCs [11,56]. AS of p53 gives rise to at least 12 isoforms. These isoforms all retain the mutation hotspot sequence (exons 5–8). The canonical p53 protein (Figure 5B) is named p53α and is normally the most abundant isoform and contains all seven functional domains. The two N-terminal transactivation domains, a proline-rich domain (PXXP), a DNA-binding domain, an oligomerization domain (OD), a nuclear localization signaling domain and a negative-regulation domain [57]). The isoforms are divided into three main groups variants, α, β or γ, based on the splicing changes at the N terminal. Isoforms in group alpha have the N terminal basic domain. Isoforms in groups beta and gamma lack this domain and use an alternate exon nine splice variants, exon 9a in group beta and exon 9b in group gamma (Figure 5) [58]. The other isoforms are the result of a variety of AS mechanisms, including alternative promoter usage and alternative initiation of translation sequences. These groups each contain their own truncated variants that arise due to internal promoters Δ40p53. Δ133p53 and Δ160p53 [57].

The isoform detected at the highest level in HNSCC was p53β [58]. Unlike the canonical p53, p53β preferentially binds to the promoter for the proapoptotic *Bax* and is unable to efficiently induce the expression of the p53 regulator MDM2. This allows it to induce apoptosis in a p53 independent manner [59]. Overexpressed p53β cooperates with full-length p53 and contributes to cellular senescence. This increase in p53β was observed in vivo in senescent colon adenomas [58]. The truncated Δ40p53 isoforms are created by AS in intron 2, and the resulting isoform lacks the transactivation domain. These isoforms have a dominant-negative effect on the activity of full-length p53 [60]

The shorter isoforms Δ133p53 and Δ160p53 lack the transactivation domain, the proline-rich domain and a part of the DNA-binding domain (Figure 5). These isoforms can interact directly with the canonical p53 and regulate its transcriptional activity [58]. A mouse model was developed expressing a p53 protein with a deletion of the first 122 amino acids. This model was used to study the role of the Δ133p53 isoforms [61]. This p53 promoted hyperproliferation in cancer and inflammation in the mice studied [61]. Other studies undertaken using mice xenograft models have shown that Δ133p53α could stimulate cell migration and angiogenesis [62]. The Δ133p53α isoform is also able to prevent p53-mediated replicative senescence, G_1 cell-cycle arrest and apoptosis [63]. In summary, p53β promotes replicative senescence, and the action of this isoform is opposite to that of Δ133p53α, which promotes proliferation. Therefore, the ratio of p53β/Δ133p53α can be used to measure cancer risk. A decrease in the ratio would favor cancer development and progression [58].

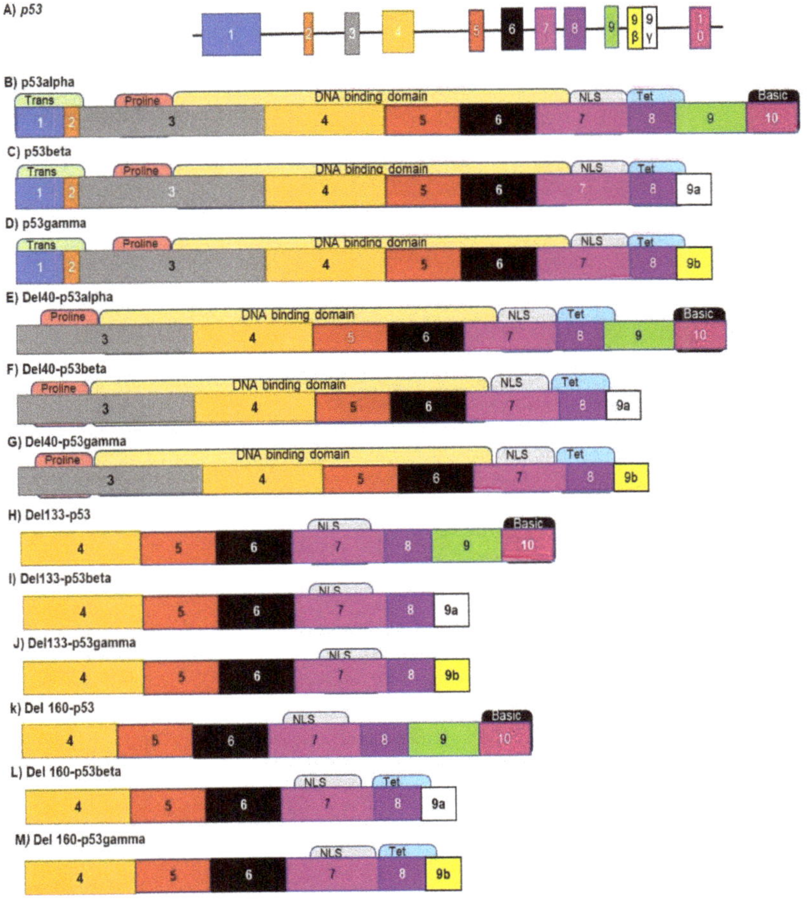

Figure 5. (**A–M**) Splicing isoforms of p53: Alternate splicing of p53 gives rise to at least 12 isoforms. The canonical isoforms (**B**) contain all 10 exons and all 6 domains. The main classification of these isoforms relies on differences in the C terminal. The alpha group contains the basic domain encoded by exon 10. The Beta family of isoforms contains the 9a exon, and the gamma exon contains the 9b exon. The extent of the deletions at the N terminal can further divide these into separate groups full length, Del40, Del 133 and Del 160 [64]. The blocks in this figure represent the different exons making up each isoform. The size of the boxes indicates the relative size of the exons.

3.5. PITX2

The homeobox gene paired-like homeodomain transcription factor 2 (PITX2) is one of the bicoid transcription factors and is spliced to give rise to four isoforms (PITX2A, PITX2B, PITX2C, PITX2D) (Figure 6). These transcription factors play a role in controlling the transcription of procollagen lysyl hydroxyl, an enzyme responsible for the formation of many body structures during development [65]. PITX2 mutations are responsible for the Axenfeld–Rieger type I condition, a disorder that affects the development of teeth, hair, and abdominal structures [66]. The expression of the PITX2 gene can control the Wnt pathway and interferes with the activation of transcription and the β - catenin cell adhesion mediator. PITX2 is also required for the induction of Cyclins A1 and D2 by recruiting coactivators [67,68]. The different isoforms of PITX2 (A, B and C) induce the transcription of different target genes. Isoform D acts as a negative regulator of the other isoforms by suppressing their transcriptional activity [69]. The isoforms A, B and C are expressed at higher levels in various cancers, where different isoforms stimulate the expression of different members of the TGFβ family [70].

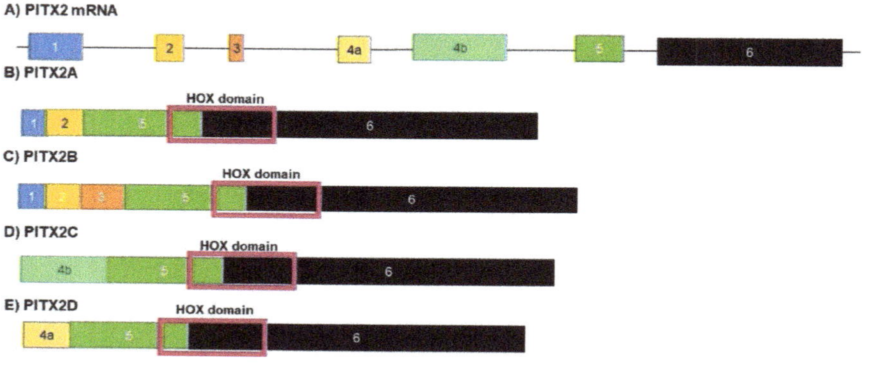

Figure 6. Isoforms of PITX2. Alternate splicing of the PITX2 mRNA gives rise to four protein isoforms. Three of these isoforms (**A–C**) have a similar function, each inducing the transcription of different genes that stimulate growth and proliferation. The final isoform, isoform (**D**), acts as a negative regulator of the other three isoforms. The blocks in this figure represent the different exons making up each isoform. The size of the boxes indicates the relative size of the exons (**E**).

3.6. Aberrant Expression of Splicing Factors and Associated Proteins in HNSCC

The expression of HPV proteins is also dependent on the hosts splicing factors. The most important of these splicing factors are hnRNPA1 and hnRNPA2, which control the expression level of the E6 protein, which is directly responsible for the increased levels of HPV-related HNSCC. The viral E7 and E6 proteins are produced due to AS of viral mRNA. The use of the 5′-splice site SD226 and 3′-splice site SA409 produce E7 mRNAs. Un-spliced viral mRNAs produce E6 mRNA, which is promoted by hnRNPA1 by inhibiting the use of the SA409 splice site, decreasing the levels of E7, and increasing those of E6. The levels of E6 are also decreased through splicing induced by hnRNPA2. This splicing factor inhibits the use of the SA-409 splice site and promotes the use of a downstream 3′ splice site named SA742 [71].

4. Non-Coding RNAs in HNSCC

Recent studies have indicated that non-coding RNAs (ncRNAs) play an important role in the development and progression of HNSCC. These RNAs regulate the expression of coding genes. MicroRNAs (miRNAs) can either promote or inhibit the expression of target genes by binding directly to their target mRNA. They then affect the stability of the mRNA [72]. This is why the aberrant regulation of miRNAs is an important contributing factor in the development of this disease [73]. LncRNAs may control gene expression by promoting transcription, silencing transcription or by promoting or inhibiting translation [74]. Not only do these ncRNAs regulate the expression of protein-coding genes, but they also regulate the expression of other ncRNAs, and since these molecules act by binding to target mRNA, they also compete for the target binding sites on these mRNA targets. Both these types of ncRNA can also be used to fulfill the role of biomarkers for cancer diagnosis and prognosis, as they are found in the body fluids [75].

4.1. MicroRNA Profile in HNSCC

A number of miRNAs have been identified as playing an important role in the development and progression or prevention of HNSCC by acting as either oncogenes or as tumor suppressors [72,76–79]. AS can generate mRNA with different MicroRNA response elements (MREs) that can alter the ability of miRNA to target them. Different miRNAs can easily be generated through the use of alternate promoters and alternate termination sequences to generate miRNAs with different 5′ and 3′ UTRs. The sequence of miRNAs can also be altered by alternate polyadenylation [80]. An early study that examined miRNA profile changes in HNSCC found that the expression of 20 miRNAs was different in HNSCC samples when compared to normal tissue [76], while a later study using more sensitive deep sequencing found 365 miRNAs with significantly different expression levels in HNSCC samples [81]. Further characterization of these miRNAs that are differentially expressed in HNSCC revealed that 49 of these miRNAs were associated in some way with p53. Sixteen of these miRNAs were also associated with lower survival rates in HNSCC patients [82].

MiRNAs whose expression changes in HNSCC cell lines and patients' samples that play a tumor suppressor role include miR-200 [83], mi-R375 [84], miR-26a [85], miR-7 [86], miR-107 [87] miR-218 [88] and members of the let-7 micro-RNA family [89]. In addition to this, multiple miRNAs were reported to be downregulated in HNSCC. These include miR-206 [90], miR-10a-5p, miR-125a-5p, miR-144-3p, miR-195-5p and miR-203 [91]. MiR-200 knockdown results in the development of aggressive cancer, while increased levels of, Mi-RNA-200 inhibits cell growth [83]. Another miRNA that acts as a tumor suppressor in multiple cancers, including HNSCC, is mi-R375; however, it was found to act as an oncogene in cancers, such as lung cancer. It was also found that the expression ratio of miR21 to mi-R375 in tumors compared to normal tissue is a good indicator of patient survival. The lower this ratio is, the worse the survival outcome [84]. MiR-26a acts as a tumor suppressor by inhibiting cell migration and metastasis as well as lowering the expression of the enhancer of zeste homolog 2 (EZH2). This results in decreased cell growth [85]. Many of the other tumor suppressor miRNAs function by inhibiting the expression of genes that promote cell proliferation. MiR-7 inhibits EGFR expression [86], miR-107 inhibits Akt, Stat3 and Rho GTPases via Protein kinase Cε (PKCε) [87]. Other tumor suppressor miRNAs function by inhibiting cell migration, invasion, and metastasis by inhibiting signaling cascades. For example, miR-218 inhibits the focal adhesion pathway, preventing cell migration [88].

The changes in miRNA expression in HPV + HNSCC have also been studied. Specific effects of HPV infection in the development of HNSCC rely on the dysregulation of miRNA expression levels and changes in the location of cellular miRNA. MiR-363 is overexpressed in HPV positive HNSCC, where it functions in cell cycle regulation and reduces cell growth and invasion [92,93]. Analysis of the transcription levels of miR-106a and miR-92a did not reveal any variations in expression between HPV+ and HPV- HNSCC cell lines [93], yet in the presence of HPV-16, MiR-155 has been shown to be downregulated [93]. Studies have shown that in HPV+ HNSCC cells, miR-181a and miR-29a were downregulated in comparison to HPV- HNSCC cells [93]. MiR-29a interacts with and stabilizes p53 [94]. Since HPV-16 E6 increases the rate of p53 degradation [20], MiR-29a deregulation in conjunction with E6 expression could further decrease p53 levels following chronic HPV infection [93].

MiRNAs that were found to function as oncogenes include miR-21 [95], miR-375 [96] and miR-184 [97]. Some of the miRNAs that promote HNSCC development and progression that function by the inhibition of apoptosis include miR-21 [95]. Additionally, many of these miRNAs whose expression is increased in HNSCC are also associated with decreased HNSCC survival; an example of this is miR-21 [96]. MiRNAs, who were found to be expressed at higher levels in HNSCC, but whose effects are not known include miR-133b, miR-455-5p and miR-196 [98], miR-26a, miR-21 [95], miR-106b-3p, miR-2, miR-19a, miR-33a and miR-31 [97].

4.2. LncRNAs in HNSCC

Multiple studies have identified numerous lncRNAs whose expression is altered in many cancers [97]. As in many other cancers, the lncRNA HOX antisense intergenic RNA (*HOTAIR*) is deregulated in HNSCC. This lncRNA is overexpressed in poorly differentiated HNSCCs, and higher expression is associated with more advanced stages of the disease [99]. Those lncRNAs whose expression is increased in HNSCC include nuclear paraspeckle assembly transcript 1 (NEAT1) [100], HOXA transcript at the distal tip (HOTTIP), urothelial cancer associated 1 (UCA1) [101], lncRNA-regulator of reprogramming (ROR) [102] and H19 [103]. The expression of other lncRNAs is downregulated in HNSCC, and this lower expression is associated with a poorer prognosis. This is a possible indication that they play an antitumor function. These include AC026166.2-001, RP11-169D4.1-001, growth-arrest-specific 5 (GAS5) [100], LET [104], X-inactive specific transcript (XIST) [105], maternally-expressed 3 (MEG3) [106], and lnc-JPHl-7 [107].

5. The Contribution of Genomic Mutations to HNSCC

The genomic changes that have been observed in HNSCC include chromosome amplification, chromosome deletion and mutations. Mutations in the genome of HNSCC patients are commonly observed and are known to contribute to cancer development and progression. This has been observed in both HPV+ and HPV- tumors [108]. Some of the most common genes that are found to be mutated in HNSCC include genes that play a role in cell cycle regulation and progression. The gene cyclin-dependent kinase inhibitor 2a (*CDKN2A)* was found to be mutated in up to 87% of HPV- HNSCC tumors. However, mutations in this gene are not common in HPV+ HNSCC [109]. Other groups of genes that are found to be mutated in HNSCC include receptor tyrosine kinases and mitogen-associated protein kinases, growth factors and growth factor receptors [109].

5.1. Mutations in P53 and Associated miRNAs

Mutations in TP53 are linked to the poor overall survival of patients when compared to patients with wild-type p53 [110]. Most of the mutations observed in the p53 gene occur in the DNA-binding region, commonly referred to as the mutation hotspot. As a result of this, mutation analysis of the genomic DNA sequence is usually confined to exons 5–8 or 5–9 [111]. A genomic DNA sequence analysis of exons 5–9 was performed in a study of 166 HNSCC patients. Mutations in p53 were found in 65 tumors (39%) [11]. Another study of 32 HNSCC tumors showed that only 8 (25% were identified as having mutations in p53. Once again, this study focused on exons 5–8 [112]. Another later study indicated that in HNSCC, mutations in exons 5–9 were reported in 22% of all tumors [111]. Mutations are also found in the non-coding regions of p53. One mutation is found between exon 6 and 7 (63 bp downstream of exon 6). The resulting mutant protein was found more commonly in samples from cancer patients. It is expected that this mutation results in the stabilization and accumulation of wild-type p53 [113]. Studies estimating the frequency of mutations in p53 in HNSCC samples have found a wide variety of results. These include 78% of HPV-tumors [114]; 60% in freshly frozen HNSCC samples regardless of HPV status [115], to frequencies as low as 30% in some HNSCC samples regardless of HPV infection [108].

In HNSCC, the TP53 tumor suppressor gene is most often mutated in tumors negative for HPV [110]. The tumor suppressor activity of p53 is accomplished through its activity as a transcription factor. Some of the genes whose transcription is initiated by p53 are miRNAs and mRNA coding for proteins involved in cell stress control, apoptosis, and DNA damage repair [110]. One example of this is the expression of miR-377-3p, the primary regulator of Sestrin 1 (SESN1), which encodes a Sestrin family member and is also known as p53-regulated protein PA26 [110]. The expression of SESN1 correlates with the expression of genes that control autophagy. It triggers TP53 expression [110] and helps to stimulate the response to DNA damage and oxidative stress in cells [110]. MiR-377-3p is a downregulator of SESN1, which directly targets 30 untranslated genome regions [110]. The lower expression of miR-377-3p is an indicator of a poor outcome for patients.

5.2. Mutation in PIK3CA

The alpha catalytic subunit of phosphatidylinositol-4,5-bisphosphate 3-kinase (*PIK3CA*) has been identified as being the most affected gene in HNSCC. In HPV+ tumors, the mutation hotspots for this gene occur in the area of the gene that codes for the protein's helical domain [109]. This domain conducts inhibitory signals to the kinase catalytic domain. This kinase activity activates downstream signaling in response to growth factors leading to cell growth. Therefore, mutations that affect the activity of this domain result in uncontrolled growth and cancer [116]. In HPV- tumors, the mutations are not localized to any single region and occur throughout the gene [109].

6. Diagnostic and Therapeutic Applications

The differences in mRNA splicing, the resulting changes in expression profiles of various protein isoform and the changes in the transcription levels of ncRNA observed in HNSCC compared to healthy tissue could be used to develop new diagnostic or prognostic biomarkers and may also be utilized as targets for the development of new therapies. There have been multiple clinical trials evaluating alterations in splicing as therapeutic targets, and they have been reviewed in detail elsewhere [117,118].

Drugs have already been developed that target AS in cancer. Some of these drugs are shown in Table 1. One of the therapeutic approaches targeting splicing events for the treatment of cancer is the search for small molecular inhibitors of splicing. This is most commonly done using a bioprospecting approach involving the search for natural products derived from bacteria or plants. Many of the most successful of these compounds inhibit splicing factors, such as SF3B (Table 1). This results in the inhibition of spliceosome assembly [118]. This includes compounds, such as pladienolides, isolated from *Streptomyces platensis*. This compound displays cytotoxic effects and the ability to induce cell cycle arrest. However, this compound is not stable, but a stable derivative molecule was developed based on pladienolide known as E7107 [119]. Although SF3B consists of 7 subunits, named SF3B1, SF3B2, SF3B3, SF3B4, SF3B5, SF3B14, and PHF5A, all these small molecules target SF3B1 [120]. Other than SF3B, many of these compounds inhibit the spliceosome by binding to other snRNPs (Table 1). However, despite the promise of these drugs, cancer cell lines have been shown to develop resistance to them. For instance, after continuous exposure to pladienolide B, human colorectal cell lines developed resistance to the drug. This resistance results from point mutations arising in *SF3B1* (*SF3B1*R1074H), which reduce the binding affinity of these compounds to SF3B [120].

Another means to target specific splicing events is through the use of specific oligonucleotides that are able to hybridize to specific regions of mRNA and by targeting specific sequences and regulate splicing to favor one isoform over another (Table 1) [121]. These oligonucleotides that target specific regions of mRNA need to have the antisense sequence of the mRNA to facilitate binding. Therefore, they have been named antisense oligonucleotides (ASOs). Another type of oligonucleotide-based therapy is known as splice-switching oligonucleotides (SSOs), and many functions by blocking the sites on the mRNA where silencers or enhancers can bind. These exonic splicing enhancer (ESE) sites or intronic splicing silencers (ISSs) can lead to the incorporation of different exons and introns [122]. Despite the promise of these oligonucleotides as therapeutic interventions targeting aberrant splicing, it is important to note that none of these oligonucleotides have been approved by the FDA for the treatment of cancer [121].

Although cancer screening panels do exist that detect splicing-factor mutations, there is as yet none that detect changes in AS events. Additionally, changes in the AS of many genes are shared across many different types of cancers. These changes and the accompanying changes in molecules downstream of these splice variants can be used to provide diagnostic and prognostic biomarkers for many cancers and then be used in specific combinations to stratify individual cancer types. Studies have shown that SESN1 mRNA, UHRF1BP11 mRNA and miRNA-377-3p are important biomarkers for predicting prognosis for patients with HPV- HNSCC [110]. This can help to stratify patients and possibly introduce new clinical strategies for managing HNSCC patients [110]. Further analysis indicated that the mRNA for ubiquitin-like containing PHD and RING finger domains 1-binding protein 1 (UHRF1BP1), and p53-regulated protein PA26 (SESN) are both associated with mutated TP53 [110].

Practical use of these isoform profiles as diagnostic or prognostic markers can be achieved using tissue biopsies from patients. The development of labeled riboprobes specific to the alternately spliced mRNA or labeled antibodies raised against isoform-specific protein regions would allow for the detection of AS variants in these tissue sections. Previously the quantification of the staining intensity in samples, such as these, was based on subjective ranking by an experienced histologist. Recently, automated systems have been developed that can automatically scan and analyze slides prepared in the above manner. An example of such a system is the TissueFAXS (TissueGnostics®, Vienna, Austria). This system has previously been shown to accurately detect the transcription and expression levels of different members of the neuron growth factor (NGF) and the neurotrophin receptor (NTR) families. The expression of different members of the NTR family has been found to be mutually exclusive in different cells, with one member being more commonly expressed in HNSCC (NTRK1, the high-affinity NTR) at higher levels than other members (p75NTR, the low-affinity NTR). Through the use of riboprobes and antibodies specific to these two family members, the TissueFAXS system was able to accurately detect and quantify the levels of these different family members in HNSCC samples [123]. An example of the workflow using such a system is given in Figure 7.

Table 1. Different classes of drugs that target alternate splicing in cancer.

Class of Compound	Compound	Target Mechanism	Effect on Splicing	
Small molecules	Pladienolides	Abolish the conformation rearrangement of SF3B1	Interfere with canonical splicing cell cycle arrest	[124–126]
	Spliceostatins	SF3B1		[127]
	Brr2	U5 snRNP Interfere with the RNA helicase activity	Stall canonical RNA splicing	[128,129]
	Pseudouridine 5 fluorouracil;	U2 snRNA	Induces change in the structure and stability of the branch site, altering base site recognition	
	Sulfonamides	U2AF-related splicing factor	Selective degradation of the U2AF splicing factor	[130]
	Hinokilflavone	U2 snRNA	Blocks spliceosome assembly	[131]
	Jerantimine A	SF3b complex	Inhibits proliferation in human cancer cell lines	[132]
	Amiprophosmethyl	mRNA	mRNA splicing and folding	[132]

Table 1. *Cont.*

Class of Compound	Compound	Target Mechanism	Effect on Splicing	
Protein inhibitor	TG-003; TG-693	CLK family interfere with ATP binding	Reduced phosphorylation of SRSF family members	[133]
	SRPIN340	SRPK family ATP binding competitor	Altered cellular localization of SRSFs	[134]
	Cpd-1/2/3	SRPK and CLK family	ATP-binding competitor	[122]
Oligonucleotide	ASO-MDM4	MDM4 transcript mRNA degradation	Reduce expression of MDM4 mRNA	[135]
	AZD9150	STAT3 transcript	Reduce expression of STAT3 mRNA	[136]
	AZD4785	KRAS transcript	Reduce expression of KRAS mRNA	[137]

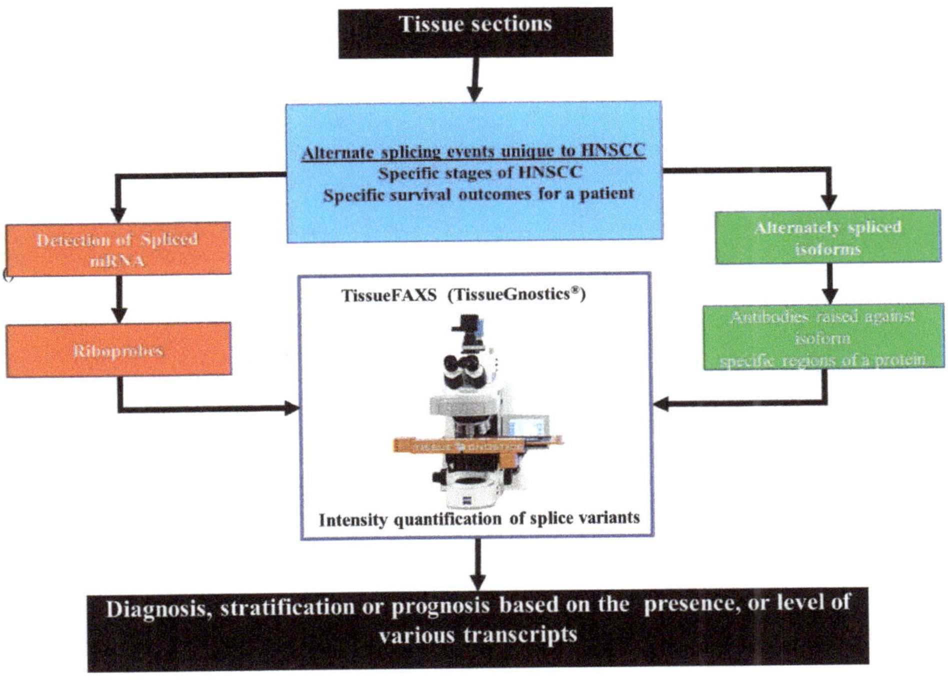

Figure 7. An example of a workflow to detect alternative splicing events using an automated detection and quantification system. The differences in the profiles of splicing events can be used to diagnose HNSCC, detect if the HNSCC is related to human papillomavirus (HPV) infection or stratify patients based on cancer stage and severity. Either riboprobes specific for mRNA transcripts or antibodies that are raised against portions of different protein isoforms that are unique to specific isoforms can be used to detect variants. An automated quantification system, such as the TissueFAXS Plus from TissueGnostics, can be used to rapidly and accurately detect and quantify the levels of these variants without the possible inaccuracies introduced by a subjective assessment by a histologist.

7. Conclusions

As the sixth most diagnosed cancer in the world, HNSCC is a large public health burden. The use of changes in the expression profile of alternate protein isoforms or ncRNAs profiles as biomarkers could prove to be a useful diagnostic tool. In addition to this, the isoforms that contribute to cancer development and progression could serve as useful targets for the development of new therapies. These new diagnostic tools and therapeutic targets could assist in the development of personalized healthcare and more precise patient stratification. However, more research is required to establish if these different profiles can truly be used as diagnostic or prognostic tools. The main problem with these markers is that entire profiles of the splicing changes must be evaluated as many different tumors will not show changes in the splicing of the same genes. For example, only a certain percentage of tumors will show alterations in the splicing of one gene, while another percentage of cancer cells will not show any changes in the splicing of the same gene. For an accurate diagnosis and patient stratification, a large amount of data concerning the changes in the populations of various isoforms or mutations will need to be analyzed in order to come to the correct conclusion. This large amount of data could be expensive to obtain for individual patients and may be time-intensive to analyze. As technology advances, this cost will decrease. Additionally, as artificial intelligence and machine learning technology advance, so the time needed to analyze data will decrease.

These molecular differences between healthy and HNSCC tissue allow patients to be defined by their prognosis and optimize the management of the disease through specific treatments or, in the worst circumstances, palliative care. Multiple methods of targeting alternative splicing are being developed. However, problems have been encountered with many of these approaches. Small molecules that can target the spliceosome, inhibiting the formation of specific splice variants, can lead to resistance to these drugs. However, the development of new drugs derived from naturally occurring compounds that can inhibit splicing may provide novel therapies that cancer cells do cannot readily develop resistance to. Despite the hurdles and the large amount of further study required to develop new biomarkers and therapeutic targets based on genetic drivers of HNSCC. These molecular profiles still offer a promising target for improving treatment outcomes and creating new diagnostic and prognostic tests for HNSCC.

Author Contributions: Conceptualization, Z.D., A.B., S.M. and M.M.; writing—original draft preparation, M.A. and R.H.; writing—review and editing, Z.D., S.M., A.B., M.M. and R.H.; visualization, Z.D., M.A., R.H.; supervision, Z.D and A.C.; project administration, Z.D and A.C.; funding acquisition, Z.D. All authors have read and agreed to the published version of the manuscript.

Funding: This research was funded by The South African Medical Research Council.

Institutional Review Board Statement: Not applicable.

Informed Consent Statement: Not applicable.

Conflicts of Interest: The authors declare no conflict of interest.

References

1. Johnson, N.W.; Amarasinghe, H.K. Epidemiology and Aetiology of Head and Neck Cancers. In *Head and Neck Cancer: Multimodality Management*; Bernier, J., Ed.; Springer International Publishing: Cham, Switzerland, 2016; pp. 1–57. [CrossRef]
2. Rettig, E.M.; D'Souza, G. Epidemiology of head and neck cancer. *Surg. Oncol. Clin. N. Am.* **2015**, *24*, 379–396. [CrossRef] [PubMed]
3. Siegel, R.L.; Miller, K.D.; Jemal, A. Cancer statistics, 2018. *CA Cancer J. Clin.* **2018**, *68*, 7–30. [CrossRef] [PubMed]
4. Zhang, L.; Zhang, W.; Wang, Y.F.; Liu, B.; Zhang, W.F.; Zhao, Y.F.; Kulkarni, A.B.; Sun, Z.J. Dual induction of apoptotic and autophagic cell death by targeting survivin in head neck squamous cell carcinoma. *Cell Death Dis.* **2015**, *6*, e1771. [CrossRef] [PubMed]
5. Polanska, H.; Raudenska, M.; Gumulec, J.; Sztalmachova, M.; Adam, V.; Kizek, R.; Masarik, M. Clinical significance of head and neck squamous cell cancer biomarkers. *Oral Oncol.* **2014**, *50*, 168–177. [CrossRef] [PubMed]
6. Economopoulou, P.; Kotsantis, I.; Psyrri, A. Special Issue about Head and Neck Cancers: HPV Positive Cancers. *Int. J. Mol. Sci.* **2020**, *21*, 3388. [CrossRef]
7. Siegel, R.; Naishadham, D.; Jemal, A. Cancer statistics, 2012. *CA Cancer J. Clin.* **2012**, *62*, 10–29. [CrossRef]
8. Cohen, E.E.; LaMonte, S.J.; Erb, N.L.; Beckman, K.L.; Sadeghi, N.; Hutcheson, K.A.; Stubblefield, M.D.; Abbott, D.M.; Fisher, P.S.; Stein, K.D.; et al. American Cancer Society Head and Neck Cancer Survivorship Care Guideline. *CA Cancer J. Clin.* **2016**, *66*, 203–239. [CrossRef]

9. Rothenberg, S.M.; Ellisen, L.W. The molecular pathogenesis of head and neck squamous cell carcinoma. *J. Clin. Investig.* **2012**, *122*, 1951–1957. [CrossRef]
10. La Vecchia, C.; Tavani, A.; Franceschi, S.; Levi, F.; Corrao, G.; Negri, E. Epidemiology and prevention of oral cancer. *Oral Oncol.* **1997**, *33*, 302–312. [CrossRef]
11. Gillison, M.L.; Koch, W.M.; Capone, R.B.; Spafford, M.; Westra, W.H.; Wu, L.; Zahurak, M.L.; Daniel, R.W.; Viglione, M.; Symer, D.E.; et al. Evidence for a causal association between human papillomavirus and a subset of head and neck cancers. *J. Natl. Cancer Inst.* **2000**, *92*, 709–720. [CrossRef] [PubMed]
12. Liu, J.; Pan, S.; Hsieh, M.H.; Ng, N.; Sun, F.; Wang, T.; Kasibhatla, S.; Schuller, A.G.; Li, A.G.; Cheng, D.; et al. Targeting Wnt-driven cancer through the inhibition of Porcupine by LGK974. *Proc. Natl. Acad. Sci. USA* **2013**, *110*, 20224–20229. [CrossRef] [PubMed]
13. Jamali, Z.; Asl Aminabadi, N.; Attaran, R.; Pournagiazar, F.; Oskouei, S.G.; Ahmadpour, F. MicroRNAs as prognostic molecular signatures in human head and neck squamous cell carcinoma: A systematic review and meta-analysis. *Oral Oncol.* **2015**, *51*, 321–331. [CrossRef] [PubMed]
14. Marur, S.; Forastiere, A.A. Head and neck squamous cell carcinoma: Update on epidemiology, diagnosis, and treatment. In *Mayo Clinic Proceedings*; Elsevier: Amsterdam, The Netherlands, 2016; pp. 386–396.
15. Zhang, X.; Feng, H.; Li, D.; Liu, S.; Amizuka, N.; Li, M. Identification of Differentially Expressed Genes Induced by Aberrant Methylation in Oral Squamous Cell Carcinomas Using Integrated Bioinformatic Analysis. *Int. J. Mol. Sci.* **2018**, *19*, 1698. [CrossRef] [PubMed]
16. Zhao, X.; Sun, S.; Zeng, X.; Cui, L. Expression profiles analysis identifies a novel three-mRNA signature to predict overall survival in oral squamous cell carcinoma. *Am. J. Cancer Res.* **2018**, *8*, 450–461. [PubMed]
17. Zhao, X.; Si, S.; Li, X.; Sun, W.; Cui, L. Identification and validation of an alternative splicing-based prognostic signature for head and neck squamous cell carcinoma. *J. Cancer* **2020**, *11*, 4571–4580. [CrossRef]
18. Dahlgren, L.; Mellin, H.; Wangsa, D.; Heselmeyer-Haddad, K.; Björnestål, L.; Lindholm, J.; Munck-Wikland, E.; Auer, G.; Ried, T.; Dalianis, T. Comparative genomic hybridization analysis of tonsillar cancer reveals a different pattern of genomic imbalances in human papillomavirus-positive and -negative tumors. *Int. J. Cancer* **2003**, *107*, 244–249. [CrossRef]
19. Braakhuis, B.J.; Snijders, P.J.; Keune, W.J.; Meijer, C.J.; Ruijter-Schippers, H.J.; Leemans, C.R.; Brakenhoff, R.H. Genetic patterns in head and neck cancers that contain or lack transcriptionally active human papillomavirus. *J. Natl. Cancer Inst.* **2004**, *96*, 998–1006. [CrossRef]
20. Zur Hausen, H. Papillomaviruses and cancer: From basic studies to clinical application. *Nat. Rev. Cancer* **2002**, *2*, 342–350. [CrossRef]
21. Leemans, C.R.; Braakhuis, B.J.; Brakenhoff, R.H. The molecular biology of head and neck cancer. *Nat. Rev. Cancer* **2011**, *11*, 9–22. [CrossRef]
22. Slebos, R.J.; Yi, Y.; Ely, K.; Carter, J.; Evjen, A.; Zhang, X.; Shyr, Y.; Murphy, B.M.; Cmelak, A.J.; Burkey, B.B.; et al. Gene expression differences associated with human papillomavirus status in head and neck squamous cell carcinoma. *Clin. Cancer Res. Off. J. Am. Assoc. Cancer Res.* **2006**, *12*, 701–709. [CrossRef] [PubMed]
23. Moskovitz, J.M.; Moy, J.; Seiwert, T.Y.; Ferris, R.L. Immunotherapy for Head and Neck Squamous Cell Carcinoma: A Review of Current and Emerging Therapeutic Options. *Oncologist* **2017**, *22*, 680–693. [CrossRef]
24. Münger, K.; Howley, P.M. Human papillomavirus immortalization and transformation functions. *Virus Res.* **2002**, *89*, 213–228. [CrossRef]
25. Gillison, M.L.; Castellsagué, X.; Chaturvedi, A.; Goodman, M.T.; Snijders, P.; Tommasino, M.; Arbyn, M.; Franceschi, S. Eurogin Roadmap: Comparative epidemiology of HPV infection and associated cancers of the head and neck and cervix. *Int. J. Cancer* **2014**, *134*, 497–507. [CrossRef] [PubMed]
26. Murphy, N.; Ring, M.; Heffron, C.C.; King, B.; Killalea, A.G.; Hughes, C.; Martin, C.M.; McGuinness, E.; Sheils, O.; O'Leary, J.J. p16INK4A, CDC6, and MCM5: Predictive biomarkers in cervical preinvasive neoplasia and cervical cancer. *J. Clin. Pathol.* **2005**, *58*, 525–534. [CrossRef]
27. Vidal, L.; Gillison, M.L. Human papillomavirus in HNSCC: Recognition of a distinct disease type. *Hematol. Oncol. Clin. N. Am.* **2008**, *22*, 1125–1142. [CrossRef] [PubMed]
28. Tran, N.; McLean, T.; Zhang, X.; Zhao, C.J.; Thomson, J.M.; O'Brien, C.; Rose, B. MicroRNA expression profiles in head and neck cancer cell lines. *Biochem. Biophys. Res. Commun.* **2007**, *358*, 12–17. [CrossRef]
29. Li, W.; Thompson, C.H.; Cossart, Y.E.; O'Brien, C.J.; McNeil, E.B.; Scolyer, R.A.; Rose, B.R. The expression of key cell cycle markers and presence of human papillomavirus in squamous cell carcinoma of the tonsil. *Head Neck* **2004**, *26*, 1–9. [CrossRef]
30. Santin, A.D.; Zhan, F.; Bignotti, E.; Siegel, E.R.; Cané, S.; Bellone, S.; Palmieri, M.; Anfossi, S.; Thomas, M.; Burnett, A.; et al. Gene expression profiles of primary HPV16- and HPV18-infected early stage cervical cancers and normal cervical epithelium: Identification of novel candidate molecular markers for cervical cancer diagnosis and therapy. *Virology* **2005**, *331*, 269–291. [CrossRef] [PubMed]
31. Martinez, I.; Wang, J.; Hobson, K.F.; Ferris, R.L.; Khan, S.A. Identification of differentially expressed genes in HPV-positive and HPV-negative oropharyngeal squamous cell carcinomas. *Eur J. Cancer* **2007**, *43*, 415–432. [CrossRef]
32. Rorke, E.A.; Zhang, D.; Choo, C.K.; Eckert, R.L.; Jacobberger, J.W. TGF-beta-mediated cell cycle arrest of HPV16-immortalized human ectocervical cells correlates with decreased E6/E7 mRNA and increased p53 and p21(WAF-1) expression. *Exp. Cell Res.* **2000**, *259*, 149–157. [CrossRef]

33. Jones, D.L.; Münger, K. Analysis of the p53-mediated G1 growth arrest pathway in cells expressing the human papillomavirus type 16 E7 oncoprotein. *J. Virol.* **1997**, *71*, 2905–2912. [CrossRef]
34. Murphy, N.; Ring, M.; Heffron, C.C.; Martin, C.M.; McGuinness, E.; Sheils, O.; O'Leary, J.J. Quantitation of CDC6 and MCM5 mRNA in cervical intraepithelial neoplasia and invasive squamous cell carcinoma of the cervix. *Mod. Pathol.* **2005**, *18*, 844–849. [CrossRef]
35. Cunningham, L.L., Jr.; Pagano, G.M.; Li, M.; Tandon, R.; Holm, S.W.; White, D.K.; Lele, S.M. Overexpression of p16INK4 is a reliable marker of human papillomavirus–induced oral high-grade squamous dysplasia. *Oral Med. Oral Pathol. Oral Radiol. Endodontology* **2006**, *102*, 77–81. [CrossRef]
36. Ruutu, M.; Peitsaro, P.; Johansson, B.; Syrjänen, S. Transcriptional profiling of a human papillomavirus 33-positive squamous epithelial cell line which acquired a selective growth advantage after viral integration. *Int. J. Cancer* **2002**, *100*, 318–326. [CrossRef] [PubMed]
37. Herdman, M.T.; Pett, M.R.; Roberts, I.; Alazawi, W.O.; Teschendorff, A.E.; Zhang, X.Y.; Stanley, M.A.; Coleman, N. Interferon-beta treatment of cervical keratinocytes naturally infected with human papillomavirus 16 episomes promotes rapid reduction in episome numbers and emergence of latent integrants. *Carcinogenesis* **2006**, *27*, 2341–2353. [CrossRef]
38. Thyrell, L.; Sangfelt, O.; Zhivotovsky, B.; Pokrovskaja, K.; Wang, Y.; Einhorn, S.; Grandér, D. The HPV-16 E7 oncogene sensitizes malignant cells to IFN-alpha-induced apoptosis. *J. Interferon Cytokine Res.* **2005**, *25*, 63–72. [CrossRef]
39. Xing, L.; Zhang, X.; Tong, D. Systematic Profile Analysis of Prognostic Alternative Messenger RNA Splicing Signatures and Splicing Factors in Head and Neck Squamous Cell Carcinoma. *DNA Cell Biol.* **2019**, *38*, 627–638. [CrossRef]
40. Li, M.; Wu, F.; Zheng, Q.; Wu, Y.; Wu, Y. Identification of Potential Diagnostic and Prognostic Values of P4HA1 Expression in Lung Cancer, Breast Cancer, and Head and Neck Cancer. *DNA Cell Biol.* **2020**, *39*, 909–917. [CrossRef]
41. Liu, S.; Hausmann, S.; Carlson, S.M.; Fuentes, M.E.; Francis, J.W.; Pillai, R.; Lofgren, S.M.; Hulea, L.; Tandoc, K.; Lu, J.; et al. METTL13 Methylation of eEF1A Increases Translational Output to Promote Tumorigenesis. *Cell* **2019**, *176*, 491–504.e21. [CrossRef] [PubMed]
42. Li, Q.; Yao, L.; Wei, Y.; Geng, S.; He, C.; Jiang, H. Role of RHOT1 on migration and proliferation of pancreatic cancer. *Am. J. Cancer Res.* **2015**, *5*, 1460–1470. [PubMed]
43. Chen, D.; Cao, L.; Wang, X. MPZL1 promotes tumor cell proliferation and migration via activation of Src kinase in ovarian cancer. *Oncol. Rep.* **2019**, *42*, 679–687. [CrossRef]
44. Liu, C.; Guo, T.; Xu, G.; Sakai, A.; Ren, S.; Fukusumi, T.; Ando, M.; Sadat, S.; Saito, Y.; Khan, Z.; et al. Characterization of Alternative Splicing Events in HPV-Negative Head and Neck Squamous Cell Carcinoma Identifies an Oncogenic DOCK5 Variant. *Clin. Cancer Res.* **2018**, *24*, 5123–5132. [CrossRef]
45. Cuenca, N.; Fernández-Sánchez, L.; Campello, L.; Maneu, V.; De la Villa, P.; Lax, P.; Pinilla, I. Cellular responses following retinal injuries and therapeutic approaches for neurodegenerative diseases. *Prog. Retin. Eye Res.* **2014**, *43*, 17–75. [CrossRef] [PubMed]
46. El-Haibi, C.P.; Bell, G.W.; Zhang, J.; Collmann, A.Y.; Wood, D.; Scherber, C.M.; Csizmadia, E.; Mariani, O.; Zhu, C.; Campagne, A.; et al. Critical role for lysyl oxidase in mesenchymal stem cell-driven breast cancer malignancy. *Proc. Natl. Acad. Sci. USA* **2012**, *109*, 17460–17465. [CrossRef]
47. Li, T.; Wu, C.; Gao, L.; Qin, F.; Wei, Q.; Yuan, J. Lysyl oxidase family members in urological tumorigenesis and fibrosis. *Oncotarget* **2018**, *9*, 20156–20164. [CrossRef] [PubMed]
48. Choi, J.; Chung, T.; Rhee, H.; Kim, Y.J.; Jeon, Y.; Yoo, J.E.; Noh, S.; Han, D.H.; Park, Y.N. Increased Expression of the Matrix-Modifying Enzyme Lysyl Oxidase-Like 2 in Aggressive Hepatocellular Carcinoma with Poor Prognosis. *Gut Liver* **2019**, *13*, 83–92. [CrossRef]
49. Wu, L.; Zhu, Y. The function and mechanisms of action of LOXL2 in cancer (Review). *Int. J. Mol. Med.* **2015**, *36*, 1200–1204. [CrossRef]
50. Cui, X.; Wang, G.; Shen, W.; Huang, Z.; He, H.; Cui, L. Lysyl oxidase-like 2 is highly expressed in colorectal cancer cells and promotes the development of colorectal cancer. *Oncol. Rep.* **2018**, *40*, 932–942. [CrossRef] [PubMed]
51. Wu, B.L.; Lv, G.Q.; Zou, H.Y.; Du, Z.P.; Wu, J.Y.; Zhang, P.X.; Xu, L.Y.; Li, E.M. Exploration of potential roles of a new LOXL2 splicing variant using network knowledge in esophageal squamous cell carcinoma. *Sci. World J.* **2014**, *2014*, 431792. [CrossRef] [PubMed]
52. Lv, G.Q.; Zou, H.Y.; Liao, L.D.; Cao, H.H.; Zeng, F.M.; Wu, B.L.; Xie, J.J.; Fang, W.K.; Xu, L.Y.; Li, E.M. Identification of a novel lysyl oxidase-like 2 alternative splicing isoform, LOXL2 Δe13, in esophageal squamous cell carcinoma. *Biochem. Cell Biol. Biochim. Biol. Cell.* **2014**, *92*, 379–389. [CrossRef]
53. Park, P.G.; Jo, S.J.; Kim, M.J.; Kim, H.J.; Lee, J.H.; Park, C.K.; Kim, H.; Lee, K.Y.; Kim, H.; Park, J.H.; et al. Role of LOXL2 in the epithelial-mesenchymal transition and colorectal cancer metastasis. *Oncotarget* **2017**, *8*, 80325–80335. [CrossRef]
54. Liu, C.; Guo, T.; Sakai, A.; Ren, S.; Fukusumi, T.; Ando, M.; Sadat, S.; Saito, Y.; Califano, J.A. A novel splice variant of LOXL2 promotes progression of human papillomavirus-negative head and neck squamous cell carcinoma. *Cancer* **2020**, *126*, 737–748. [CrossRef]
55. Xu, L.; Go, E.P.; Finney, J.; Moon, H.; Lantz, M.; Rebecchi, K.; Desaire, H.; Mure, M. Post-translational modifications of recombinant human lysyl oxidase-like 2 (rhLOXL2) secreted from Drosophila S2 cells. *J. Biol. Chem.* **2013**, *288*, 5357–5363. [CrossRef] [PubMed]
56. Sidransky, D. Molecular genetics of head and neck cancer. *Curr. Opin. Oncol.* **1995**, *7*, 229–233. [CrossRef]

57. Surget, S.; Khoury, M.P.; Bourdon, J.-C. Uncovering the role of p53 splice variants in human malignancy: A clinical perspective. *OncoTargets Ther.* **2013**, *7*, 57–68. [CrossRef]
58. Fujita, K.; Mondal, A.M.; Horikawa, I.; Nguyen, G.H.; Kumamoto, K.; Sohn, J.J.; Bowman, E.D.; Mathe, E.A.; Schetter, A.J.; Pine, S.R.; et al. p53 isoforms Delta133p53 and p53beta are endogenous regulators of replicative cellular senescence. *Nat. Cell Biol* **2009**, *11*, 1135–1142. [CrossRef]
59. Bourdon, J.C.; Fernandes, K.; Murray-Zmijewski, F.; Liu, G.; Diot, A.; Xirodimas, D.P.; Saville, M.K.; Lane, D.P. p53 isoforms can regulate p53 transcriptional activity. *Genes Dev.* **2005**, *19*, 2122–2137. [CrossRef] [PubMed]
60. Courtois, S.; Verhaegh, G.; North, S.; Luciani, M.G.; Lassus, P.; Hibner, U.; Oren, M.; Hainaut, P. DeltaN-p53, a natural isoform of p53 lacking the first transactivation domain, counteracts growth suppression by wild-type p53. *Oncogene* **2002**, *21*, 6722–6728. [CrossRef]
61. Slatter, T.L.; Hung, N.; Campbell, H.; Rubio, C.; Mehta, R.; Renshaw, P.; Williams, G.; Wilson, M.; Engelmann, A.; Jeffs, A.; et al. Hyperproliferation, cancer, and inflammation in mice expressing a Δ133p53-like isoform. *Blood* **2011**, *117*, 5166–5177. [CrossRef] [PubMed]
62. Ungewitter, E.; Scrable, H. Delta40p53 controls the switch from pluripotency to differentiation by regulating IGF signaling in ESCs. *Genes Dev.* **2010**, *24*, 2408–2419. [CrossRef]
63. Bernard, H.; Garmy-Susini, B.; Ainaoui, N.; Van Den Berghe, L.; Peurichard, A.; Javerzat, S.; Bikfalvi, A.; Lane, D.P.; Bourdon, J.C.; Prats, A.C. The p53 isoform, Δ133p53α, stimulates angiogenesis and tumour progression. *Oncogene* **2013**, *32*, 2150–2160. [CrossRef] [PubMed]
64. Hofstetter, G.; Berger, A.; Fiegl, H.; Slade, N.; Zorić, A.; Holzer, B.; Schuster, E.; Mobus, V.J.; Reimer, D.; Daxenbichler, G.; et al. Alternative splicing of p53 and p73: The novel p53 splice variant p53delta is an independent prognostic marker in ovarian cancer. *Oncogene* **2010**, *29*, 1997–2004. [CrossRef]
65. Quentien, M.H.; Barlier, A.; Franc, J.L.; Pellegrini, I.; Brue, T.; Enjalbert, A. Pituitary transcription factors: From congenital deficiencies to gene therapy. *J. Neuroendocrinol.* **2006**, *18*, 633–642. [CrossRef] [PubMed]
66. Lin, C.R.; Kioussi, C.; O'Connell, S.; Briata, P.; Szeto, D.; Liu, F.; Izpisúa-Belmonte, J.C.; Rosenfeld, M.G. Pitx2 regulates lung asymmetry, cardiac positioning and pituitary and tooth morphogenesis. *Nature* **1999**, *401*, 279–282. [CrossRef] [PubMed]
67. Kioussi, C.; Briata, P.; Baek, S.H.; Rose, D.W.; Hamblet, N.S.; Herman, T.; Ohgi, K.A.; Lin, C.; Gleiberman, A.; Wang, J.; et al. Identification of a Wnt/Dvl/beta-Catenin –> Pitx2 pathway mediating cell-type-specific proliferation during development. *Cell* **2002**, *111*, 673–685. [CrossRef]
68. Liu, Y.; Huang, Y.; Zhu, G.Z. Cyclin A1 is a transcriptional target of PITX2 and overexpressed in papillary thyroid carcinoma. *Mol. Cell. Biochem.* **2013**, *384*, 221–227. [CrossRef] [PubMed]
69. Cox, C.J.; Espinoza, H.M.; McWilliams, B.; Chappell, K.; Morton, L.; Hjalt, T.A.; Semina, E.V.; Amendt, B.A. Differential regulation of gene expression by PITX2 isoforms. *J. Biol. Chem.* **2002**, *277*, 25001–25010. [CrossRef]
70. Basu, M.; Bhattacharya, R.; Ray, U.; Mukhopadhyay, S.; Chatterjee, U.; Roy, S.S. Invasion of ovarian cancer cells is induced byPITX2-mediated activation of TGF-β and Activin-A. *Mol. Cancer* **2015**, *14*, 162. [CrossRef]
71. Zheng, Y.; Jönsson, J.; Hao, C.; Chaghervand, S.S.; Cui, X.; Kajitani, N.; Gong, L.; Wu, C.; Schwartz, S. hnRNP A1 and hnRNP A2 inhibit splicing to HPV16 splice site SA409 through a UAG-containing sequence in the E7 coding region. *J. Virol.* **2020**. [CrossRef]
72. Wu, B.H.; Xiong, X.P.; Jia, J.; Zhang, W.F. MicroRNAs: New actors in the oral cancer scene. *Oral Oncol.* **2011**, *47*, 314–319. [CrossRef]
73. Sannigrahi, M.; Sharma, R.; Panda, N.; Khullar, M. Role of non-coding RNAs in head and neck squamous cell carcinoma: A narrative review. *Oral Dis.* **2018**, *24*, 1417–1427. [CrossRef]
74. Ransohoff, J.D.; Wei, Y.; Khavari, P.A. The functions and unique features of long intergenic non-coding RNA. *Nat. Rev. Mol. Cell Biol.* **2018**, *19*, 143–157. [CrossRef]
75. Denaro, N.; Merlano, M.C.; Russi, E.G.; Lo Nigro, C. Non coding RNAs in head and neck squamous cell carcinoma (HNSCC): A clinical perspective. *Anticancer Res.* **2014**, *34*, 6887–6896. [PubMed]
76. Ramdas, L.; Giri, U.; Ashorn, C.L.; Coombes, K.R.; El-Naggar, A.; Ang, K.K.; Story, M.D. miRNA expression profiles in head and neck squamous cell carcinoma and adjacent normal tissue. *Head Neck* **2009**, *31*, 642–654. [CrossRef] [PubMed]
77. Wong, T.S.; Liu, X.B.; Wong, B.Y.; Ng, R.W.; Yuen, A.P.; Wei, W.I. Mature miR-184 as Potential Oncogenic microRNA of Squamous Cell Carcinoma of Tongue. *Clin. Cancer Res.* **2008**, *14*, 2588–2592. [CrossRef] [PubMed]
78. Childs, G.; Fazzari, M.; Kung, G.; Kawachi, N.; Brandwein-Gensler, M.; McLemore, M.; Chen, Q.; Burk, R.D.; Smith, R.V.; Prystowsky, M.B.; et al. Low-level expression of microRNAs let-7d and miR-205 are prognostic markers of head and neck squamous cell carcinoma. *Am. J. Pathol.* **2009**, *174*, 736–745. [CrossRef]
79. Hui, A.B.; Lenarduzzi, M.; Krushel, T.; Waldron, L.; Pintilie, M.; Shi, W.; Perez-Ordonez, B.; Jurisica, I.; O'Sullivan, B.; Waldron, J.; et al. Comprehensive MicroRNA profiling for head and neck squamous cell carcinomas. *Clin. Cancer Res.* **2010**, *16*, 1129–1139. [CrossRef]
80. Hata, A.; Lieberman, J. Dysregulation of microRNA biogenesis and gene silencing in cancer. *Sci. Signal.* **2015**, *8*, re3. [CrossRef] [PubMed]
81. Yoon, A.J.; Wang, S.; Kutler, D.I.; Carvajal, R.D.; Philipone, E.; Wang, T.; Peters, S.M.; LaRoche, D.; Hernandez, B.Y.; McDowell, B.D.; et al. MicroRNA-based risk scoring system to identify early-stage oral squamous cell carcinoma patients at high-risk for cancer-specific mortality. *Head Neck* **2020**, *42*, 1699–1712. [CrossRef] [PubMed]

82. Ganci, F.; Sacconi, A.; Ben-Moshe, N.B.; Manciocco, V.; Sperduti, I.; Strigari, L.; Covello, R.; Benevolo, M.; Pescarmona, E.; Domany, E.; et al. Expression of TP53 mutation-associated microRNAs predicts clinical outcome in head and neck squamous cell carcinoma patients. *Ann. Oncol.* **2013**, *24*, 3082–3088. [CrossRef]
83. Kita, Y.; Vincent, K.; Natsugoe, S.; Berindan-Neagoe, I.; Calin, G.A. Epigenetically regulated microRNAs and their prospect in cancer diagnosis. *Expert Rev. Mol. Diagn.* **2014**, *14*, 673–683. [CrossRef]
84. Harris, T.; Jimenez, L.; Kawachi, N.; Fan, J.B.; Chen, J.; Belbin, T.; Ramnauth, A.; Loudig, O.; Keller, C.E.; Smith, R.; et al. Low-level expression of miR-375 correlates with poor outcome and metastasis while altering the invasive properties of head and neck squamous cell carcinomas. *Am. J. Pathol.* **2012**, *180*, 917–928. [CrossRef] [PubMed]
85. Yu, L.; Lu, J.; Zhang, B.; Liu, X.; Wang, L.; Li, S.Y.; Peng, X.H.; Xu, X.; Tian, W.D.; Li, X.P. miR-26a inhibits invasion and metastasis of nasopharyngeal cancer by targeting EZH2. *Oncol. Lett.* **2013**, *5*, 1223–1228. [CrossRef] [PubMed]
86. Kalinowski, F.C.; Giles, K.M.; Candy, P.A.; Ali, A.; Ganda, C.; Epis, M.R.; Webster, R.J.; Leedman, P.J. Regulation of epidermal growth factor receptor signaling and erlotinib sensitivity in head and neck cancer cells by miR-7. *PLoS ONE* **2012**, *7*, e47067. [CrossRef] [PubMed]
87. Datta, J.; Smith, A.; Lang, J.C.; Islam, M.; Dutt, D.; Teknos, T.N.; Pan, Q. microRNA-107 functions as a candidate tumor-suppressor gene in head and neck squamous cell carcinoma by downregulation of protein kinase Cϵ. *Oncogene* **2012**, *31*, 4045–4053. [CrossRef] [PubMed]
88. Kinoshita, T.; Nohata, N.; Hanazawa, T.; Kikkawa, N.; Yamamoto, N.; Yoshino, H.; Itesako, T.; Enokida, H.; Nakagawa, M.; Okamoto, Y.; et al. Tumour-suppressive microRNA-29s inhibit cancer cell migration and invasion by targeting laminin-integrin signalling in head and neck squamous cell carcinoma. *Br. J. Cancer* **2013**, *109*, 2636–2645. [CrossRef] [PubMed]
89. Alajez, N.M.; Shi, W.; Wong, D.; Lenarduzzi, M.; Waldron, J.; Weinreb, I.; Liu, F.F. Lin28b promotes head and neck cancer progression via modulation of the insulin-like growth factor survival pathway. *Oncotarget* **2012**, *3*, 1641–1652. [CrossRef]
90. Zhang, T.; Han, G.; Wang, Y.; Chen, K.; Sun, Y. MicroRNA expression profiles in supraglottic carcinoma. *Oncol. Rep.* **2014**, *31*, 2029–2034. [CrossRef]
91. Lu, Z.M.; Lin, Y.F.; Jiang, L.; Chen, L.S.; Luo, X.N.; Song, X.H.; Chen, S.H.; Zhang, S.Y. Micro-ribonucleic acid expression profiling and bioinformatic target gene analyses in laryngeal carcinoma. *OncoTargets Ther.* **2014**, *7*, 525–533. [CrossRef]
92. Chapman, B.V.; Wald, A.I.; Akhtar, P.; Munko, A.C.; Xu, J.; Gibson, S.P.; Grandis, J.R.; Ferris, R.L.; Khan, S.A. MicroRNA-363 targets myosin 1B to reduce cellular migration in head and neck cancer. *BMC Cancer* **2015**, *15*, 861. [CrossRef]
93. Wald, A.I.; Hoskins, E.E.; Wells, S.I.; Ferris, R.L.; Khan, S.A. Alteration of microRNA profiles in squamous cell carcinoma of the head and neck cell lines by human papillomavirus. *Head Neck* **2011**, *33*, 504–512. [CrossRef] [PubMed]
94. Park, S.Y.; Lee, J.H.; Ha, M.; Nam, J.W.; Kim, V.N. miR-29 miRNAs activate p53 by targeting p85 alpha and CDC42. *Nat. Struct. Mol. Biol.* **2009**, *16*, 23–29. [CrossRef] [PubMed]
95. Li, J.; Huang, H.; Sun, L.; Yang, M.; Pan, C.; Chen, W.; Wu, D.; Lin, Z.; Zeng, C.; Yao, Y.; et al. MiR-21 indicates poor prognosis in tongue squamous cell carcinomas as an apoptosis inhibitor. *Clin. Cancer Res.* **2009**, *15*, 3998–4008. [CrossRef]
96. Avissar, M.; Christensen, B.C.; Kelsey, K.T.; Marsit, C.J. MicroRNA expression ratio is predictive of head and neck squamous cell carcinoma. *Clin. Cancer Res.* **2009**, *15*, 2850–2855. [CrossRef] [PubMed]
97. Liu, C.J.; Lin, S.C.; Yang, C.C.; Cheng, H.W.; Chang, K.W. Exploiting salivary miR-31 as a clinical biomarker of oral squamous cell carcinoma. *Head Neck* **2012**, *34*, 219–224. [CrossRef] [PubMed]
98. Saito, K.; Inagaki, K.; Kamimoto, T.; Ito, Y.; Sugita, T.; Nakajo, S.; Hirasawa, A.; Iwamaru, A.; Ishikura, T.; Hanaoka, H.; et al. MicroRNA-196a is a putative diagnostic biomarker and therapeutic target for laryngeal cancer. *PLoS ONE* **2013**, *8*, e71480. [CrossRef]
99. Fu, W.M.; Lu, Y.F.; Hu, B.G.; Liang, W.C.; Zhu, X.; Yang, H.D.; Li, G.; Zhang, J.F. Long noncoding RNA Hotair mediated angiogenesis in nasopharyngeal carcinoma by direct and indirect signaling pathways. *Oncotarget* **2016**, *7*, 4712–4723. [CrossRef]
100. Zou, A.E.; Ku, J.; Honda, T.K.; Yu, V.; Kuo, S.Z.; Zheng, H.; Xuan, Y.; Saad, M.A.; Hinton, A.; Brumund, K.T.; et al. Transcriptome sequencing uncovers novel long noncoding and small nucleolar RNAs dysregulated in head and neck squamous cell carcinoma. *RNA* **2015**, *21*, 1122–1134. [CrossRef]
101. Tang, H.; Wu, Z.; Zhang, J.; Su, B. Salivary lncRNA as a potential marker for oral squamous cell carcinoma diagnosis. *Mol. Med. Rep.* **2013**, *7*, 761–766. [CrossRef]
102. Li, L.; Gu, M.; You, B.; Shi, S.; Shan, Y.; Bao, L.; You, Y. Long non-coding RNA ROR promotes proliferation, migration and chemoresistance of nasopharyngeal carcinoma. *Cancer Sci.* **2016**, *107*, 1215–1222. [CrossRef]
103. Wu, T.; Qu, L.; He, G.; Tian, L.; Li, L.; Zhou, H.; Jin, Q.; Ren, J.; Wang, Y.; Wang, J.; et al. Regulation of laryngeal squamous cell cancer progression by the lncRNA H19/miR-148a-3p/DNMT1 axis. *Oncotarget* **2016**, *7*, 11553–11566. [CrossRef]
104. Sun, Q.; Liu, H.; Li, L.; Zhang, S.; Liu, K.; Liu, Y.; Yang, C. Long noncoding RNA-LET, which is repressed by EZH2, inhibits cell proliferation and induces apoptosis of nasopharyngeal carcinoma cell. *Med. Oncol.* **2015**, *32*, 226. [CrossRef] [PubMed]
105. Song, P.; Ye, L.F.; Zhang, C.; Peng, T.; Zhou, X.H. Long non-coding RNA XIST exerts oncogenic functions in human nasopharyngeal carcinoma by targeting miR-34a-5p. *Gene* **2016**, *592*, 8–14. [CrossRef]
106. Jia, L.F.; Wei, S.B.; Gan, Y.H.; Guo, Y.; Gong, K.; Mitchelson, K.; Cheng, J.; Yu, G.Y. Expression, regulation and roles of miR-26a and MEG3 in tongue squamous cell carcinoma. *Int. J. Cancer* **2014**, *135*, 2282–2293. [CrossRef] [PubMed]
107. Zou, A.E.; Zheng, H.; Saad, M.A.; Rahimy, M.; Ku, J.; Kuo, S.Z.; Honda, T.K.; Wang-Rodriguez, J.; Xuan, Y.; Korrapati, A.; et al. The non-coding landscape of head and neck squamous cell carcinoma. *Oncotarget* **2016**, *7*, 51211–51222. [CrossRef] [PubMed]

108. Stransky, N.; Egloff, A.M.; Tward, A.D.; Kostic, A.D.; Cibulskis, K.; Sivachenko, A.; Kryukov, G.V.; Lawrence, M.S.; Sougnez, C.; McKenna, A.; et al. The mutational landscape of head and neck squamous cell carcinoma. *Science* **2011**, *333*, 1157–1160. [CrossRef]
109. Chung, C.H.; Guthrie, V.B.; Masica, D.L.; Tokheim, C.; Kang, H.; Richmon, J.; Agrawal, N.; Fakhry, C.; Quon, H.; Subramaniam, R.M.; et al. Genomic alterations in head and neck squamous cell carcinoma determined by cancer gene-targeted sequencing. *Ann. Oncol.* **2015**, *26*, 1216–1223. [CrossRef] [PubMed]
110. El Baroudi, M.; Machiels, J.P.; Schmitz, S. Expression of SESN1, UHRF1BP1, and miR-377-3p as prognostic markers in mutated TP53 squamous cell carcinoma of the head and neck. *Cancer Biol. Ther.* **2017**, *18*, 775–782. [CrossRef] [PubMed]
111. Balz, V.; Scheckenbach, K.; Götte, K.; Bockmühl, U.; Petersen, I.; Bier, H. Is the p53 inactivation frequency in squamous cell carcinomas of the head and neck underestimated? Analysis of p53 exons 2-11 and human papillomavirus 16/18 E6 transcripts in 123 unselected tumor specimens. *Cancer Res.* **2003**, *63*, 1188–1191.
112. Sisk, E.A.; Soltys, S.G.; Zhu, S.; Fisher, S.G.; Carey, T.E.; Bradford, C.R. Human papillomavirus and p53 mutational status as prognostic factors in head and neck carcinoma. *Head Neck* **2002**, *24*, 841–849. [CrossRef]
113. Avigad, S.; Barel, D.; Blau, O.; Malka, A.; Zoldan, M.; Mor, C.; Fogel, M.; Cohen, I.J.; Stark, B.; Goshen, Y.; et al. A novel germ line p53 mutation in intron 6 in diverse childhood malignancies. *Oncogene* **1997**, *14*, 1541–1545. [CrossRef]
114. Agrawal, N.; Frederick, M.J.; Pickering, C.R.; Bettegowda, C.; Chang, K.; Li, R.J.; Fakhry, C.; Xie, T.X.; Zhang, J.; Wang, J.; et al. Exome sequencing of head and neck squamous cell carcinoma reveals inactivating mutations in NOTCH1. *Science* **2011**, *333*, 1154–1157. [CrossRef] [PubMed]
115. Pickering, C.R.; Zhang, J.; Yoo, S.Y.; Bengtsson, L.; Moorthy, S.; Neskey, D.M.; Zhao, M.; Ortega Alves, M.V.; Chang, K.; Drummond, J.; et al. Integrative genomic characterization of oral squamous cell carcinoma identifies frequent somatic drivers. *Cancer Discov.* **2013**, *3*, 770–781. [CrossRef]
116. Ghalamkari, S.; Alavi, S.; Mianesaz, H.; Khosravian, F.; Bahreini, A.; Salehi, M. A novel carcinogenic PI3Kα mutation suggesting the role of helical domain in transmitting nSH2 regulatory signals to kinase domain. *Life Sci.* **2021**, *269*, 118759. [CrossRef]
117. Agrawal, A.A.; Yu, L.; Smith, P.G.; Buonamici, S. Targeting splicing abnormalities in cancer. *Curr. Opin. Genet. Dev.* **2018**, *48*, 67–74. [CrossRef]
118. Lee, S.C.; Abdel-Wahab, O. Therapeutic targeting of splicing in cancer. *Nat. Med.* **2016**, *22*, 976–986. [CrossRef]
119. Kotake, Y.; Sagane, K.; Owa, T.; Mimori-Kiyosue, Y.; Shimizu, H.; Uesugi, M.; Ishihama, Y.; Iwata, M.; Mizui, Y. Splicing factor SF3b as a target of the antitumor natural product pladienolide. *Nat. Chem. Biol.* **2007**, *3*, 570–575. [CrossRef]
120. Yokoi, A.; Kotake, Y.; Takahashi, K.; Kadowaki, T.; Matsumoto, Y.; Minoshima, Y.; Sugi, N.H.; Sagane, K.; Hamaguchi, M.; Iwata, M.; et al. Biological validation that SF3b is a target of the antitumor macrolide pladienolide. *FEBS J.* **2011**, *278*, 4870–4880. [CrossRef]
121. Castanotto, D.; Stein, C.A. Antisense oligonucleotides in cancer. *Curr. Opin. Oncol.* **2014**, *26*, 584–589. [CrossRef]
122. McClorey, G.; Wood, M.J. An overview of the clinical application of antisense oligonucleotides for RNA-targeting therapies. *Curr. Opin. Pharmacol.* **2015**, *24*, 52–58. [CrossRef] [PubMed]
123. Dudás, J.; Dietl, W.; Romani, A.; Reinold, S.; Glueckert, R.; Schrott-Fischer, A.; Dejaco, D.; Chacko, L.J.; Tuertscher, R.; Schartinger, V.H.; et al. Nerve Growth Factor (NGF)-Receptor Survival Axis in Head and Neck Squamous Cell Carcinoma. *Int. J. Mol. Sci.* **2018**, *19*, 1771. [CrossRef] [PubMed]
124. Vanzyl, E.J.; Sayed, H.; Blackmore, A.B.; Rick, K.R.C.; Fernando, P.; McKay, B.C. The spliceosome inhibitors isoginkgetin and pladienolide B induce ATF3-dependent cell death. *PLoS ONE* **2020**, *15*, e0224953. [CrossRef] [PubMed]
125. Sakai, Y.; Yoshida, T.; Ochiai, K.; Uosaki, Y.; Saitoh, Y.; Tanaka, F.; Akiyama, T.; Akinaga, S.; Mizukami, T. GEX1 compounds, novel antitumor antibiotics related to herboxidiene, produced by Streptomyces sp. I. Taxonomy, production, isolation, physicochemical properties and biological activities. *J. Antibiot.* **2002**, *55*, 855–862. [CrossRef]
126. Hong, D.S.; Kurzrock, R.; Naing, A.; Wheler, J.J.; Falchook, G.S.; Schiffman, J.S.; Faulkner, N.; Pilat, M.J.; O'Brien, J.; LoRusso, P. A phase I, open-label, single-arm, dose-escalation study of E7107, a precursor messenger ribonucleic acid (pre-mRNA) splicesome inhibitor administered intravenously on days 1 and 8 every 21 days to patients with solid tumors. *Investig. New Drugs* **2014**, *32*, 436–444. [CrossRef] [PubMed]
127. Nakajima, H.; Sato, B.; Fujita, T.; Takase, S.; Terano, H.; Okuhara, M. New antitumor substances, FR901463, FR901464 and FR901465. I. Taxonomy, fermentation, isolation, physico-chemical properties and biological activities. *J. Antibiot.* **1996**, *49*, 1196–1203. [CrossRef] [PubMed]
128. Hahn, D.; Kudla, G.; Tollervey, D.; Beggs, J.D. Brr2p-mediated conformational rearrangements in the spliceosome during activation and substrate repositioning. *Genes Dev.* **2012**, *26*, 2408–2421. [CrossRef]
129. Iwatani-Yoshihara, M.; Ito, M.; Klein, M.G.; Yamamoto, T.; Yonemori, K.; Tanaka, T.; Miwa, M.; Morishita, D.; Endo, S.; Tjhen, R.; et al. Discovery of Allosteric Inhibitors Targeting the Spliceosomal RNA Helicase Brr2. *J. Med. Chem.* **2017**, *60*, 5759–5771. [CrossRef]
130. Wang, B.B.; Brendel, V. Molecular characterization and phylogeny of U2AF35 homologs in plants. *Plant. Physiol.* **2006**, *140*, 624–636. [CrossRef]
131. Pozzi, B.; Bragado, L.; Will, C.L.; Mammi, P.; Risso, G.; Urlaub, H.; Lührmann, R.; Srebrow, A. SUMO conjugation to spliceosomal proteins is required for efficient pre-mRNA splicing. *Nucleic Acids Res.* **2017**, *45*, 6729–6745. [CrossRef]
132. Bonnal, S.; Vigevani, L.; Valcárcel, J. The spliceosome as a target of novel antitumour drugs. *Nat. Rev. Drug Discov.* **2012**, *11*, 847–859. [CrossRef]

133. Gammons, M.V.; Lucas, R.; Dean, R.; Coupland, S.E.; Oltean, S.; Bates, D.O. Targeting SRPK1 to control VEGF-mediated tumour angiogenesis in metastatic melanoma. *Br. J. Cancer* **2014**, *111*, 477–485. [CrossRef] [PubMed]
134. Siqueira, R.P.; Barbosa Éde, A.; Polêto, M.D.; Righetto, G.L.; Seraphim, T.V.; Salgado, R.L.; Ferreira, J.G.; Barros, M.V.; de Oliveira, L.L.; Laranjeira, A.B.; et al. Potential Antileukemia Effect and Structural Analyses of SRPK Inhibition by N-(2-(Piperidin-1-yl)-5-(Trifluoromethyl)Phenyl)Isonicotinamide (SRPIN340). *PLoS ONE* **2015**, *10*, e0134882. [CrossRef] [PubMed]
135. Dewaele, M.; Tabaglio, T.; Willekens, K.; Bezzi, M.; Teo, S.X.; Low, D.H.; Koh, C.M.; Rambow, F.; Fiers, M.; Rogiers, A.; et al. Antisense oligonucleotide-mediated MDM4 exon 6 skipping impairs tumor growth. *J. Clin. Investig.* **2016**, *126*, 68–84. [CrossRef] [PubMed]
136. Hong, D.; Kurzrock, R.; Kim, Y.; Woessner, R.; Younes, A.; Nemunaitis, J.; Fowler, N.; Zhou, T.; Schmidt, J.; Jo, M.; et al. AZD9150, a next-generation antisense oligonucleotide inhibitor of STAT3 with early evidence of clinical activity in lymphoma and lung cancer. *Sci. Transl. Med.* **2015**, *7*, 314ra185. [CrossRef]
137. Ross, S.J.; Revenko, A.S.; Hanson, L.L.; Ellston, R.; Staniszewska, A.; Whalley, N.; Pandey, S.K.; Revill, M.; Rooney, C.; Buckett, L.K.; et al. Targeting KRAS-dependent tumors with AZD4785, a high-affinity therapeutic antisense oligonucleotide inhibitor of KRAS. *Sci. Transl. Med.* **2017**, *9*. [CrossRef]

MDPI
St. Alban-Anlage 66
4052 Basel
Switzerland
Tel. +41 61 683 77 34
Fax +41 61 302 89 18
www.mdpi.com

Genes Editorial Office
E-mail: genes@mdpi.com
www.mdpi.com/journal/genes

www.ingramcontent.com/pod-product-compliance
Lightning Source LLC
Chambersburg PA
CBHW061557210926
51850CB00009B/2133
* 9 7 8 3 0 3 6 5 7 1 1 6 4 *